Weather and Climate: The M.P. Singh Volume, Part I

Edited by
Sethu Raman
Maithili Sharan

2005

Birkhäuser Verlag
Basel · Boston · Berlin

Reprint from Pure and Applied Geophysics
(PAGEOPH), Volume 162 (2005) No. 8/9

Editor(s)

Maithili Sharan
Professor, Centre for Atmospheric Sciences
Indian Institute of Technology, Delhi
Hauz Khas New Delhi 11016
India

e-mail: mathilis@cas.iitd.ernet.in

Sethu Raman
Professor and State Climatologist
North Carolina State University
1005 Capability Dr., Suite 213
Campus Box 7236
Raleigh, NC 27695
U.S.A.

e-mail: raman@ncsu.edu

A CIP catalogue record for this book is available from the Library of Congress,
Washington D.C., USA

Bibliographic information published by Die Deutsche Bibliothek:
Die Deutsche Bibliothek lists this publication in the Deutsche Nationalbibliographie; detailed
bibliographic data is available in the internet at <http://dnb.ddb.de>

ISBN 3-7643-7296-6 Birkhäuser Verlag, Basel – Boston – Berlin

© 2005 Birkhäuser Verlag, P.O.Box 133, CH-4010 Basel, Switzerland
Part of Springer Science+Business Media
Printed on acid-free paper produced from chlorine-free pulp TCF ∞
Printed in Germany

ISBN-10: 3-7643-7296-6
ISBN-13: 978-3-7643-7296-5

9 8 7 6 5 4 3 2 1

PURE AND APPLIED GEOPHYSICS
Vol. 162, No. 8–9, 2005

Contents

B. Weather

Pure appl. geophys. 162 (2005) 1397–1400
0033–4553/05/091397–4
DOI 10.1007/s00024-005-2675-x

© Birkhäuser Verlag, Basel, 2005

Pure and Applied Geophysics

Weather and Climate: The M.P. Singh Volume
Part I

Preface

Weather and climate are of concern to virtually all countries worldwide. For many countries the economy depends largely on agriculture, which is significantly affected by variations in weather and climate. Many of the South Asian countries, for example, are prone to natural hazards such as tropical cyclones, droughts, and floods. In fact, the Indian economy is dubbed by many as a "Monsoon Gamble." Rapid industrialization and urbanization add to further deterioration of our environment. Changes in weather and climate can be traced to this environmental degradation and changes in land-use patterns brought on by rapid industrialization. Atmospheric and oceanic processes assume significance in understanding all aspects of weather and climate.

A number of field experiments have been conducted to gain an understanding of the physics of the atmosphere and the oceans. Significant advances have taken place in the understanding of atmospheric and oceanic processes and in the development of atmospheric/ oceanic models. Modern technology such as weather and ocean sensing satellites, fast communication systems, improved measurement systems and efficient computational techniques are now used in updating the information base. With increased computational power, finer resolution global / regional models have been developed for weather and climate. This has led to an increase in the predictability of weather-related phenomena at various scales. This special volume on Weather and Climate has provided contributions to the latest developments in this field.

In this volume, peer-reviewed papers related to mathematical techniques/ modeling, numerical simulations, atmospheric and oceanic processes and field experiments are included to gain insight into the weather and climate system. Papers are divided broadly into three groups, namely climate, weather and air quality. The first eight papers are related to climate, the next eleven papers pertain to weather, including land surface processes, and the last seven papers deal with air quality studies. This volume has been divided in to two parts. Part I contains the papers related to climate and weather whereas papers pertaining to boundary layer processes and air quality are included in Part II.

The paper by Marchuk *et al.* summarizes the "splitting methods" of numerical modeling and illustrates their application to a high resolution model of the Indian Ocean. The article from Mitra *et al.* describes results from a series of numerical experiments designed to determine the fidelity of predictions of major features of the Indian monsoon season in the period 1987 to 2002, using a state-of-the-art Florida State University coupled ocean-atmospheric general circulation model. The coupled model is also shown to capture the anomalous dry conditions of the monsoon 2002 season. Pielke *et al.* have discussed in their paper the various aspects of the 2002 Colorado drought to analyze the question whether it is a routine or unprecedented drought. The study of Sanjeeva Rao and Sikka examines several aspects of the intra-seasonal oscillations to understand its mechanisms and how it modulates the performance of the monsoon using data from BOBMEX and ARMEX experiments. The aim of the study undertaken by Mohanty *et al.* is to understand the climate diagnostics of the Asian summer monsoon and the role of equatorial convection on the summer monsoon activity over India. Le Treut and Bellon show that the direct impact of radiative heating on the development of monsoon may depend in a very complex manner on a combination of strongly interacting radiative and dynamical feedbacks.

Khandekar *et al.* have presented a review of the present status of global warming science. They have focused on evolution of the earth's atmosphere, and the greenhouse effect. Beniston's paper provides a concise overview of the peculiarities of mountain climates, pointing out the sensitivity of the atmosphere to these systems with a special reference to the European Alps.

There are several papers on weather in this volume dealing with radiation, tornados, tropical cyclones, and land use. The paper by Ramanathan and Ramana discusses the presence of absorbing aerosols over Kathmandu, Nepal and Kanpur, India and their implications on radiation balance. The article by Rao *et al.* presents the structures of mesocirculations that generated tornadoes associated with Tropical Cyclones Frances (1998) in Texas and in Louisiana. The study (Subramaniam *et al.*) addresses the thermal, salinity and circulation responses at sea surface due to intense tropical cyclones in 1999 and 2001 in the northern tropical Indian Ocean, based on satellite measurements and model simulations. Dube *et al.* in their paper describe a depth-averaged storm surge model for simulating the surges along the Orissa coast. This study emphasizes the impact of the Mahanadi River on overall surge development along the Orissa coast.

In the second part of this volume, scientific papers related to the atmospheric boundary layer and air quality studies are included. The land surface processes and surface characteristics play a dominant role in the development of boundary layer and in turn on weather and climate. For weather prediction and climate modeling, the understanding of the behavior and predictability of the atmospheric boundary layer is crucial. The paper by Raman *et al.* is a numerical study on convection initiation caused by differences in soil types. Arya provides an overview on micrometeorology and atmospheric boundary layer discussing, recent developments in the field in relation to weather and climate. Orr *et al.* presents detailed features of mesoscale flows over Greenland, specifically, to understand the influence of high orography on stable flows over and around Greenland. Hozumi and Ueda consider air flow over large mountains in an idealized situation of uniform approach flow, and in the absence of local thermally driven flows on the slopes, e.g. cooling as in the case of Greenland (Orr *et al.*) and heating as the southern slopes of the Himalayas. The effect of gentle slopes on strong and weak wind nocturnal boundary layer (NBL) is investigated by Gopalakrishnan *et al.* This study indicates that the wind profiles, temperature profiles and surface layer turbulence characteristics are sensitive to the imposed geostrophic wind when small slopes are present, especially for light winds. Using techniques of nonlinear analysis, Shi *et al.* examine in their paper the behavior of NBL by analyzing a set of four partial differential equations and this paper explores the role of initial conditions in the final solution and carries out an eigen value analysis to determine whether a pure limit cycle exists. The paper by Vasudev Murthy *et al.* presents a complete asymptotic analysis of a simple model for the evolution of the nocturnal temperature distribution on bare soil in calm and clear conditions. This simple model is able to explain the occurrence of minima in the vertical temperature profile during the calm/clear night and thus describes the structure and dynamics of the Ramdas layer.

Pollutants emitted into the atmosphere from various sources such as industrial chimneys, vehicular exhaust, power plants, nuclear and chemical plants etc. degrade the environment and in turn influence weather and climate. Sharan and Modani describe an analytical model for the dispersion of pollutants in a finite layer in low wind conditions. The proposed model is validated with the data from the Hanford diffusion experiment in stable conditions and IIT diffusion experiment in unstable conditions. The paper by Rao addresses the analysis and quantification of various types of uncertainties associated with the prediction of concentrations from atmospheric dispersion models. Gego *et al.* address the important problem of consistency between various measurement networks with a particular reference to the particulate nitrate, sulfate and ammonium concentrations. Neophytou and Britter provide a simple model for the variation of smoke and oxygen concentrations in an inclined two-dimensional tunnel that is open at the top.

Following the collapse of the New York World Trade Center Towers (WTC) on September 11, 2001, several studies were initiated to monitor air quality to better understand the impact of emissions from the disaster. The next three papers deal with the various aspects. Urban heat island and roughness length effects, sea breeze structure and behavior are among the features examined by Child and Raman. Gilliam *et al.* critically examine a dispersion modeling system (CALMET-CAL-PUFF) used to simulate the WTC emissions transport. The first paper focuses on the performance of the diagnostic meteorological model, CALMET during various synoptic flow events whereas in the second, the dispersion patterns of a simulated WTC plume are presented for a three-month period from September 11, 2001.

Earlier research related to meteorology in India was primarily confined to the India Meteorological Department (IMD), the country's National Weather Service. IMD devoted significant concentration on the operational needs of the country. However, efforts on the fundamental research for the understanding of atmospheric and oceanic features were limited in scope. Dr. M. P. Singh, Professor of Applied Mathematics, Indian Institute of Technology (IIT), Delhi (India) took an initiative and played a key role in the development of research and development activities in the areas of atmospheric and oceanic sciences at IIT Delhi. At his initiative, a Centre for Atmospheric Sciences was set-up at IIT Delhi which was cosponsored by the India Meteorological Department on a cost sharing basis. India is among the very few countries where such a joint venture sponsored by the National Weather Service to promote weather research in an academic institute has been estabilished. To recognize the contributions of Professor M.P. Singh in research and developmental activities in the area of atmospheric and oceanic sciences, we honor him by dedicating this issue of Pure and Applied Geophysics as "M.P. Singh volume on Weather and Climate."

We would like to place on record our gratitude to all the authors who have contributed to this special issue. Their association with this special issue will benefit immensely the scientific community at large, especially younger scientists. Late Professor G. V. Rao played a key role in the planning of this special issue. He fell victim to a swimming accident on the Mexican Pacific Coast while he was performing research related to the North American Monsoon Experiment. We wish to thank all the reviewers for their valuable comments/suggestions. We also thank Dr. Renata Dmowska, Topical Editor, Pure and Applied Geophysics for her assistance in issuing out this topical edition in honor of Professor M.P. Singh.

Finally, we express our sincere thanks to the editorial staff of Pure and Applied Geophysics for the meticulous care they have taken in preparing both parts of this special edition.

Editors:

Maithili Sharan
Professor, Centre for Atmospheric Sciences,
Indian Institute of Technology, Delhi
Hauz Khas, New Delhi 110016, India.

Sethu Raman
Professor and State Climatologist,
North Carolina State University,
Raleigh, North Carolina 27695, U.S.A.

Pure appl. geophys. 162 (2005) 1401–1403
0033–4553/05/091401–3
DOI 10.1007/s00024-005-2676-9

❘Pure and Applied Geophysics

Brief Biography of Professor M.P. Singh

Maithili Sharan[1] and Sethu Raman[2]

Professor Singh was born in Etah, India in 1931. He received his Master's degree from Lucknow University, was a Fulbright Scholar, and received his Ph.D. degree from the University of Maryland (U.S.A.) in 1960, after which he worked as Postdoctoral Research Associate at Cornell University and Goddard Space Flight Center, U.S.A.

Earlier research related to meteorology in India was primarily confined to collaborations with the India Meteorological Department (IMD), the country's National Weather Service. IMD devoted significant major time focusing on the operational needs of the country. However, efforts on fundamental research for the understanding of atmospheric and oceanic features were limited in scope. Dr. M. P. Singh, Professor of Applied Mathematics, Indian Institute of Technology (IIT), Delhi (India) took an initiative and played a key role in the development of research and development activities in the areas of atmospheric and oceanic sciences at IIT Delhi. At his initiative, a Centre for Atmospheric Sciences was set-up at IIT Delhi

[1]Centre for Atmospheric Sciences, Indian Institute of Technology, Delhi, Hauz Khas, New Delhi 110016, India.

[2]State Climate Office of North Carolina. North Carolina State University, Raleigh, North Carolina 27695-7236, U.S.A.

which is cosponsored by the Indian Meteorological Department on a cost sharing basis. India is one of very few countries to establish such a joint venture sponsored by the National Weather Service to promote fundamental research in an academic Institute.

The Centre for Atmospheric Sciences has attained prominence over the years and is now widely recognized as a leading institution on research in atmospheric and marine sciences. Professor Singh initiated several international collaboration programs between the Centre and various well known Universities and Research Institutions in USA, UK, France, the then Soviet Union, and Japan. He was one of the original contributors to the proposal to acquire a Super-Computer facility in India for conducting research in Medium Range Weather Forecasting for which the National Centre for Medium Range Weather forecasts was founded in the 1980s. Professor Singh has played a key role in developing multidisciplinary programs in the country. He represented India in the scientific delegation to the then Soviet Union in 1988 and in the Indo-US Subcommission on Science and Technology in 1985 and 1987. Professor Singh was a member of the Indian delegation to attend the prestigious Science and Technology initiative meeting in 1983, the so-called Blue Ribbon panel on Meteorology arranged by the late Indian Prime Minister, Mrs. Indira Gandhi and the late US President Mr. Ronald Reagan to achieve targets in mission-oriented thrust areas identified by the two Governments. He was a member of ICSU (International Council for Scientific Union) International Committee for the International Decade for Natural Disaster Reduction during 1990–92.

Professor Singh is one of the founding faculty members of the prestigious Indian Institute of Technology, Delhi (IITD). IITs have acquired international reputation and recognition. Professor Singh has played a pioneering role in promoting basic research to study monsoon phenomena; he organized the international conference on *Monsoon Dynamics* in December 1977, which was attended by leading world experts and the proceedings were published in the form of a book entitled "Monsoon Dynamics" printed by the Cambridge University Press. After the unfortunate Bhopal Gas leak, he started highlighting the importance of development of appropriate *Atmospheric Dispersion Models* relevant to the tropical environment. He was the first to provide scientific analysis of the dispersion of the deadly MIC leak from the Union Carbide premises in Bhopal, which killed thousands of persons and caused permanent physical impairment to hundreds of thousands of the inhabitants; this work was published in the prestigious "J. of Hazardous Materials". Professor Singh has made significant contributions in atmospheric boundary layer studies and air quality modeling related to the tropics.

To highlight the importance of studies in the field of atmospheric boundary layer and dispersion processes, he organized several international conferences on *Air Quality* in India, Italy and Brunei. He was one of the Guest Editors of the two issues of Atmospheric environment which covered the conferences organized by him in India in 1988 and 1993; he was also one of the Guest Editors of the special issue of

Pure and Applied Geophysics entitled "Air Quality" which addressed the international conference on *Air Quality Dispersion* organized by Professor Singh in Brunei in 1999. As one of the Directors of the workshops on air quality modeling and climate and cloud physics organized by the International Center for Theoretical Physics in Italy in 1987 and 1990, he ably provided leadership to the scientists from the Third World countries interested in acquiring knowledge in Air Quality, Weather and Climate.

He is the co-editor of a book titled "Dynamics of Atmospheric Flow" published under the "Advances in Fluid Mechanic Series, UK." He was on the Editorial Board of the journal "Atmospheric Environment" (1988–1994); "Non-linear World (1992–1996) and "Advances in Fluid Mechanic Series" issued by Computation Mechanics Publications, UK (1994-1998). Additional conference / workshops organized by Professor Singh in emerging areas include:

a) A workshop on "Environmental Management in Oil Industry: A course for Corporate Managers" at IIT Delhi, July 1989.
b) Indo-Japan Symposium on "Air Pollution Scientific Research and its Prevention Technology" at Hotel Maurya Sheraton, Delhi, October 1989.
c) A regional Symposium on "Acid Rain in Asia" at IIT Delhi; sponsored by the World Bank, Feb., 1993.

Professor Singh has supervised 27 Ph.D., students who are now highly placed academically, several of them in the United States. He has stimulated numerous Indian researchers to initiate research activities in the field of Air Quality, Weather and Climate; he has approximately 100 research publications in prestigious international journals such as Journal of Atmospheric Sciences, Journal of Applied Meteorology, Boundary Layer Meteorology, Atmospheric Environment, Philosophical Transactions of the Royal Society and is the Fellow of the Indian National Science Academy and National Academy of Sciences, India.

Professor Singh is currently the Director of the prestigious Ansal Institute of Technology (AIT), Gurgaon, devoted to promote international academic collaboration. The Institute has been set up in the private sector, unlike most of the reputed academic institutions in the country, all of which are funded by the Government.

A. Climate

Pure appl. geophys. 162 (2005) 1407–1429
0033–4553/05/091407–23
DOI 10.1007/s00024-005-2677-8

▌**Pure and Applied Geophysics**

Splitting Numerical Technique with Application to the High Resolution Simulation of the Indian Ocean Circulation

G.I. MARCHUK,[1] A.S. RUSAKOV, V.B. ZALESNY,[1] and N.A. DIANSKY[1]

Abstract—The aim of this paper is twofold : To present an efficient numerical technique for the simulation of the ocean general circulation (OGC) and to apply it to the simulation of the Indian Ocean dynamics with high spatial resolution. To solve model equations we use the splitting method by physical processes and space coordinates. We select the main parts of the model operator and then perform their numerical treatment independently of one another. We describe the general methodology and some special aspects of this approach. Numerical treatment of the monsoon circulation is performed on the basis of the sigma-coordinate primitive equation model, which was developed at the Institute of Numerical Mathematics (Moscow, Russia). We present and briefly analyze the results of the numerical experiment with high spatial resolution $1/8°$ along latitude, $1/12°$ along longitude, and with 21 vertical sigma levels.

Key words: Ocean dynamics, monsoon circulation, numerical methods, sigma coordinate, splitting technique.

1. Introduction

The monsoon atmosphere circulation is a key process of the natural environment in the Indo-Asian region (SINGH *et al.*, 1990; SHUKLA and PAOLINO, 1983). The peculiarities of the monsoon circulation affect all aspects of life in the countries situated on the coast of the Indian Ocean. The monsoon precipitation regime above India, the time at which the rainy season begins, rainfall intensity, and the duration of precipitation are the main indicators of the forecast which are necessary for agriculture and industry of India (RAJEEVAN, 2003).

A great amount of heat and moisture, which enters the atmosphere from the surface of the Indian Ocean, defines the peculiarities of the summer southwest monsoon (SINGH *et al.*, 1983). The circulation of the Indian Ocean redistributes heat in the upper ocean layer and consumes the wind energy of the atmosphere and

The work was supported by the Russian Foundation for the Basic Research (03-05-64354, 02-05-64909) and by the Russian Academy of Sciences (10002-251/OMN-03/026-020/240603-807).

[1] Institute of Numerical Mathematics RAS, Moscow, Russia

therefore it is of vital importance in the formation and maintenance of the monsoon regime.

To know the structure and the variability of such characteristics of the ocean as sea-surface temperature, heat storage in the upper mixed layer, and currents is necessary for predicting the monsoon circulation. The adequate simulation of the interaction of the atmosphere and the Indian Ocean calls for the development of the system of oceanic observational data assimilation (BARNIER *et al.*, 1994; WENZEL *et al.*, 2001), in the on-line operation as well. When solving the above problem it is impossible to do without modern hydrodynamical models of the circulation of the Indian Ocean.

The aim of the present work is to develop efficient numerical methods of predicting the ocean dynamics. These methods are used for calculating the monsoon circulation in the Indian Ocean, which is characterized by the unique seasonable cycle and complex spatial and temporal variability (SHANKAR *et al.*, 2002). A dramatic peculiarity of the north Indian Ocean is that its currents are radically changed under the action of variable winds of summer and winter monsoons. The observational data show that most currents in the north Indian Ocean reverse their direction from winter to summer (SHANKAR *et al.*, 2002). For the adequate simulation of the complex dynamics of the Indian Ocean and the peculiarities of its eddy structure it is necessary to use models with high spatial resolution, which are physically complete and numerically efficient. We dwell on two aspects of numerical simulation of the ocean dynamics. These are the development of an efficient numerical technique and its application to the simulation of the complex dynamics of the Indian Ocean with high spatial resolution.

OGC models are extremely complex, developing systems. They are based on nonlinear differential equations describing the evolution of three-dimensional velocity, temperature, salinity fields as well as pressure and density. Two main parts can be singled out in the operator of the system of ocean dynamics equations. The first one is the classical established basis, viz. a subsystem describing the dynamics of rotating fluid in the framework of approximations traditional in oceanology (BRYAN, 1969; GILL, 1982; MARCHUK and SARKISYAN, 1988). The second one includes physical parameterizations of various kinds, which change as we gain a better understanding of natural phenomena (GRIFFIES *et al.*, 2000). On this basis we use the decomposition of the problem operator i.e., the splitting method by physical processes as a building block to construct the model and develop efficient numerical methods for solving it. On physical grounds we select the main parts of the operator and then perform their numerical treatment independently of one another. Here we present this line of investigation. We give considerable attention to the description of general methodology of the model construction and the methods of solving the classical part of the OGC equations, we do not dwell on subgrid parameterization. The approach proposed is applied to the solution of the problem of the dynamics of the Indian Ocean. We present and briefly analyze the results of the numerical simulation of the seasonal cycle of the Indian Ocean circulation with high spatial resolution 1/8° along latitude, 1/12° along longitude, and with 21 vertical levels.

Numerical treatment of the monsoon circulation is performed on the basis of one version of the model of ocean dynamics, which was developed at the Institute of Numerical Mathematics (ZALESNY, 1996; DIANSKY et al., 2002). The model is based on primitive equations in the Boussinesq, hydrostatics, and "rigid lid" approximations, which are written at the bottom following the σ-coordinate system. In the model the horizontal components of the velocity vector, potential temperature, and salinity are prognostic variables, while the vertical velocity and pressure are diagnostic ones.

The main peculiarity of the model, which distinguishes it from the other ocean models (see the review by GRIFFIES et al., 2000), is that the numerical technique is based on the splitting method by physical processes and space coordinates (MARCHUK, 1980, 1988). To this end, ocean model equations are written in special symmetrized form. The form of the equations is chosen so that it is convenient to represent the operator of the differential problem as a sum of simpler operators, each being nonnegative in the norm defined by the law of conservation of total energy. This enables one to split the operator of the complete problem into a set of simpler operators and construct spatial approximations of the corresponding groups of terms (in different equations) so that the "energy" relation (the conservation law) which holds for the original differential problem should hold for all the splitted discrete problems.

Splitting of model equations is performed at several levels. The macro-level of splitting is splitting of three-dimensional equations by physical processes. At higher levels the process of splitting selects the simplest locally one-dimensional (with respect to the space) equations. For example, the transport-diffusion equation for the tracer is solved along separate coordinates.

The outline of this paper is as follows. In Section 2 we discuss the key features of our approach to the construction and implementation of the OGC model, which is based on the splitting method. The essence of the method is illustrated by simple examples. In Section 3 we formulate the OGC equations, describe the architecture of the splitted model, transformations of the equations at separate splitting stages: their symmetrization and regularization. Particular emphasis is placed upon the choice of the special symmetrized form of ocean dynamics equations in the σ-coordinate system. We provide the form of equations, which allows one to diminish the error of approximation of horizontal pressure gradients on the given profile of vertical density stratification. In Section 4 we discuss the performance and results of the numerical experiment on simulation of the monsoon circulation in the north Indian Ocean with high spatial resolution 1/8°×1/12°×21 (steps along latitude, longitude, and the number of σ-levels along the vertical, respectively). In Section 5 we formulate the main conclusions.

2. *The Splitting Method as a Methodological Basis for the Construction of a Numerical Model of a Complicated Physical Process*

The key points of the approach proposed are as follows.

- The methodological basis for the construction of numerical models of different complexity levels is the splitting method.
- The splitting method can be considered not only as a cost-effective method of integrating the complex OGC problem with respect to the time but as the basis for the construction of the hierarchical model system as well.
- In the framework of the unified approach there can be constructed a particular model of ocean dynamics of a different complexity: from the point of view of its physical completeness, dimension, and spatial resolution.
- The splitting method is defined for solving systems of equations with nonnegative operators. This property is established *a priori* for the differential problem considered. We find an integral invariant or a conservation law which holds in the model in the absence of external sources and internal energy sinks.
- When using the splitting method the form of a differential problem is of great importance. The most convenient form of equations is their symmetrized form. By the symmetrized form we mean the form of equations, which satisfies the conditions:

  — the symmetrized form gives the form of the adjoint operator, which is close to the original one,
  — this form leads to the finite difference approximation retaining the main properties typical of original differential operators (symmetry, skew-symmetry, nonnegativeness),
  — from the form naturally follows the splitting of the problem operator into the sum of simple nonnegative operators.

- The key point of the construction of a splitted hierarchical model system and the method of its solution is the decomposition of the original problem into the set of simple subproblems with nonnegative operators.

The choice of this splitting is frequently nontrivial and not unique (MARCHUK *et al.*, 1987). The splitting process reduces to the choice of a set of separate problems of simpler structure. The established conservation law holds for every selected problem. Several levels of different depth can be selected in splitting. The splitting macro-level is based on splitting by physical processes. The simplest one-dimensional (with respect to space) problems can be selected at higher levels.

- On splitting the problem at a macro-level the transformation of the problem at some stage can be required. This can be, for example, *filtration* (simplification) of equations and *regularization* i.e., the inclusion of additional terms which can improve the numerical algorithm.

- Software and algorithms for solving splitted problems.

  On regularizing the splitting stages the question of the choice of a method for solving the problem at the selected stage arises. When choosing a space-approximation technique the property of the selected problem should be taken into account. Different problems (at some stages) can call for different approximation techniques and solvers. In general, the joint model can combine finite-difference schemes and finite-element ones; some problems can be approximated with a higher order of accuracy and so on.
- Module principle and the model software.

The natural property of the splitted model is its module principle: a separate problem — a separate module. The joint model can be "composed" of the different number of modules. The computational characteristics of the model can be improved by changing separate computational modules. Mathematical aspects of the splitting method and its application to the solution of a wide class of physical problems are presented in SAMARSKII (1962), YANENKO (1967), and MARCHUK, (1980, 1988).

The essence of the method is the following. Suppose there is the nonstationary problem

$$\frac{\partial \varphi}{\partial t} + A\varphi = f, \qquad t \in (0, T],$$
$$\varphi = \varphi^0 \qquad t = 0 \tag{2.1}$$

where A is a nonnegative operator which can be represented as superposition of simpler operators $A_i (i = 1, 2, 3, ..., I)$:

$$A = \sum_{i=1}^{I} A_i \quad A_i \geq 0, \ \forall i.$$

To solve (2.1) we use the following method. We reduce the solution of the original problem with the complex operator A to the solution of a set of problems with simpler operators A_i. For example, if $A = A_1 + A_2$, we can use the following two-cycle splitting scheme (MARCHUK, 1988; MARCHUK and SARKISYAN, 1988) to solve the problem (2.1):

$$\left(E + \frac{\tau}{2}A_1\right)\varphi^{j-1/2} = \left(E - \frac{\tau}{2}A_1\right)\varphi^{j-1},$$
$$\left(E + \frac{\tau}{2}A_2\right)\varphi^{j} = \left(E - \frac{\tau}{2}A_2\right)\varphi^{j-1/2}, \tag{2.2}$$
$$\bar{\varphi} = \varphi^{j} + 2\tau f^{j},$$
$$\left(E + \frac{\tau}{2}A_2\right)\varphi^{j+1/2} = \left(E - \frac{\tau}{2}A_2\right)\bar{\varphi},$$
$$\left(E + \frac{\tau}{2}A_1\right)\varphi^{j+1} = \left(E - \frac{\tau}{2}A_1\right)\varphi^{j+1/2}, \quad j = 1, 2, ..., J - 1, \quad \varphi^0 = \varphi(t = 0), \tau = T/J.$$

The scheme (2.2) is absolutely stable and approximates (2.1) with the second order of accuracy with respect to time, provided $\frac{\tau}{2}\|A_i\| < 1$.

To solve (2.1) we can also use the simple implicit splitting scheme:

$$\bar{\varphi} = \varphi^j + \tau f^j,$$
$$(E + \tau A_1)\varphi^{j+1/2} = \bar{\varphi}, \tag{2.3}$$
$$(E + \tau A_2)\varphi^{j+1} = \varphi^{j+1/2}.$$

The scheme (2.3) is absolutely stable and approximates (2.1) with the first order of accuracy with respect to time. It is more cost-effective than (2.2) but less accurate with respect to time.

The splitting method can be used more widely; on its basis we can develop a numerical model of a complicated process. We can improve the original model by including additional splitting stages into (2.2) or (2.3). We can change the original model. For example, on splitting the problem into a chain of subproblems we can change (simplify and/or regularize) the problem at some stage.

We present a simple example. Assume that when solving some problem we use the procedure of filtering out high-frequency harmonics along the x-coordinate from the solution. To this end, at each time step j we recalculate the vector solution φ^j by the formula

$$\bar{\varphi}_i = \left(\varphi^j_{i+1} + 2\varphi^j_i + \varphi^j_{i-1}\right)/4.$$

Writing this formula as

$$(\bar{\varphi}_i - \varphi^j_i)/\tau = \frac{h^2}{4\tau}\left(\varphi^j_{i+1} - 2\varphi^j_i + \varphi^j_{i-1}\right)/h^2,$$

we see that the filtering procedure can be considered as the inclusion of an additional splitting stage. At the additional stage we solve the diffusion equation by the explicit scheme:

$$\frac{\partial \varphi}{\partial t} = \mu^\tau \varphi_{xx},$$

where μ^τ is the coefficient of computational viscosity:

$$\mu^\tau = \frac{h^2}{4\tau}.$$

It is not difficult to show that if there is viscosity with the coefficient no less than μ^τ at one of the splitting stages implemented implicitly, then the numerical scheme is absolutely stable.

3. Mathematical Model of Ocean Dynamics

We present the mathematical formulation of the ocean dynamics problem. In the spherical coordinates (λ, θ, z) we have (BRYAN, 1969; ZALESNY, 1996)

$$\frac{du}{dt} - (l - m \cdot \cos\theta \cdot u)v = -\frac{m}{\rho_0}\frac{\partial p}{\partial\lambda} + \frac{\partial}{\partial z}\nu_u\frac{\partial u}{\partial z} + F^u,$$

$$\frac{dv}{dt} + (l - m \cdot \cos\theta \cdot u)u = -\frac{n}{\rho_0}\frac{\partial p}{\partial\theta} + \frac{\partial}{\partial z}\nu_v\frac{\partial v}{\partial z} + F^v,$$

$$\frac{\partial p}{\partial z} = g\rho_w, \tag{3.1}$$

$$m\left[\frac{\partial u}{\partial\lambda} + \frac{\partial}{\partial\theta}\left(\frac{n}{m}v\right)\right] + \frac{\partial w}{\partial z} = 0,$$

$$\frac{dT}{dt} = \frac{\partial}{\partial z}\nu_T\frac{\partial T}{\partial z} + F^T,$$

$$\frac{dS}{dt} = \frac{\partial}{\partial z}\nu_s\frac{\partial S}{\partial z} + F^S,$$

$$\rho_w = \rho_w(T, S, p), \quad \text{in} \quad D(\lambda, \theta, z)$$

where

$$\frac{d}{dt} = \frac{\partial}{\partial t} + mu\frac{\partial}{\partial\lambda} + nv\frac{\partial}{\partial\theta} + w\frac{\partial}{\partial z},$$

$$F^* = m^2\frac{\partial}{\partial\lambda}\mu_*\frac{\partial^*}{\partial\lambda} + mn\frac{\partial}{\partial\theta}\mu_*\frac{n}{m}\frac{\partial^*}{\partial\theta}.$$

The system of equations (3.1) is considered on the time interval $(0,t]$ in the three-dimensional domain D. The domain D is bounded by the boundary ∂D which consists of the undisturbed sea surface $z = 0$, the lateral (coastal) surface $\sum$, and the bottom relief $H(\lambda, \theta)$.

The corresponding boundary and initial conditions are added to the above system of equations. In particular, along the vertical coordinate we have for $z = 0$:

$$\nu_u\frac{\partial u}{\partial z} = -\frac{\tau_1}{\rho_0}, \quad \nu_v\frac{\partial v}{\partial z} = -\frac{\tau_2}{\rho_0}, \quad w = 0, \tag{3.2}$$

$$\nu_T\frac{\partial T}{\partial z} = D_T(T_S - T) + Q_T, \quad \nu_S\frac{\partial S}{\partial z} = D_S(S_0 - S) + Q_S,$$

for $z = H(\lambda, \theta)$:

$$w = m\frac{\partial H}{\partial\lambda}u + n\frac{\partial H}{\partial\theta}v. \tag{3.3}$$

On the lateral surface $\sum$ we give the no-slip condition and the conditions of no heat and salt fluxes at the bottom and on $\sum$.

We give initial conditions for $t = 0$:

$$u = u^0, \quad v = v^0, \quad T = T^0, \quad S = S^0. \tag{3.4}$$

Here λ is the longitude, $\theta = 90 + \psi$, where ψ is the latitude, z is the vertical downward coordinate, (u,v,w) is the velocity field, T is potential temperature, S is salinity, p is pressure, ρ_w is seawater density which is the known function of potential temperature, salinity, and pressure ; the terms $F^u, \ldots, F^S$ describe the horizontal turbulent transport; v_u, v_v, v_T, v_S are the coefficients of vertical turbulent diffusivity; $\mu_u, \mu_v, \mu_T, \mu_S$ are the corresponding coefficients of horizontal diffusivity, l is the Coriolis parameter: $l = -2\Omega \cos \theta, \quad m = \frac{1}{r \sin \theta}, \quad n = \frac{1}{r}$, r is the radius of the earth.

The procedure of constructing the numerical model and the algorithm for its numerical treatment includes several successive steps (ZALESNY, 1996). We outline the procedure.

3.1 The Ocean Dynamics Equation in the σ-coordinate System

The first step of the transformations consists in introducing the σ-coordinate system. The σ-transformation introduced into atmospheric models by Phillips more than 40 years ago (PHILLIPS, 1957) has found wide application in solving the problems of meteorology and oceanology (WASHINGTON and PARKINSON, 1986; SINGH *et al.*, 1995; HAIDVOGEL and BECKMANN, 1999). Using this approach, we rewrite (3.1)–(3.3) in the new system $(\lambda_1, \theta_1, \sigma) : \lambda_1 = \lambda, \theta_1 = \theta, \sigma = z/H(\lambda, \theta)$. In this case we have

$$\frac{\partial}{\partial \lambda} = \frac{\partial}{\partial \lambda_1} - \frac{1}{H} \frac{\partial H}{\partial \lambda_1} \frac{\partial}{\partial \sigma}, \quad \frac{\partial}{\partial \theta} = \frac{\partial}{\partial \theta_1} - \frac{1}{H} \frac{\partial H}{\partial \theta_1} \frac{\partial}{\partial \sigma}, \quad \frac{\partial}{\partial z} = \frac{1}{H} \frac{\partial}{\partial \sigma}.$$

The operators of turbulent transport and boundary conditions are written accordingly.

In the σ-system the continuity equation takes the form

$$m \left[\frac{\partial Hu}{\partial \lambda_1} + \frac{\partial}{\partial \theta_1} \left(\frac{n}{m} Hv \right) \right] + \frac{\partial w_1}{\partial \sigma} = 0, \tag{3.5}$$

where w_1 is the new vertical velocity

$$w_1 = w - m\sigma \left[\frac{\partial H}{\partial \lambda_1} u + \frac{n}{m} \frac{\partial H}{\partial \theta_1} v \right]. \tag{3.6}$$

The boundary condition for the new vertical velocity at the bottom, for $\sigma = 1$, is the same as at the surface:

$$w_1 = 0. \tag{3.7}$$

3.2 The Total Energy Conservation Law and Macro-splitting of the Problem

We write the hydrostatic equation as

$$\frac{\partial p}{\partial z} = g(\rho + \delta\rho),$$ (3.8)

$$\delta\rho = \rho_w - \rho,$$

where ρ is potential density. Unlike the density ρ_w, it does not depend on pressure. If all the terms describing turbulent transport processes are neglected as well as the last term $\delta\rho$ in (3.8), the total energy conservation law holds:

$$\frac{\partial}{\partial t}\int_D \left[H\rho_0\frac{u^2 + v^2}{2} - \sigma gH^2\rho\right]dD = 0.$$ (3.9)

The second step of the transformations consists in splitting the system of equations (3.1)–(3.3) by physical processes. Allowing for the conservation law (3.9), we single out three energy-independent splitting stages.

As the first subsystem we select equations describing the transport diffusion of momentum, taking into account metric terms. Dropping the subscripts on the variables λ_1, θ_1, we have

$$\frac{du}{dt} + m \cdot \cos\theta \cdot u \cdot v = \frac{1}{H^2}\frac{\partial}{\partial\sigma}v_u\frac{\partial u}{\partial\sigma} + \mathbf{F}_1^u,$$ (3.10)

$$\frac{dv}{dt} - m \cdot \cos\theta \cdot u \cdot u = \frac{1}{H^2}\frac{\partial}{\partial\sigma}v_v\frac{\partial v}{\partial\sigma} + \mathbf{F}_1^v.$$

As the second subsystem we select the equation of turbulent heat and salt exchange. We have

$$\frac{\partial T}{\partial t} = \frac{1}{H^2}\frac{\partial}{\partial\sigma}v_T\frac{\partial T}{\partial\sigma} + F_1^T,$$ (3.11)

$$\frac{\partial S}{\partial t} = \frac{1}{H^2}\frac{\partial}{\partial\sigma}v_s\frac{\partial S}{\partial\sigma} + F_1^S.$$

In (3.10), (3.11) F_1^* are the terms describing the turbulent exchange in the σ-coordinate system. At the third stage we have equations of adjustment of velocity and density fields. We consider this stage in greater detail below.

3.3 Symmetrization of Systems of Equations

The next step of the transformations consists in symmetrizing the obtained systems of equations.

We present a simple example of symmetrization of the transport equation for the tracer φ in the nondivergent flow field. Three forms of the transport equation are known: conventional, divergent, and semidivergent. Using the semidivergent form, we have

$$\frac{H}{m}\frac{\partial\varphi}{\partial t} + \frac{1}{2}\left[Hu\frac{\partial\varphi}{\partial\lambda} + \frac{\partial}{\partial\lambda}(Hu\varphi) + \frac{n}{m}Hv\frac{\partial\varphi}{\partial\theta} + \frac{\partial}{\partial\theta}\left(\frac{n}{m}Hv\varphi\right) + \frac{w_1}{m}\frac{\partial\varphi}{\partial\sigma} + \frac{\partial}{\partial\sigma}\left(\frac{w_1}{m}\varphi\right)\right] = 0.$$

(3.12)

From a computational standpoint, the semidivergent form (3.12) has the following useful properties:

- this form admits simple finite-difference approximation retaining the skew-symmetry property (MARCHUK, 1980),
- using this form, it is easy to obtain the decomposition of the operator of the problem into the sum of three simple nonnegative transport operators along the coordinates λ, θ, σ,
- the operator of the adjoint equation coincides with the original one.

This raises up the question: In what form is it convenient to write the equations at the stage of adjustment of velocity and potential density fields? With the above example in mind, we discuss this question in more detail.

3.4 Symmetrization of Equations of Adjustment of Velocity and Density Fields

Neglecting $\delta\rho$ in the hydrostatic equation for simplicity, at the adjustment stage we have

$$\frac{\partial u}{\partial t} - lv + \frac{m}{\rho_0}\left[\frac{\partial p}{\partial\lambda} - \underbrace{g\frac{\partial H\sigma}{\partial\lambda}\rho}_{1}\right] = 0,$$

$$\frac{\partial v}{\partial t} + lu + \frac{n}{\rho_0}\left[\frac{\partial p}{\partial\theta} - \underbrace{g\frac{\partial H\sigma}{\partial\theta}\rho}_{2}\right] = 0,$$

$$\frac{1}{\rho_0}\left[\frac{\partial p}{\partial\sigma} - \underbrace{g\frac{\partial H\sigma}{\partial\sigma}\rho}_{3}\right] = 0,$$ (3.13)

$$m\left[\frac{\partial Hu}{\partial\lambda} + \frac{\partial}{\partial\theta}\left(\frac{n}{m}Hv\right)\right] + \frac{\partial w_1}{\partial\sigma} = 0,$$

$$\frac{H}{m}\frac{\partial\rho}{\partial t} + \underbrace{\frac{\partial}{\partial\lambda}(Hu\rho)}_{1} + \underbrace{\frac{\partial}{\partial\theta}\left(\frac{n}{m}Hv\rho\right)}_{2} + \underbrace{\frac{\partial}{\partial\sigma}\left(\frac{w_1}{m}\rho\right)}_{3} = 0$$

The adjustment equations (3.13) are written in terms of potential density. This can be done if we assume that potential density is a sufficiently smooth function of potential temperature and salinity.

To the system of equations (3.13) are added the corresponding no-normal flow conditions on the lateral boundary dD as well as the kinematical condition along the vertical:

$$w_1 = 0 \quad \text{for} \quad \sigma = 0, \, \sigma = 1. \tag{3.14}$$

If we take the inner product of the system (3.13) and the vector

$$(\rho_0 Hu, \, \rho_0 Hv, \, \rho_0 w_1, \, p, \, -gH\sigma),$$

then with allowance for the boundary conditions the law of conservation of total energy (3.9) holds. It should be noted that the form of the terms depending on density in the first three equations of (3.13) is consistent with the divergent form of the equation for potential density. The terms marked by identical numbers are in pairs energetically neutral.

There exist several different forms of adjustment equations (3.13), which are due to different representations of the terms depending on potential density.

Now we write the equations in more general form. Assume $f = f(\sigma H)$ is some known smooth function of the vertical coordinate $z \equiv \sigma H$ and $f' \equiv \frac{df}{d(\sigma H)} \neq 0$ is its derivative. We write the adjustment equations as

$$\frac{\partial u}{\partial t} - lv + \frac{m}{\rho_0} \frac{\partial \tilde{p}}{\partial \lambda} = \frac{mg}{2\rho_0} \left[\frac{\rho}{f'} \frac{\partial f}{\partial \lambda} - f \frac{\partial}{\partial \lambda} \left(\frac{\rho}{f'} \right) \right],$$

$$\frac{\partial v}{\partial t} - lu + \frac{n}{\rho_0} \frac{\partial \tilde{p}}{\partial \theta} = \frac{ng}{2\rho_0} \left[\frac{\rho}{f'} \frac{\partial f}{\partial \theta} - f \frac{\partial}{\partial \theta} \left(\frac{\rho}{f'} \right) \right],$$

$$\frac{1}{\rho_0} \frac{\partial \tilde{p}}{\partial \sigma} = \frac{g}{2\rho_0} \left[\frac{\rho}{f'} \frac{\partial f}{\partial \sigma} - f \frac{\partial}{\partial \sigma} \left(\frac{\rho}{f'} \right) \right], \tag{3.15}$$

$$m \left[\frac{\partial Hu}{\partial \lambda} + \frac{\partial}{\partial \theta} \left(\frac{n}{m} Hv \right) \right] + \frac{\partial w_1}{\partial \sigma} = 0,$$

$$\frac{H}{m} \frac{\partial}{\partial t} \left(\frac{\rho}{f'} \right) + \frac{1}{2} \left[Hu \frac{\partial}{\partial \lambda} \left(\frac{\rho}{f'} \right) + \frac{\partial}{\partial \lambda} \left(Hu \frac{\rho}{f'} \right) + \frac{n}{m} Hv \frac{\partial}{\partial \theta} \left(\frac{\rho}{f'} \right) + \frac{\partial}{\partial \theta} \left(\frac{n}{m} Hv \frac{\rho}{f'} \right) \right.$$

$$+ \frac{w_1}{m} \frac{\partial}{\partial \sigma} \left(\frac{\rho}{f'} \right) + \frac{\partial}{\partial \sigma} \left(\frac{w_1 \, \rho}{m \, f'} \right) \right] = \frac{\rho}{f} \left[Hu \frac{\partial}{\partial \lambda} \left(\frac{f}{f'} - \sigma H \right) \right.$$

$$\left. + \frac{n}{m} Hv \frac{\partial}{\partial \theta} \left(\frac{f}{f'} - \sigma H \right) + \frac{w_1}{m} \frac{\partial}{\partial \sigma} \left(\frac{f}{f'} - \sigma H \right) \right]$$

where

$$\tilde{p} = p - \frac{g}{2} \frac{\rho}{f'} f.$$

Choosing the form of the function f, we can obtain different forms of model equations.

One of the difficulties in using σ-models of ocean dynamics is associated with the presence of the truncation error of horizontal pressure gradients

(WASHINGTON and PARKINSON, 1986; GRIFFIES *et al.*, 2000). In our case, these are the right-hand sides of the first three equations (3.15) (the terms depending on potential density).

In the geopotential z-system, if density does not depend on the horizontal coordinates λ and θ, motion does not occur. In the σ-system, due to the approximation error of pressure gradients along the surface $\sigma = $ const nonzero velocities occur. With pronounced density stratification along the vertical and with large gradients of the bottom relief, these fictitious velocities can be significant. Choosing the function f in the special way, we can reduce this effect.

As an illustration we give the following examples. Let potential density depend only on the vertical coordinate $z \equiv \sigma H$ and satisfy

$$\rho = \sigma H.$$

It is easily seen that if we choose f also as

$$f = \sigma H,$$

the right-hand side of the first three equations (3.15) vanishes. It means that the horizontal pressure gradients do not generate in σ-coordinate system artificial velocities.

Assume now that potential density satisfies the condition

$$\rho = \rho_0 e^{\alpha \sigma H}.$$

In this case, we can eliminate errors in the pressure gradients choosing f as

$$f^2 = \rho_0 e^{\alpha \sigma H}.$$

Finally, presume that potential density is an arbitrary function $\tilde{\rho}(\sigma H)$ or it can be approximated by this function with high accuracy. Then choosing the function f as

$$f^2 = 2 \int \tilde{\rho}(\sigma H) d(\sigma H),$$

we can see that for $\rho = \tilde{\rho}(\sigma H)$ the terms in the right-hand sides of (3.15) vanish. Note that with usual finite-difference approximation of the equation of motion this property is satisfied using the staggered grid C in the spatial variables λ, θ, σ.

3.5 Splitting of Adjustment Equations

The adjustment stage is the most tedious stage of calculations. To increase the computational efficiency at this stage we can use further splitting of equations. We demonstrate the splitting procedure at the adjustment stage using, as an example, the symmetrized forms of equations which are consistent with the semidivergent form of the equations for density, i.e., we put $f = \sigma H$ in (3.15).

In this case, at the first internal splitting stage we have

$$\frac{\partial u}{\partial t} = -\frac{mg}{2\rho_0}\left[\sigma H \frac{\partial \rho}{\partial \lambda} - \frac{\partial H\sigma}{\partial \lambda}\rho\right],$$

$$\frac{\partial v}{\partial t} = 0, \tag{3.16}$$

$$\frac{H}{m}\frac{\partial \rho}{\partial t} + \frac{1}{2}\left[\frac{\partial}{\partial \lambda}(Hu\rho) + Hu\frac{\partial}{\partial \lambda}\rho\right] = 0.$$

At the second internal splitting stage

$$\frac{\partial u}{\partial t} = 0,$$

$$\frac{\partial v}{\partial t} = -\frac{ng}{2\rho_0}\left[\sigma H \frac{\partial \rho}{\partial \theta} - \frac{\partial H\sigma}{\partial \theta}\rho\right], \tag{3.17}$$

$$\frac{H}{m}\frac{\partial \rho}{\partial t} + \frac{1}{2}\left[\frac{\partial}{\partial \theta}\frac{(nHv\rho)}{m} + \frac{n}{m}Hv\frac{\partial}{\partial \theta}\rho\right] = 0$$

Finally, at the third stage

$$\frac{\partial u}{\partial t} - lv + \frac{m}{\rho_0}\frac{\partial p_1}{\partial \lambda} = 0,$$

$$\frac{\partial v}{\partial t} + lv + \frac{n}{\rho_0}\frac{\partial p_1}{\partial \theta} = 0,$$

$$\frac{\partial p_1}{\partial \sigma} = -\frac{g}{2}\left[\sigma H \frac{\partial \rho}{\partial \sigma} - \frac{\partial H\sigma}{\partial \sigma}\rho\right], \tag{3.18}$$

$$m\left[\frac{\partial Hu}{\partial \lambda} + \frac{\partial}{\partial \theta}\left(\frac{n}{m}Hv\right)\right] + \frac{\partial w_1}{\partial \sigma} = 0,$$

$$\frac{H}{m}\frac{\partial \rho}{\partial t} + \frac{1}{2}\left[\frac{\partial}{\partial \sigma}\left(\frac{w_1}{m}\rho\right) + \frac{w_1}{m}\frac{\partial \rho}{\partial \sigma}\right] = 0,$$

$$p_1 = p - \frac{g}{2}\sigma H\rho.$$

Thus, at the first and second splitting stages we arrive at the solution of locally one-dimensional problems along the coordinates λ and θ, while at the third stage - at the more complex three-dimensional problem.

To equations (3.18) are added the boundary no-normal flow conditions. A peculiarity of the formulation of the initial boundary value problem for (3.18), which is typical of hydrodynamic equations, is that there are no boundary conditions for pressure. This leads to additional difficulties when solving numerically the problem, in particular, when calculating pressure along the vertical. In this case, we can use once again additional splitting involving the selection of the vertical-averaged motion component. To this end we write equations (3.18) in the equivalent form from which the solution algorithm naturally follows. We have

$$\frac{\partial u}{\partial t} - l(v - \bar{v} + \tilde{v}) + \frac{m}{\rho_0} \frac{\partial p_1'}{\partial \lambda} + R\bar{u} = -\frac{m}{\rho_0} \frac{\partial \bar{p}_1}{\partial \lambda} + R\bar{u},$$

$$\frac{\partial v}{\partial t} + l(u - \bar{u} + \tilde{u}) + \frac{n}{\rho_0} \frac{\partial p_1'}{\partial \theta} + R\bar{v} = -\frac{n}{\rho_0} \frac{\partial \bar{p}_1}{\partial \theta} + R\bar{v},$$

$$\frac{\partial H\bar{u}}{\partial \lambda} + \frac{\partial}{\partial \theta} \left(\frac{n}{m} H\bar{v}\right) = 0, \tag{3.19}$$

$$m\left[\frac{\partial H(u - \bar{u})}{\partial \lambda} + \frac{\partial}{\partial \theta} \left(\frac{n}{m} H(v - \bar{v})\right)\right] + \frac{\partial w_1}{\partial \sigma} = 0,$$

$$\frac{H}{m} \frac{\partial \rho}{\partial t} + \frac{1}{2} \left[\frac{\partial}{\partial \sigma} \left(\frac{w_1}{m} \rho\right) + \frac{w_1}{m} \frac{\partial \rho}{\partial \sigma}\right] = 0,$$

where

$$\bar{a} = \int_0^1 a \, d\sigma, (a = u, v), \quad \bar{p}_1 = p_1(0) - \frac{g}{2} \int_0^1 d\sigma \int_0^\sigma \left(\sigma H \frac{\partial \rho}{\partial \sigma} - \frac{\partial \sigma H}{\partial \sigma} \rho\right) d\sigma, \quad p_1' = p_1 - \bar{p}_1,$$

R is some nonnegative function, for example $R = \text{const} \equiv \varepsilon, \ 0 \leq \varepsilon << 1$.

Using the above representation, we develop the solution algorithm. It consists in solving the following three subsystems. The first subsystem is

$$\frac{\partial u}{\partial t} = R\bar{u},$$

$$\frac{\partial v}{\partial t} = R\bar{v}, \tag{3.20}$$

$$\frac{\partial \rho}{\partial t} = 0.$$

The second subsystem is

$$\frac{\partial u}{\partial t} - l\bar{v} + \frac{m}{\rho_0} \frac{\partial \bar{p}_1}{\partial \lambda} + R\bar{u} = 0,$$

$$\frac{\partial v}{\partial t} + l\bar{u} + \frac{n}{\rho_0} \frac{\partial \bar{p}_1}{\partial \theta} + R\bar{v} = 0, \tag{3.21}$$

$$\frac{\partial H\bar{u}}{\partial \lambda} + \frac{\partial}{\partial \theta} \left(\frac{n}{m} H\bar{v}\right) = 0,$$

$$\frac{\partial \rho}{\partial t} = 0.$$

The third subsystem is

$$\frac{\partial u}{\partial t} - l(v - \bar{v}) + \frac{m}{\rho_0} \frac{\partial p'_1}{\partial \lambda} = 0,$$

$$\frac{\partial v}{\partial t} + l(u - \bar{u}) + \frac{n}{\rho_0} \frac{\partial p'_1}{\partial \theta} = 0, \tag{3.22}$$

$$\frac{H}{m}\frac{\partial \rho}{\partial t} + \frac{1}{2}\left[\frac{\partial}{\partial \sigma}\left(\frac{w_1}{m}\rho\right) + \frac{w_1}{m}\frac{\partial \rho}{\partial \sigma}\right] = 0,$$

where

$$\frac{w_1}{m} = \int\limits_{\sigma}^{0}\left[\frac{\partial H(u - \bar{u})}{\partial \lambda} + \frac{\partial}{\partial \theta}\left(\frac{nH(v - \bar{v})}{m}\right)\right]d\sigma.$$

Note that by selecting depth-averaged motion we obtain the following result. In numerical calculations two boundary conditions along the vertical for the velocity w_1 are exactly satisfied, and correct calculations of the vertical structure of the pressure field are performed.

3.6 Regularization of Problems at Some Splitting Stages

The procedure of ε-regularization is frequently used to increase the stability of the numerical solution. The method of ε-regularization involves the addition of certain terms with small coefficients $\varepsilon < < 1$ to the original equation. The method of artificial compressibility, which is proposed by YANENKO (1967) for solving equations of viscous incompressible fluid, can serve as an example of the ε-regularization procedure. The idea of the artificial compressibility method is to replace the continuity equation by a nonstationary equation of the form

$$\varepsilon\frac{\partial p}{\partial t} + \frac{\partial u}{\partial x} + \frac{\partial v}{\partial y} = 0,\ 0 < \varepsilon < 1.$$

We apply the above approach to regularization of the problem (3.19). We drop the "rigid lid" condition at the undisturbed ocean surface and instead of it we use

$$w = -\frac{1}{g\rho_0}\frac{\partial p_0}{\partial t}, \tag{3.23}$$

where

$$p_0 = -g\rho_0\varsigma,$$

ς – is the ocean surface height.

From a mathematical standpoint, the transition from the model with the "rigid lid" approximation to the 'milder' dynamical condition can be considered as regularization of the two-dimensional incompressible fluid dynamics problem. In this case, the efficiency of calculations of the depth-averaged velocities increases.

Physically, the transition to the boundary condition (3.23) implies the introduction of the dynamics of external gravity waves into the model (GILL, 1982; MARCHUK *et al.*, 1987). In this case, in the chain of splitted subsystems, only the stage describing evolution of the vertical-averaged fields changes. In the terms $\bar{u}, \bar{v}, \bar{p}$ the equations have the form

$$\frac{\partial \bar{u}}{\partial t} - l \cdot \bar{v} + \frac{m}{\rho_0} \frac{\partial \bar{p}_1}{\partial \lambda} + R\bar{u} = 0,$$

$$\frac{\partial \bar{v}}{\partial t} + l \cdot \bar{u} + \frac{n}{\rho_0} \frac{\partial \bar{p}_1}{\partial \theta} + R\bar{v} = 0, \qquad (3.24)$$

$$\frac{1}{g\rho_0} \frac{\partial p_0}{\partial t} + m \left[\frac{\partial H\bar{u}}{\partial \lambda} + \frac{\partial}{\partial \theta} \left(\frac{n}{m} H\bar{v} \right) \right] = 0,$$

$$\frac{\partial \rho}{\partial t} = 0,$$

$$\bar{p}_1 = p_0 - \frac{g}{2} \int_0^1 d\sigma \int_0^\sigma \left(\sigma H \frac{\partial \rho}{\partial \sigma} - \frac{\partial \sigma H}{\partial \sigma} \rho \right) d\sigma.$$

To the system (3.24) on the closed coastal boundary $\sum$ are added the no-normal flow conditions.

Note

When writing the third subsystem of equations of adjustment of velocity and potential density fields (3.19) we added and subtracted the terms $Ru, Rv, R << 1$. On the one hand, this is an equivalent transformation of the differential system. On the other hand, using the splitting algorithm (3.20)–(3.22) to solve the problem, we can increase the efficiency of calculations of vertical-averaged motion. It is not difficult to show that the splitting scheme is absolutely stable and has the first order of accuracy with respect to time if we use an explicit scheme to solve (3.20) and an implicit one to solve (3.21). If the stage (3.20) is dropped in the hierarchical model system, this can also be considered as ε-regularization of the original problem. In this case, regularization involves the introduction of friction with the coefficient R, which acts on the depth-averaged flow component.

4. The Numerical Experiment on Simulation of Monsoon Circulation in the Indian Ocean

The method presented was used as the basis for constructing the numerical model of monsoon circulation in the Indian Ocean with high spatial resolution.

The computational domain covered the north Indian Ocean (10°S–30°N, 38°E–103°E). The spatial resolution was $1/8° \times 1/12° \times 21$. The bottom topography was interpolated to the computational grid from the five-minute data array ETOPO5. The minimum bottom depth in the domain was 7 m, the maximal one was about 6000 m.

When treating vertical-averaged flows the stage (3.20) was dropped. To solve the equations describing the adjustment of velocity and density fields we put $f = \sigma H$ in (3.15). Horizontal eddy diffusivity was 1.0×10^2 m^2/s, eddy viscosity was 1.0×10^3 m^2/s, and the friction coefficients R was 1.0×10^{-6} s^{-1}. Vertical mixing was parameterized using the scheme of PAKANOWSKI and PHILANDER (1981).

The aim of the experiment was to calculate the dynamically consistent seasonable cycle of velocity, temperature, and salinity fields of the Indian Ocean. The model was driven at the sea surface by wind-stress (NCEP reanalysis) and by prescribed temperature and salinity (LEVITUS et al., 1998). The model was spun up for seven years from a state of rest and January temperature and salinity (LEVITUS et al., 1998).

Figures 1–3 present the model instantaneous velocity fields at depths of 60 m, 400 m, and 1000 m. The calculated flows differ widely during the winter and summer monsoons, as the observational data show (TOMCZAK and GODFREY, 2003; SHANKAR et al., 2002).

We shall dwell briefly on the structure and seasonal variability of the main currents in the Indian Ocean and compare them with the observational data.

Equatorial currents. The numerical experiment shows that the North Equatorial Current is observed from January to March when the winter monsoon is fully established. The horizontal velocities are about 0.5–1.0 m/s near the coasts of Sri Lanka, in the sector between the equator and 6°N as well as between 60°E and 75°E. In the equatorial zone the westward Equatorial Counter Current with the velocities of order 0.5–0.8 m/s is observed. The current is located to the south of 2°S, and its intensity is diminished westward. According to the calculations this current does not extend farther than 70°E and merges with the westward countercurrent which is approximately equal in intensity. This pattern is in good agreement with the observational study (TOMCZAK and GODFREY 2003), except that the countercurrent is much weaker according to the data. During the months of the monsoon change the Indian Equatorial Jet with velocities 0.7 m/s and higher is pronounced. In the period of the summer monsoon the model also reproduces the Southwest Monsoon Current (TOMCZAK and GODFREY 2003) which reaches the velocities 0.5–0.8 m/s south and southeast of Sri Lanka.

Currents in the Arabian Sea and the Bay of Bengal. The model describes adequately the structure of the Somali Current as well as the appearance of the Great Whirl towards the end of the summer monsoon. In September the velocities reach 2 m/s at a depth of 30 m. In March and in January the current reverses its direction.

In summer the northeastward and southeastward Summer Monsoon Current (SHANKAR et al., 2002) flows across the Arabian Sea basin (Fig. 1). It divides into two branches off Oman at approximately 15N°. In one branch, which is situated close to the Oman coast, water moves clockwise and forms the West India Coastal Current (WICC) off India (SHANKAR et al., 2002). The second branch deviates from the coastal current into the open ocean and forms the Summer

January horizontal circulation at 60m

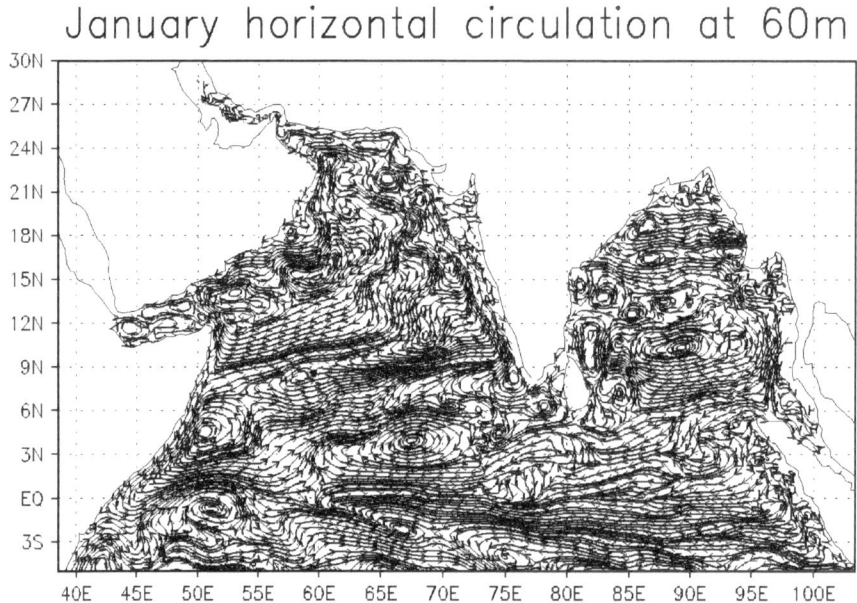

July horizontal circulation at 60m

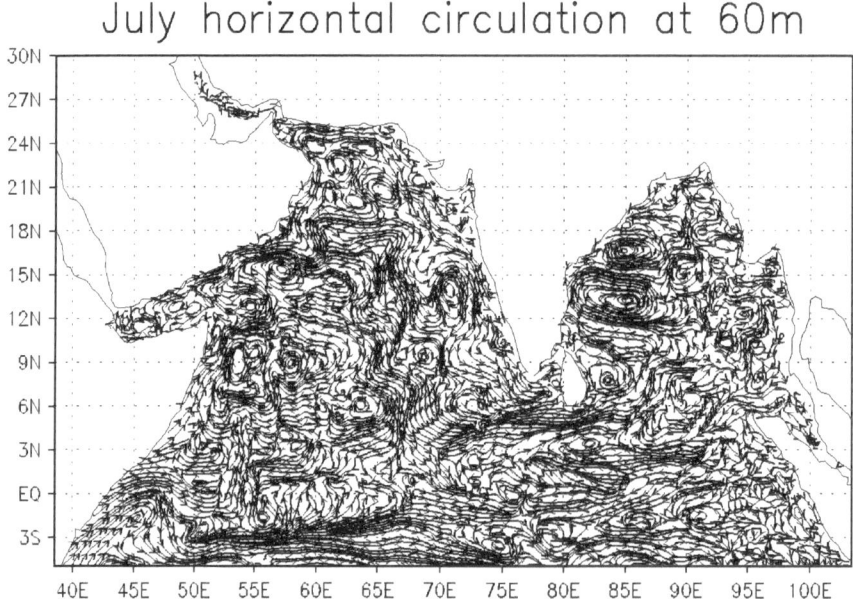

Figure 1
Model instantaneous velocity fields at depth 60 m.

Monsoon Current (SMC) (Shankar *et al.*, 2002). It should be noted that the complete pattern of currents is rather complex. There are several cyclones and anticyclones inside the Arabian Sea, between two main currents. In the period of

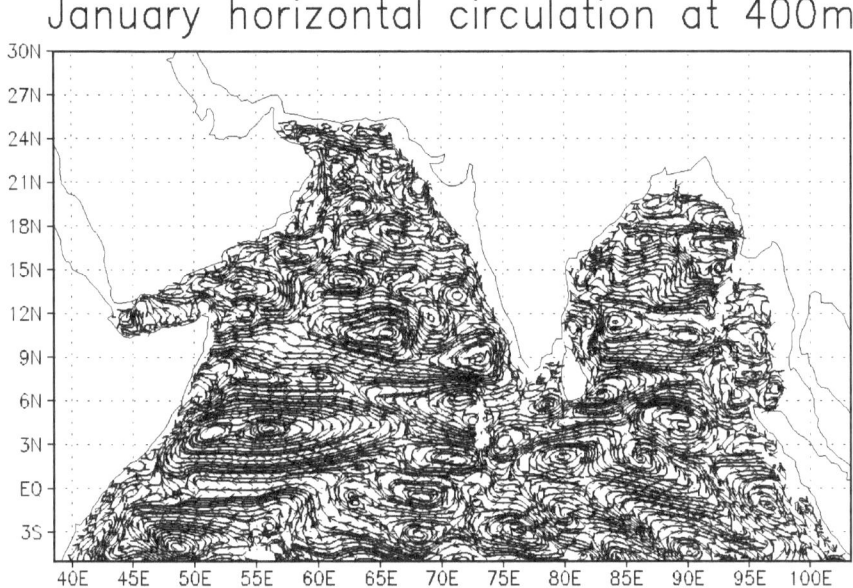

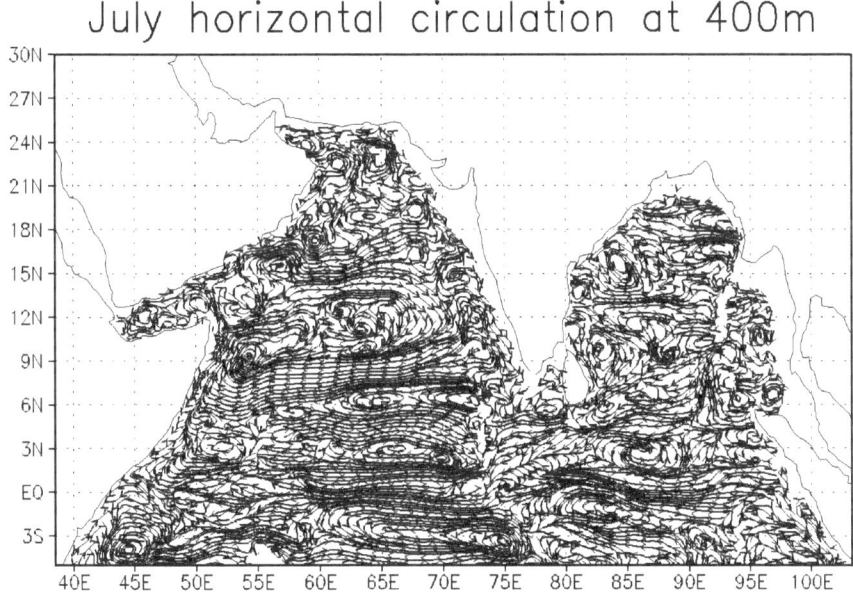

Figure 2
Model instantaneous velocity fields at depth 400 m.

the winter monsoon the structure of the velocity field is also rather complex (Fig. 1). In January the Somali Current reverses its direction. Along the entire western coastline of India there appears the coastal northwestward current in the

January horizontal circulation at 1000m

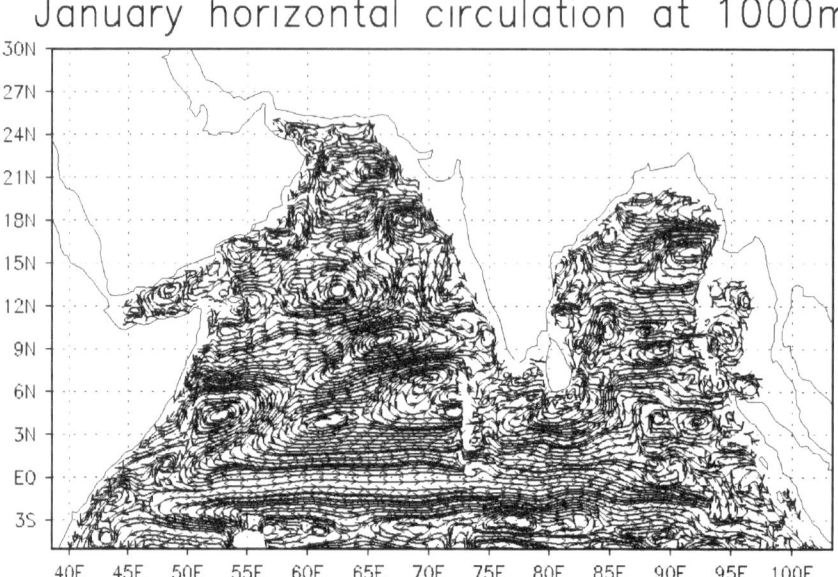

July horizontal circulation at 1000m

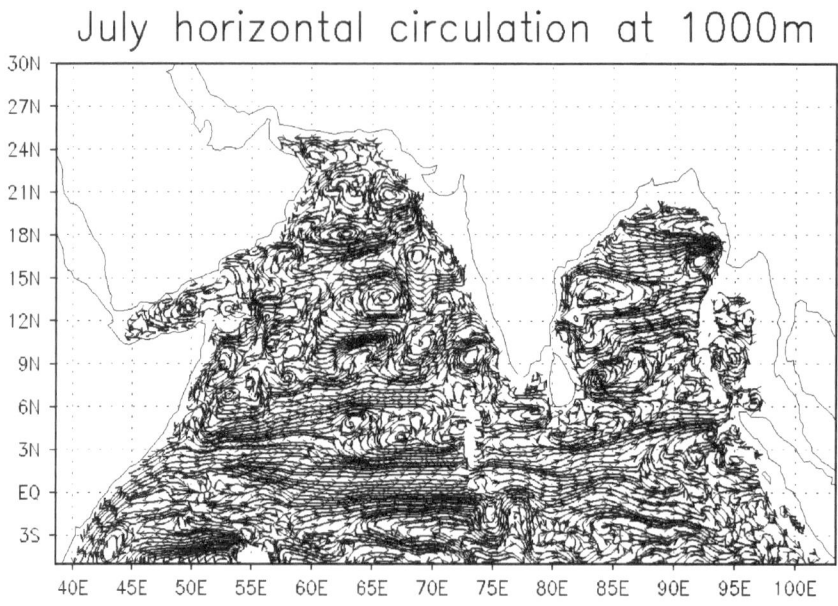

Figure 3
Model instantaneous velocity fields at depth 1000 m

opposite direction as compared to the summer period. Along most of the coastline of the Arabian Sea the currents also reverse their direction as compared to the summer season. There are many local eddies and countercurrents.

The anticyclonic circulation prevails in the Bay of Bengal during most of the year. In March-April the local, rather stable anticyclonic vorticity develops northeast of Sri Lanka. Its horizontal size is about 500 km, its thickness is of order 200 m, the velocities reach 0.8 m/s. By September the eddy intensity is diminished; at the end of October-November the circulation reverses its direction and becomes cyclonic. Several eddies are formed; their total large-scale structure may be interpreted as the East Indian Winter Jet (TOMCZAK and GODFREY 2003). The Jet velocities are of order 0.4–0.7 m/s. The structure of currents is close to the scheme constructed by the observational data (TOMCZAK and GODFREY 2003) though the pattern differs from the scheme by certain details. At 75°–77°E, south of Sri Lanka, the East Indian Winter Jet meets with the countercurrent entering into the Bay of Bengal. This countercurrent is formed farther west in the open ocean by the eastward current because of its division into two branches. The first branch turns southward and merges with the Equatorial Jet (TOMCZAK and GODFREY 2003). The second one first flows northeastward along the eastern periphery of the above cyclone, flows around it and turns southward forming the countercurrent off Sri Lanka. Water returns into the Arabian Sea, following two paths. First, as the narrow jet about 40 m in depth through the shallow strait between India and Sri Lanka. Second, along the western periphery of the cyclone off the east coast of Sri Lanka as in TOMCZAK and GODFREY (2003).

The vertical structure of currents in the north Indian Ocean is rather complex (Figs. 2 and 3). However, there is one common property: The currents are of distinct zonal character in deeper layers below 300 m. Currents are less intensive and better regulated. The experiment shows that even in the deep ocean the variability of currents with time is appreciable. However, it should be noted that variability is typical of open-ocean eddies rather than the large-scale currents themselves.

5. Conclusion

- In the paper we set forth a common approach to the construction of a numerical model of ocean dynamics which is based on splitting by physical processes and geometric coordinates. Model equations are split on several levels. The splitting macro-level is splitting of three-dimensional equations by physical processes. On higher levels the splitting process is to select the simplest equations which are locally one-dimensional with respect to space. The application of the above approach to the solution of the problem of the Indian Ocean circulation with high spatial resolution demonstrated that this numerical technique performs well.
- The peculiarity of the numerical model of the Indian Ocean dynamics is that the adjustment equations of potential density and velocity fields are written in the sigma-coordinate system in the generalized symmetrized form. This transforma-

tion allows one to decrease truncation errors occurring in horizontal pressure gradient terms in the sigma model and construct a stable computational procedure.

- A comparison of the results of the Indian Ocean simulation with the schemes of currents constructed on the basis of observational data (SHANKAR *et al.*, 2002; TOMCZAK and GODFREY 2003) shows that the model reproduces the monsoon circulation reasonably well. The high resolution enables us to reproduce not only the large-scale structure of monsoon currents, but to describe local peculiarities of its space-time variability as well. The calculations show the high eddy activity of the Indian Ocean. Numerous cyclones and anticyclones are observed in the open ocean, coastal areas, and the deep ocean. Ocean eddies can modify the structure of basin scale currents. With high spatial resolution which is accompanied by the high eddy activity, requirements for the observational data are greater. For the detailed assessment of model calculations and forecasts it is essential to have comprehensive spatial observations. Satellite observations can supply the information regarding the ocean surface; however, to have such information about deep ocean layer is an unresolved problem. The development of observing systems such as the profiling floats ARGO can in part fill the gap. However, in this case, to solve an extremely complex problem of observational data processing and assimilation in the moving coordinate system (of the type of Lagrangian coordinates) is required.

REFERENCES

BARNIER, B., CAPELLA, J., and O'BRIEN, J.J. (1994), *The Use of Satellite Scatterometer Winds to Drive a Primitive Equation Model of the Indian Ocean: The Impact of Bandlike Sampling,* J. Geophys. Res. *99,* C7, 14,187–14,196.

BRYAN, K. (1969), *A Numerical Method for the Study of the Circulation of the World Ocean,* J. Comput. Physics *4,* 347–376.

DIANSKY, N.A., BAGNO, A. V., and ZALESNY, V. B. (2002), *Sigma Model of Global Ocean Circulation and Its Sensitivity to Variations in Wind Stress.* Izvestiya, Atmospheric and Oceanic Physics (Izvestiya Rossiiskoi Akademii Nauk Fizika Atmosfery I Okeana). *38,* No. 4, 477–494.

GILL, A.E. (Ed.) *Atmosphere-Ocean Dynamics* (Academic Press, New York 1982).

GRIFFIES, S.M., BOENING C., BRYAN F.O., CHASSIGNET, E.P., GERDES, R., HASUMI, H., HIRST, A., TREGUIER, A.-M., and WEBB, D. (2000), *Developments in Ocean Climate Modelling,* Ocean Modelling *2,* 123–192.

HAIDVOGEL, D.B. and BECKMANN, A., *Numerical Ocean Circulation Modelling* (Imperial College Press, 1999).

LEVITUS, S., BOYER, T.P., CONKRIGHT, M.E., O'BRIEN, T., ANTONOV, J., STEPHENS, C., STATHOPLOS, L., JOHNSON, and D., GELFELD R. (1998), NOAA Atlas NESDIS 18, *World Ocean Database* 1998: Vol. 1: Introduction (U.S. Gov. Printing Office, Washington, D.C.) 346 pp.

MARCHUK, G.I., *Methods of Computational Mathematics* (Nauka, Moscow 1980).

MARCHUK, G.I., *Splitting-up Methods* (Nauka, Moscow 1988).

MARCHUK, G.I., DYMNIKOV, V.P., and ZALESNY, V.B., *Mathematical Models in Geophysical Hydrodynamics and Numerical Methods of their Realization* (Gidrometeoizdat, Leningrad 1987).

MARCHUK, G.I. and SARKISYAN, A.S., *Mathematical Modelling of Ocean Circulation* (Springer-Verlag, Berlin, Heidelberg, New York, London, Paris, Tokyo 1988).

PHILLIPS, N.A. (1957), *A Coordinate System Having Some Special Advantages for Numerical Forecasting*, J. Meteorology *14*, 184–185.

RAJEEVAN, M. (2003), *Prediction of Indian Summer Monsoon: Status Problems and Prospects*, Curr. Science *11*, 1451–1457.

SAMARSKII, À.À. (1962), *On Convergence of the Fractional-step Method for Heat Equation*, Zh. Vych. Mat. Mat. Fiz. *2*, 6, 1117–1121.

SHANKAR, D., VINAYACHANDRAN, P.N., UNNIKRISHNAN, A.S., and SHETYE, S.R. (2002), *The Monsoon Currents in the North Indian Ocean*, Progr. Oceanogr. *52*(1), 63–119.

SHUKLA, J. and PAOLINO, D.A. (1983), *The Southern Oscillation and Long-range Forecasting of the Summer Monsoon Rainfall in Peninsular India*, Mausam *33*, 399–404.

SINGH, M.P., MOHANTY U.C., and DUBE S.K. (1983), *A Study of Heat and Moisture Budget over the Arabian Sea and their Role in the Onset and Maintenance of Summer Monsoon*, J. Met. Soc. Japan *61*, 218.

SINGH, M.P., RAMAN, S., TEMPLEMAN, B., TEMPLEMAN, S., HOLT T., MURTHY, A.B., AGARWAAL P., NIGAM, S., PRABHU A., and AMEENULLAH S. (1990), *Structure of the Indian Southwesterly Pre-monsoon and Monsoon Boundary Layers: Observations and Numerical Simulation*, Atmosp. Environ. *24A*, 723–734.

SINGH, M.P., SHARAN, M., and YADAV, A.K. (1995), *Comparison of Various Sigma Schemes for Estimating Dispersion of Air Pollutants in Low Winds*, Atmosp. Environ. *29*, 2501–2509.

TOMCZAK, M., and GODFREY, S.J., *Regional Oceanography: An Introduction* (Pergamon 2003).

WASHINGTON, W.M., and PARKINSON, C.L., *Three-dimensional Climate Modeling* (University Science Books, Mill Valley, California; Oxford University Press, Oxford, New York 1986).

WENZEL, M., SHROETER, J., and OLBERS, D. (2001), *The Annual Cycle of the Global Ocean Circulation as Determined by 4D VAR Data Assimilation*, Progress in Oceanog. *48*, 73–119.

YANENKO, N.N., *Fractional-step Method of Solving Multidimensional Problems of Mathematical Physics* (Nauka, Novosibirsk 1967).

ZALESNY, V.B. (1996), *Numerical Simulation and Analysis of the Sensitivity of Large-scale Ocean Dynamics*, Russ. J. Numer. Anal. Math. Modelling *11*, 6, 421–443.

(Received September 30, 2003, accepted January 21, 2004)
Published Online First: May 25, 2005

 To access this journal online:
http://www.birkhauser.ch

Pure appl. geophys. 162 (2005) 1431–1454
0033–4553/05/091431–24
DOI 10.1007/s00024-005-2678-7

© Birkhäuser Verlag, Basel, 2005

❙Pure and Applied Geophysics

Seasonal Prediction for the Indian Monsoon Region with FSU Ocean-atmosphere Coupled Model: Model Mean and 2002 Anomalous Drought

A.K. Mitra,[1,2] L. Stefanova,[1] T.S.V. Vijaya Kumar,[1] and T.N. Krishnamurti[1]

Abstract—Seasonal climate prediction for the Indian summer monsoon season is critical for strategic planning of the region. The mean features of the Indian summer monsoon and its variability, produced by versions of the 'Florida State University Coupled Ocean-Atmosphere General Circulation Model' (FSUCGCM) hindcasts, are investigated for the period 1987 to 2002. The coupled system has full global ocean and atmospheric models with coupled assimilation. Four member models were created by choosing different combinations of parameterizations of the physical processes in the atmospheric model component. Lower level wind flow patterns and rainfall associated with the summer monsoon season are examined from this fully coupled model seasonal integrations. By comparing with observations, the mean monsoon condition simulated by this coupled model for the June, July and August periods is seen to be reasonably realistic. The overall spatial low-level wind flow patterns and the precipitation distributions over the Indian continent and adjoining oceanic regions are comparable with the respective analyses. The anomalous below normal large-scale precipitation and the associated anomalous low-level wind circulation pattern for the summer monsoon season of 2002 was predicted by the model three months in advance. For the Indian summer monsoon, the ensemble mean is able to reproduce the mean features better compared to individual member models.

Key words: Seasonal forecasting, coupled ocean-atmosphere model, monsoon system, numerical modeling, ensemble prediction.

1. Introduction

Summer monsoon season rainfall is vital for agricultural operations in India. Long periods of drought or flood during any summer monsoon season can have a devastating effect on human life and the country's economy. Reliable seasonal to interannual prediction of summer monsoon rainfall is required for the Indian subcontinent. Apart from agriculture, interest in seasonal prediction has grown from

---

[1] Department of Meteorology, The Florida State University, Tallahassee, Florida, 32306-4520, U.S.A. E-mail: tnk@io.met.fsu.edu

[2] Permanent affiliation NCMRWF, New Delhi, India

the commerce and industry related sectors also. In recent years, advances have been made to understand, model and predict the climate of monsoon a season in advance. Studies indicate that for tropics the seasonal mean circulation and rainfall over some regions of the globe are determined mainly by SSTs. The current state of prediction over the monsoon region is not much better than the prediction of long-term mean climate, and the skill of prediction of seasonal anomalies is very low. Indian summer monsoon rainfall is the manifestation of the dynamically coupled ocean-atmosphere system of the Indian-Pacific region. Processes in this region are continually evolving from year to year. Prediction of reasonably accurate SSTs is possible from coupled ocean-atmosphere models. Realistic simulations/forecasts of seasonal climate are best done by coupled ocean-atmosphere models since they do not rely on a prescription of future SSTs (PALMER and CO-AUTHORS, 2004). In recent years, coordinated multi-institutional scientific research projects on seasonal prediction have been launched. PROVOST (Prediction of climate variations on seasonal to interannual timescales), DSP (The Dynamical Seasonal Prediction), SMIP (Seasonal Prediction Model Intercomparison Project) of CLIVAR/WCRP, DEMETER (Development of multi-model ensemble system for seasonal to interannual prediction), PROMISE (Predictability and variability of monsoons and the agricultural and hydrological impacts of climate) and SHIVA (Studies of the hydrology, influence and variability of the Asia summer monsoon) are a few among them.

In recent years, it has become possible by global coupled ocean-atmosphere models to simulate many observed aspects of climate variability like El Nino events and monsoon interannual variability (MEEHL, *et al.*, 2000). Global coupled ocean-atmosphere general circulation models can now potentially simulate a realistic annual cycle of sea-surface temperature (SST). The annual cycle of SST in the Indian Ocean is dominated by the response to the varying heat flux. The annual cycle of surface heat flux in the Indian Ocean is due about equally to the latent and short-wave fluxes. Variations in the magnitude of the surface heat flux are more important than mixed-layer depth variations in determining the SST response to the flux of heat across the ocean surface (DEWITT and SCHNEIDER, 1999). The simulated heat flux, wind stress, upper ocean thermal structure and mixed layer depth of the ocean by the coupled model are critical to the realistic simulation of SST. Atmosphere-ocean coupling was found to play a crucial role in simulating the mean Asian summer monsoon rainfall in comparison with the results from a stand-alone AGCM (XIOUHUA *et al.*, 2002). Both local and remote air-sea interactions in the tropical Indian and Pacific Oceans contribute to better simulation of the Asian summer monsoon.

Predictability of precipitation over the Indian region is influenced by the slow SST variations related to ENSO and dependent on the regional scale fluctuations unrelated to ENSO SST (internal variability). A segment of moisture convergence in the north Indian Ocean can be attributed to the trade wind regimes of the Pacific Ocean during ENSO (FASULLO and WEBSTER, 2003). They also showed that anomalous SST gradient between the eastern equatorial and the central subtropical Pacific Ocean prior to the monsoon onset, together with its associated moisture transports, decides the monsoon pattern in relation to ENSO. The circulation in the Indian Ocean prior to and during the monsoon onset shares a strong association with the intensity of ENSO monsoon coupling. A realistic simulation of the complex interaction between nonlinear intraseasonal oscillations and the annual cycle can only produce a realistic monsoon rainfall pattern in numerical models. Most of the models have difficulty in capturing the realistic intraseasonal oscillations in monsoon. Because of the chaotic nature of the monsoon atmosphere, it is necessary to use ensembles of the global coupled climate prediction models; the so-called multi-model ensemble.

Simulating the interannual large-scale rainfall variability associated with Indian summer monsoon by AGCMs with even prescribed SSTs is a challenging task (SPERBER et al., 2001). Predicting the regional-scale rainfall variations a season ahead is even considerably more difficult. However, recent development of superensemble technique for weather and climate predictions at Florida State University (KRISH-NAMURTI et al., 2002, 2001, 2000a, 2000b) is found to be successful in producing a deterministic forecast superior to individual member models and better than the multi-model ensemble mean forecast. Use of the FSU superensemble method leads to improved forecasting skills in terms of the reduction of RMS errors and the improvement of anomaly correlation coefficients. This is also true when a probabilistic forecast is issued using the FSU superensemble technique (STEFANOVA and KRISHNAMURTI, 2002). This gives hope for improving the forecasts for the monsoon region by combined use of numerical models and statistical techniques. Before constructing the multi-model superensemble, it is necessary to evaluate the skill and capabilities of the individual member models. In this study we focus on the quality of simulations for the Indian region by the 'Florida State University Coupled Ocean-atmosphere general Circulation Model' (FSUCGCM) for the summer monsoon season.

Evolution of the sea-level pressure anomaly contrast between the Indian continent and the surrounding oceans is the governing mechanism for the monsoon flow evolution. The northward migration of the low-pressure anomaly from the Indian Ocean and the associated moisture transport by the low-level westerly jet towards India describes the evolution of the precipitation field over India and the adjoining areas. However, there exists significant interannual variability of this monsoon system. The monsoon season of 2002 was anomalously dry and the rainfall during the season was considerably below normal. In this study we have used four versions of the Florida State University Coupled Ocean-atmosphere Model, by selecting different combinations of the parameterizations of the physical processes of deep cumulus convection and radiation. By using the four versions, three months of hindcasts from 1987 to 2002 were produced every month from the respective initial conditions. Data from the coupled model forecasts and observations are used to document different mean model characteristics relevant for the simulation/prediction of Indian summer monsoon. In a mean sense, the coupled model is able to capture the large-scale features of monsoon circulation and the associated rainfall. It could also successfully capture the anomalous dry conditions of the 2002 monsoon season.

2. *FSU Coupled Model and Coupled Assimilation*

Combining the FSU atmospheric global spectral model (KRISHNAMURTI *et al.*, 1998) with a Hamburg ocean model HOPE (LATIF 1987), LAROW and KRISHNAM-URTI (1998) constructed a coupled climate model. The atmospheric part of this coupled model is a global spectral model with 14 vertical layers and is resolved by 63 horizontal waves. Earlier studies (SPERBER *et al.*, 1994) indicate that lower resolution models (T42) have problems in correctly simulating the sequences of development of synoptic-scale milestones that characterize the Indian summer monsoon flow, leading to unrealistic temporal and spatial distributions of monsoon rainfall. In our study the horizontal resolution of the model is set at T63. The atmospheric model has realistic physical processes — surface fluxes, boundary layer, deep and shallow convection, large scale condensation and radiation.

The oceanic counterpart of the coupled model is a primitive equation global model with variable meridional resolution (0.5° near the equator and decreasing to 5.0° near the northern and southern boundaries located at 70°N and 70°S, respectively). Within 10° of the equator, the meridional resolution is 0.5°. From 10° to 20° latitudes, the resolution is 1°. Near the northern and southern boundaries it becomes 5°. In the zonal direction the resolution is constant at 5.0°. The model has realistic bottom topography. The ocean model contains 17 irregularly spaced vertical levels with 10 levels located within the uppermost 300 meters.

The FSUCGCM performance as seen in SSTs of the equatorial pacific region has been documented earlier in terms of its capability to capture the 1997–1998 El Niño and the associated intraseasonal oscillation on the Madden-Julian time scale during the onset of said El Niño (KRISHNAMURTI *et al.*, 2000c; BACHIOCHI and KRISHNAM-URTI, 2000). The entire life cycle of the SST anomaly related to the 1997–1998 El Niño was captured in that 18-month long integration starting from April 1, 1997 initial conditions. A comprehensive coupled data assimilation phase has been designed for the FSUCGCM. It consists of physical initialization (KRISHNAMURTI *et al.*, 1991) with observed rainfall estimates and nudging of the SST only during this assimilation. The coupled model integrations were quite sensitive to the use of comprehensive coupled assimilation. The present modeling starts with a 10-year ocean spin-up phase where the observed winds (and related surface wind stress) drive the ocean model only. The spin-up period was from January, 1976 to May, 1986. During part of this period (1976 to 1979) the model was forced with FSU winds and climatological heating only. Then from 1980 to May 1986, additionally the model was nudged towards Reynold's SST. This ocean spin-up is followed by a coupled atmosphere ocean (and land) data assimilation phase, following LaRow and KRISHNAMURTI (1998). This coupled assimilation phase provides initial conditions for the coupled system.

From June 1986 onwards both models were integrated in coupled mode (coupled assimilation). Here the coupled assimilations continued to produce the initial conditions for the seasonal forecasts runs. ECMWF analyses were used daily for initializing the atmospheric part of the model. Physical initialization was performed once a day with observed rain to improve the air-sea fluxes. During this coupled assimilation the SSTs were relaxed towards the observed Reynold's SST once in seven days.

3. Description of Model Runs and Observed Data

Since June 1986 the above-mentioned coupled assimilation is being carried out continuously in near real time to prepare atmosphere and ocean initial conditions for the coupled system. During coupled integration, the coupling is performed every two hours (every 16th time step of the atmospheric model integration). In the present study the multimodels are constructed using two cumulus parameterization schemes (modified Kuo's scheme following KRISHNAMURTI and Bedi, 1988; and Arakawa-Schubert type scheme following GRELL, 1993) and two radiation parameterization schemes (an emissivity-absorbtivity based radiative transfer algorithm following CHANG, 1979 and a band model for radiative transfer following LACIS and HANSEN, 1974) in the atmospheric model only. Four possible permutations were used to construct the four versions of the FSU coupled climate model. We do not create any ensemble member run by prescribing different initial conditions or by changing the ocean model part. For convenience the following acronyms have been attached to different versions of the coupled model:

KOR — Kuo type convection with Chang radiation computations
KNR — Kuo type convection with Lacis and Hansen radiation computation
AOR — AS type convection with Chang radiation computations
ANR — AS type convection with Lacis and Hansen radiation computation

Starting from 31 December 1986, every 15 days (twice a month) three-month forecasts were made with the four different versions of the coupled model. It may be noted here that during the coupled assimilation, only one version (KOR) of the atmospheric model was used in coupled integrations. All the coupled model seasonal forecasts were made twice a month from identical initial conditions. Every 15 days, seasonal forecasts (3-months integration) were made from the initial conditions for a period of 16 years (December 1986 to December 2002). In any given year, 24 sets (2 per month) of initial conditions were available for making seasonal forecasts. In this study we use the forecasts made from that initial condition which is nearer to the end of a calendar month. Consequently from daily forecast data it is easier to prepare monthly mean data matching exactly with calendar months. Then we have both model and observed (analyzed) monthly mean data for comparison in terms of calendar months of different years. The coupled ocean-atmosphere model has limitations in terms of the predictability. Forecasts beyond the three-month periods might have errors larger than the standard deviation of the parameter (GOSWAMI and SHUKLA, 1991). In this study, we have restricted the forecast length to a maximum of up to three months period for all experiments of seasonal prediction. We know that climate models can respond in different ways to the predicted SSTs. Ensemble members can show model to model variability in the estimates of the seasonal mean signal of monsoon rainfall. Hence, in the following section we present mostly the results from the ensemble mean (EMN) of the four member models. For model comparison purpose we use the CMAP (XIE and ARKIN, 1996) large-scale analyzed rainfall data and the ECMWF wind analyses available from NCAR data archives.

4. Results and Discussions

For all four versions of the coupled model the mean model conditions (model climate) was computed by averaging the forecasts (forecast length of month 1, 2 and 3) from 1987 to 2002 (16 years). For the Indian summer monsoon season we collected the mean of June, July and August data and showed the seasonal mean as JJA. If it is the one month forecast for JJA it means all the one-month forecasts valid for June (initial condition of 31 May), July (initial condition of June 30) and August (initial condition of July 31), respectively were taken collectively. Similarly, when it is a two-month forecast valid for JJA, it means all the two-month forecasts valid for June (initial condition of 30 April), July (initial condition of May 31) and August (initial condition of June 30), respectively were taken. For three month forecasts valid for JJA, all the three-month forecasts valid for June (initial condition of 31 March), July (initial condition of April 30) and August (initial condition of May 31), respectively were taken. For comparison purposes the respective analyzed data were also averaged for the valid months over the same period of 1987 to 2002.

The predictability of the interannual variability of Asian summer monsoon rainfall in a model is dependent on how well that model simulates the climatological mean monsoon (SPERBER and PALMER, 1996). Models with a better rainfall climatology have less systematic errors and higher ability to simulate interannual variations. Climate drift is almost inevitable in coupled ocean-atmosphere models. Sometimes the climate drift can be larger than the interannual signal. In this section, we examine the model climate and its drift if any.

4.1 Low-level Winds

The mean low-level (850 hPa) wind for the JJA is shown in Figure 1. Top left panel shows the observed mean wind for the JJA period. The rest of the three panels show the month-1, month-2 and month-3 forecasts, respectively valid for the same period from the coupled model's ensemble mean (average of four member models). The wind direction is shown in a streamline pattern drawing. The wind speed is shown by shading at intervals of five m/s (meters per second) also starting at five m/s. The model low-level wind regime is able to capture the observed pattern very realistically. The cross-equatorial flow from the Indian Ocean of the Southern Hemisphere, turning of wind near Africa and formation of a monsoon trough from the Bay of Bengal to the central parts of India is seen in the model climate winds. The low-level Somali Jet is also reproduced in the coupled model forecasts of month-one through three. The consistency of the forecasts from month one to three is also good. There is no shift of the monsoon flow regime and pattern from one month to the other. The strength of the Somali Jet is slightly weaker as compared to the

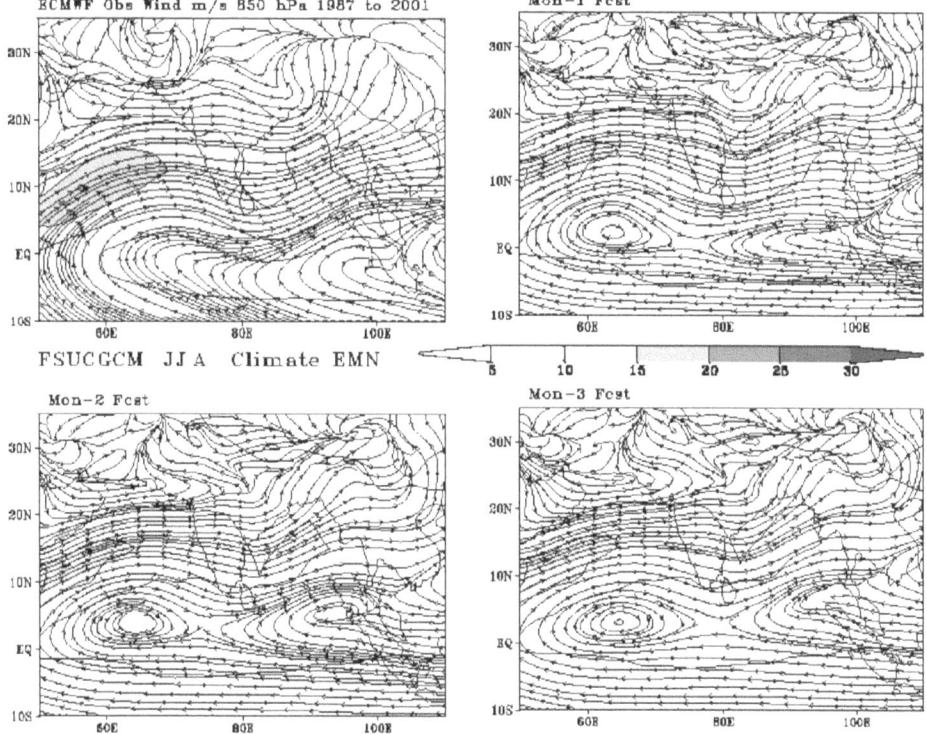

Figure 1
Mean observed and coupled model simulated low-level winds for the monsoon season (JJA) over India at 850 hPa.

observation. Majority of the numerical models have a tendency to weaken the Indian monsoon flow. Our model integrations were performed at T63 resolution. However, for comparison purposes, the model output was always bilinearly interpolated from model Gaussian grid to $2.5° \times 2.5°$ regular latitude/longitude grids. Some intensity might have been lost due to interpolation.

4.2 Summer Monsoon Rainfall

The mean observed and model summer monsoon rainfall pattern for JJA period is shown in Figure 2. The observed mean (top left panel) shows rainfall distribution in mm/day. We find an observed rain area of 15 mm/day near the west coast of India. The head of the Bay of Bengal (associated with monsoon Lows) also shows a high rain area of 20 mm/day. The model forecasts (month one through three) valid for JJA also reproduce a realistic monsoon rainfall compared to the observations. Both the west coast and the head Bay of Bengal region rainfall maxima have been captured realistically. From month one through month three, the model forecasts do not show any drift or inconsistency in general. The model has a slight tendency to intensify (as

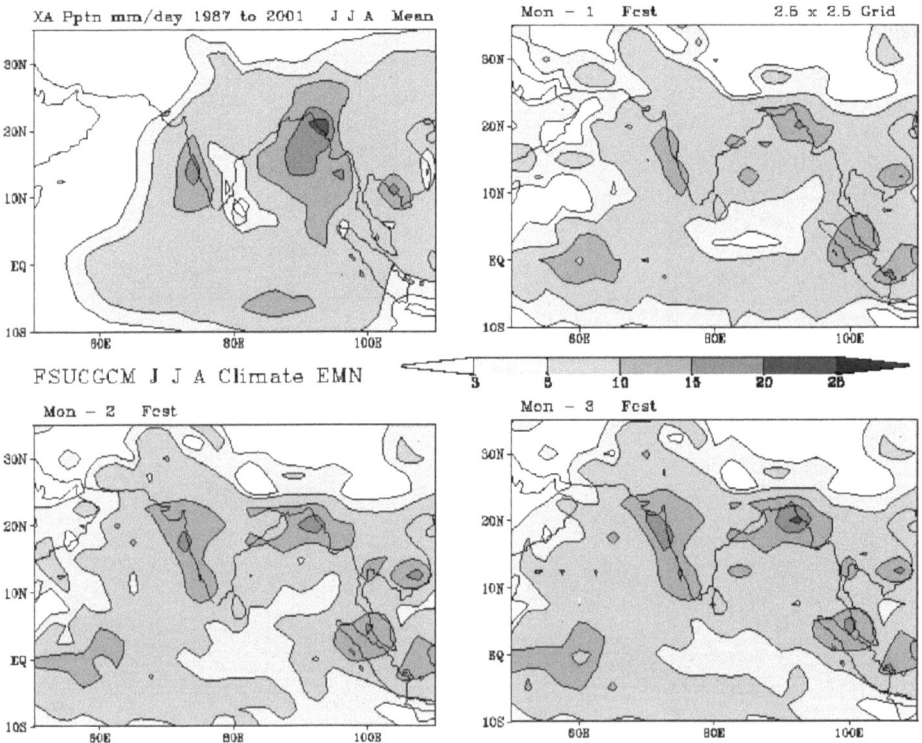

Figure 2
Mean observed and coupled model simulated rainfall for the monsoon season (JJA) over India.

seen in west coast and Bay of Bengal rainfall) the monsoon from month one through three. Figures 3 and 4 show the observed and forecasts of rainfall for the global (0 to 360 longitude; 70 S to 70 N latitude) and Indian (70 E to 120 E; Equator to 25 N latitude) regions, respectively. We see a good global hydrological (rainfall) consistency between month one, month two and month three forecasts. The Indian region in the model exhibits a realistic annual cycle of the monsoon rainfall (Fig. 4). As seen earlier in Figure 2, we see the intensity of monsoon rainfall increases from month one forecasts to month three forecasts. However, results from the month three forecasts are in best agreement with the observed rains.

The ability of coupled GCMs to simulate a reasonable seasonal cycle is a necessary condition for confidence in their prediction of seasonal means. The AMIP study showed some models simulate the observed seasonal migration of the primary rain belt; in several others this rain belt remains over the equatorial oceans in all seasons. The skill in simulation of excess/deficit summer monsoon rainfall over the Indian region is found to be much larger for models which simulate a realistic seasonal cycle of rainfall. Figure 5 shows the mean observed and model seasonal

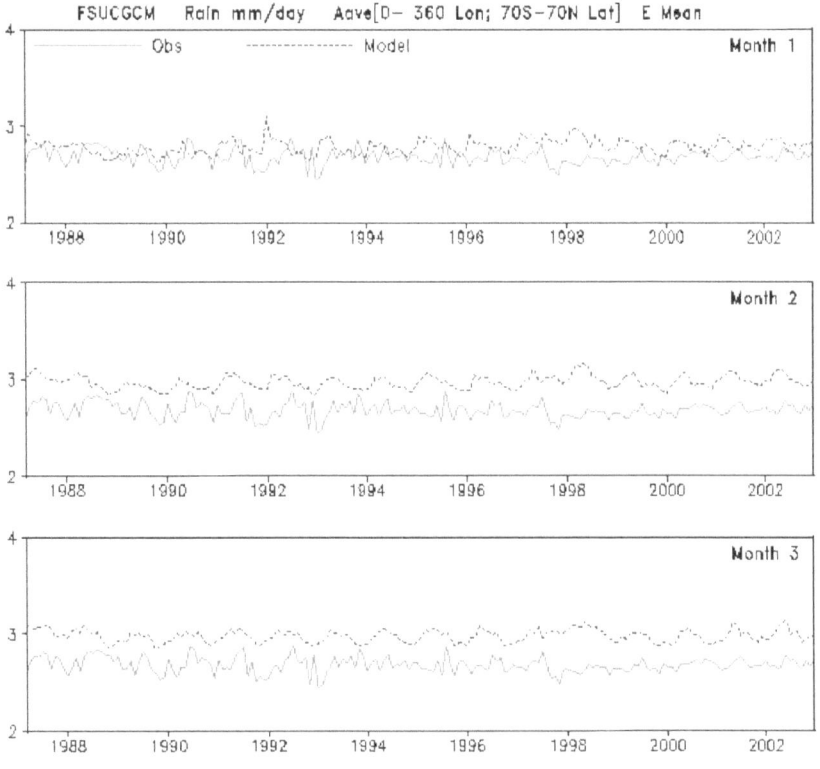

Figure 3
Observed and coupled model simulated global rainfall from 1987 to 2002 for different lengths of forecasts.

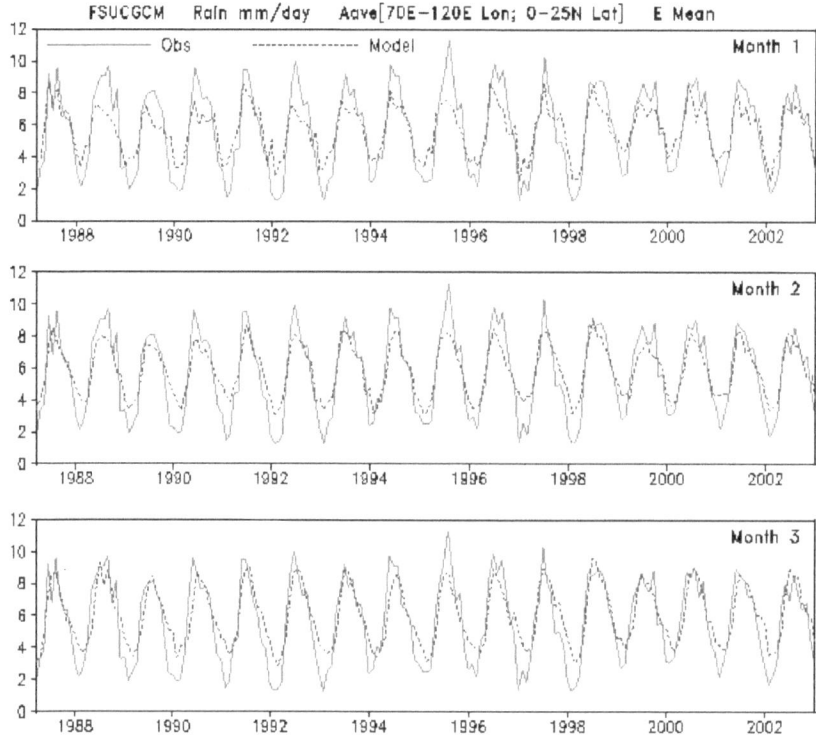

Figure 4
Observed and coupled model simulated monsoon rainfall from 1987 to 2002 for different lengths of
forecasts.

cycle of monsoon rainfall as seen at 90°E longitude (Bay of Bengal). The location and
timing of the maxima in model matches with the observation. However, the intensity
of the monsoon rain during the month one forecast is weaker as compared to the
observations. GADGIL and SAJANI (1998) in their study from AMIP model
intercomparison of monsoon precipitation concluded that the realistic simulation
of seasonal cycle for the Asia-Pacific region (70°E to 140°E longitudes) is a good
indicator of the respective model's capability to reproduce the interannual summer
monsoon rainfall variability. Figure 6 shows the seasonal cycle of the Asia-Pacific
region rainfall from coupled model and observations. The model is able to capture
the seasonal cycle in a realistic way. However, in a mean sense the month two and
three intensities are better than the month one forecasts.

The simulation of the details of the large-scale tropical rainfall pattern (intensity
and location) associated with the ITCZ is another indicator of the model's
performance. Diurnal variations, MJO etc influence the ITCZ. Proper representation
of cumulus convection, boundary layer processes, radiation and the interaction
among them are crucial for proper simulation of the ITCZ and the monsoon system.

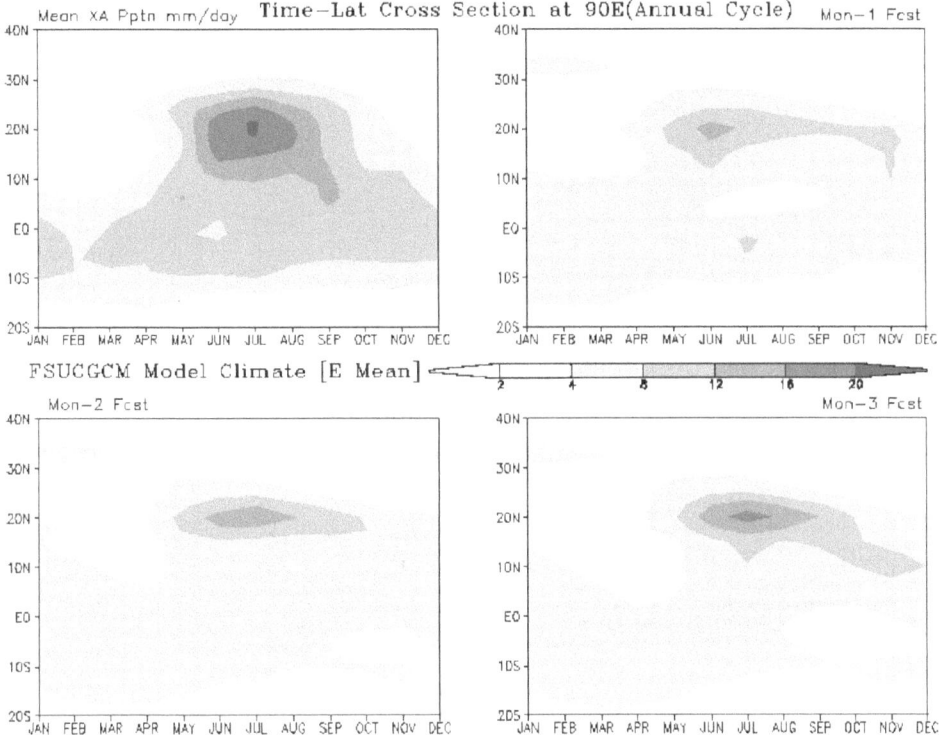

Figure 5
Seasonal cycle of the mean observed and coupled model simulated monsoon rainfall at 90° E for different lengths of forecasts.

Figure 7 shows the JJA mean observed and coupled model forecasts of rain associated with the ITCZ. The coupled model is able to reproduce a realistic ITCZ pattern. The month one, two and three forecasts are consistent and there is little shift of rainfall belts from month one to three. However, the intensity of rainfall in the east and west Pacific is slightly less compared to observations. Similar problems of reduced intensity over the west Pacific region were reported by KANG *et al.* (2002) in the CLIVAR/Monsoon intercomparison project where the climatological variations of the summer monsoon rainfall were evaluated from 10 GCMs for the Asia-Pacific region.

4.3 SST Prediction for Pacific

SST anomaly in the tropical Pacific shows large interannual variability. We examine the coupled model forecasts of the SST anomaly for the NINO3 region (150°W to 90°W; 5°S to 5°N) from 1987 to 2002 (Fig. 8). All the forecasts of month one, two and three show reasonably good SST anomaly forecasts. By month three the intensity forecasts for the SST anomaly deteriorates as compared to month one

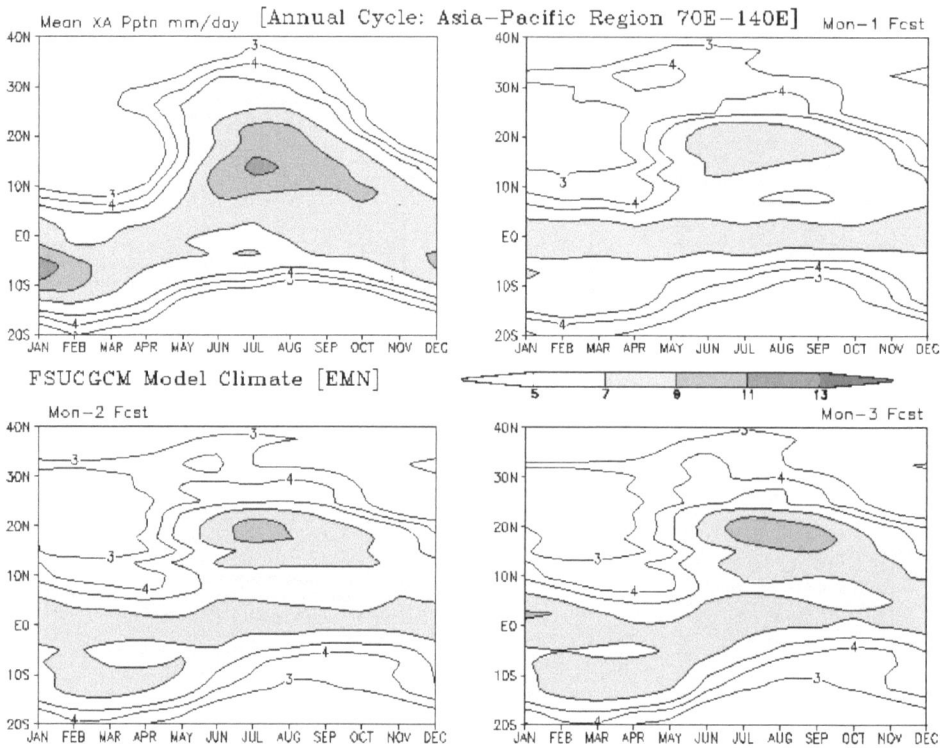

Figure 6
Seasonal cycle of the mean observed and coupled model simulated monsoon rainfall for Asia-Pacific region (for different lengths of forecasts).

forecasts. In an earlier study by KRISHNAMURTI *et al.* (2000c), the whole life cycle of the SST anomaly related to 1997–1998 El Niño was captured in the 18-month long integration starting from April 1, 1997 initial conditions. This demonstrates that the coupled model has good capability to predict the SST anomalies in seasonal time scale.

4.4 *Drought of Monsoon 2002 Season*

Predictability of the monsoon in a model can vary from year to year depending on the synoptic conditions in both the atmosphere and oceans. Due to feed back processes in a coupled model, any error in the predicted wind anomaly or the predicted SST anomaly can have detrimental effect on the final rainfall predictions. For the Indian region, the summer monsoon season of 2002 was an anomalous one with below normal observed rainfall. This severely impacted the Indian agriculture and the economy as a whole. The traditional statistical forecast from IMD indicated a normal monsoon for the year 2002. However, the dynamical seasonal forecasts

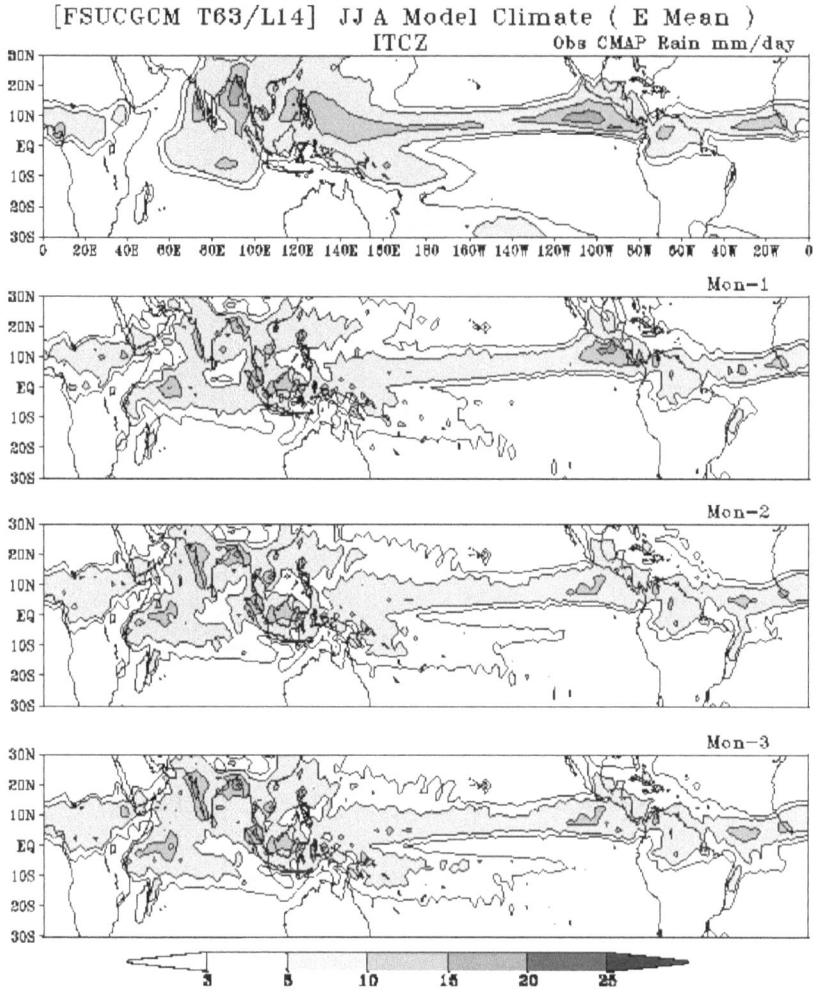

Figure 7
Mean observed and coupled model simulated ITCZ rainfall during JJA for different lengths of forecasts.

from the ECMWF were indicating below normal rainfall for JJA 2002 (with 1 month lead) for parts of India from the runs of 01 May 2002 itself. Confirming that, the ECMWF system again predicted the below normal rainfall for JAS 2002 period from 01 June 2002 initial conditions (GADGIL et al., 2003). Because of the poor performance of the seasonal prediction models, particularly for the Indian monsoon region, perhaps these forecasts were not taken seriously by many concerned organizations.

In this section we examine the performance of the FSU coupled model seasonal hindcasts for the JJA 2002 monsoon period for the Indian region. First, we look at the three month (90 days integration) forecasts starting from the 31

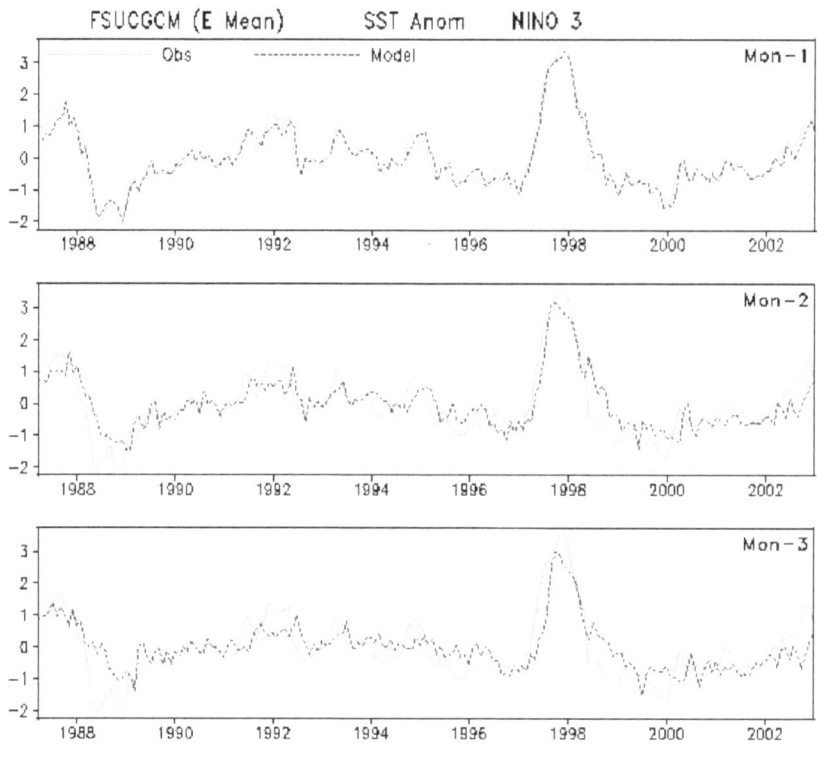

Figure 8

Observed and coupled model simulated SST anomaly for NINO3 region from 1987 to 2002 period.

May 2002 initial conditions. The anomalous low level (850 hPa) monsoon winds from the four member coupled models are shown in Figure 9. The anomaly field valid for JJA during 2002 season indicates an anticyclone over the Indian land mass. This anticyclone is unfavorable for the monsoon trough and the associated rain over the region. In most of the versions we find an anomalous low over the Arabian Sea region that will affect the moisture transport towards the west coast and the landmass. The ensemble mean of the four models indicates the anomalous bad (negative rain) monsoon flow conditions in Figure 10, where the observed anomalous wind from the ECMWF analyses are also seen for comparison with the model. Over the northern Arabian Sea region both the model and the observation show the anomalous cyclonic wind. Setting up of the anticyclone over the Indian landmass is also seen in the observation and the model mean JJA (mean of 90 individual days) forecast. Associated with the wind the respective anomaly rainfall prediction valid for JJA 2002 is examined now. Anomalous rainfall prediction from the four member coupled models are shown in Figure 11. Similar to wind, this is again the result from the same three-month (90 days integration) hindcasts starting from the 31 May 2002 initial conditions. All the

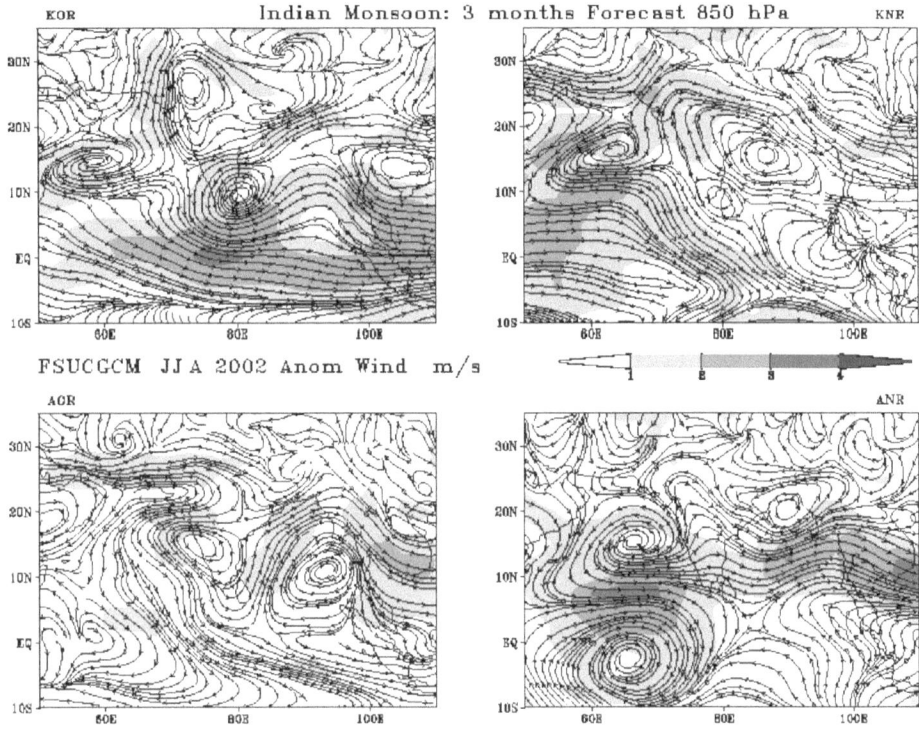

Figure 9
Three-month forecasts of anomalous low-level wind flow pattern for monsoon 2002 period (JJA) from four
member coupled models.

four member models are able to capture the negative anomaly of rainfall over the
Indian landmass. South of the equator, all four member models indicate a positive
intense anomalous rain belt. The ensemble from the four member models and the
corresponding anomalous observed rain for JJA 2002 are shown in Figure 12. We
find a very good agreement between the predicted and observed anomalous JJA
2002 rainfall. It is interesting to note that this type of anomalous wind and
rainfall patterns valid for JJA 2002 was predicted three months in advance by the
end of May 2002 itself by the FSU coupled model system. Next, we examine the
model's capability to predict the 2002 anomalous monsoon in another diagnostic
sense. Figure 13 shows the month one, month two and month three forecasts valid
for the JJA 2002 period and the corresponding observation. For example, the
month-3 forecast valid for JJA is a composite of the third month forecast only
from the initial conditions of 31 March, 30 April and 31 May of 2002,
respectively. Similarly, the month-2 forecast valid for JJA is a composite of the
second month forecast only from the initial conditions of 30 April, 31 May and 30
June of 2002. We find that all the months' forecasts are able to capture the

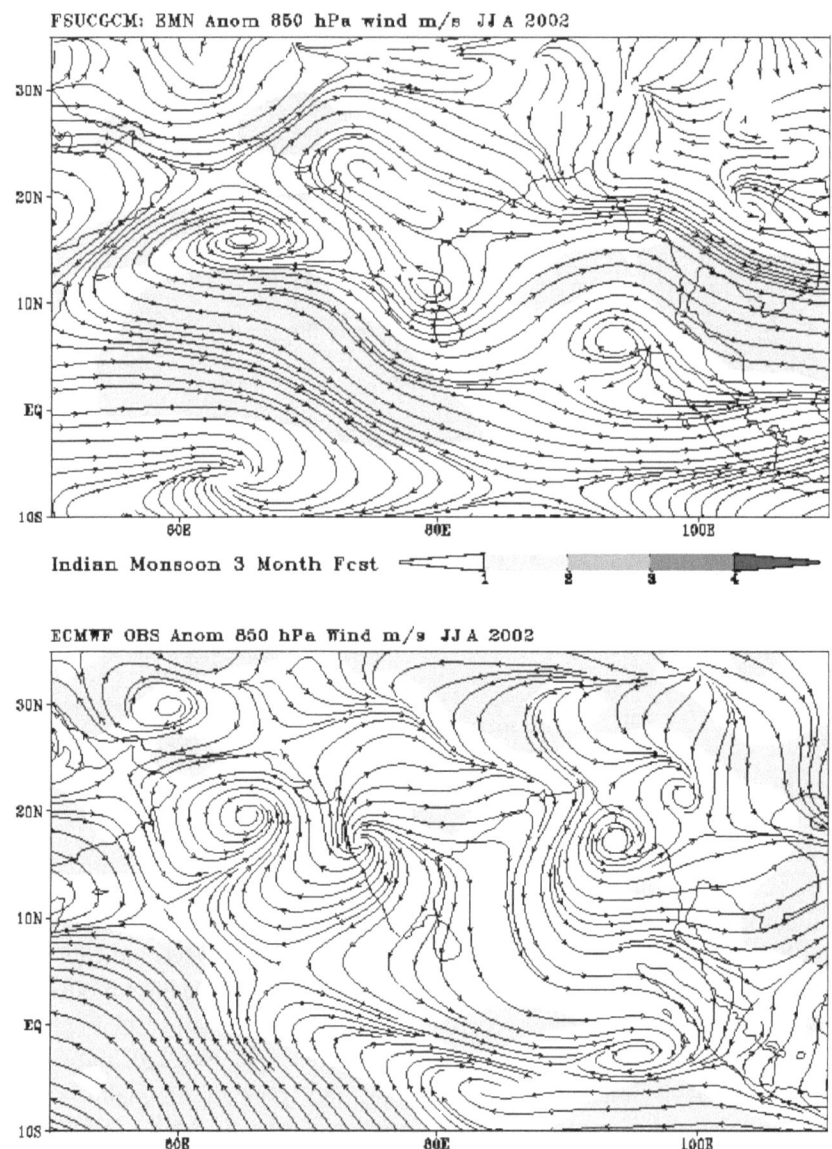

Figure 10
Three-month forecasts of anomalous low-level wind-flow pattern for monsoon 2002 period (JJA) from ensemble mean.

anomalous anticyclonic pattern of wind over India and a cyclonic pattern over the northern Arabian Sea. Even though the forecasts were made as early as 31 March (for month-3 forecast), it is noteworthy that good consistency exists among the forecasts from month one to month three. This indicates the robustness of the

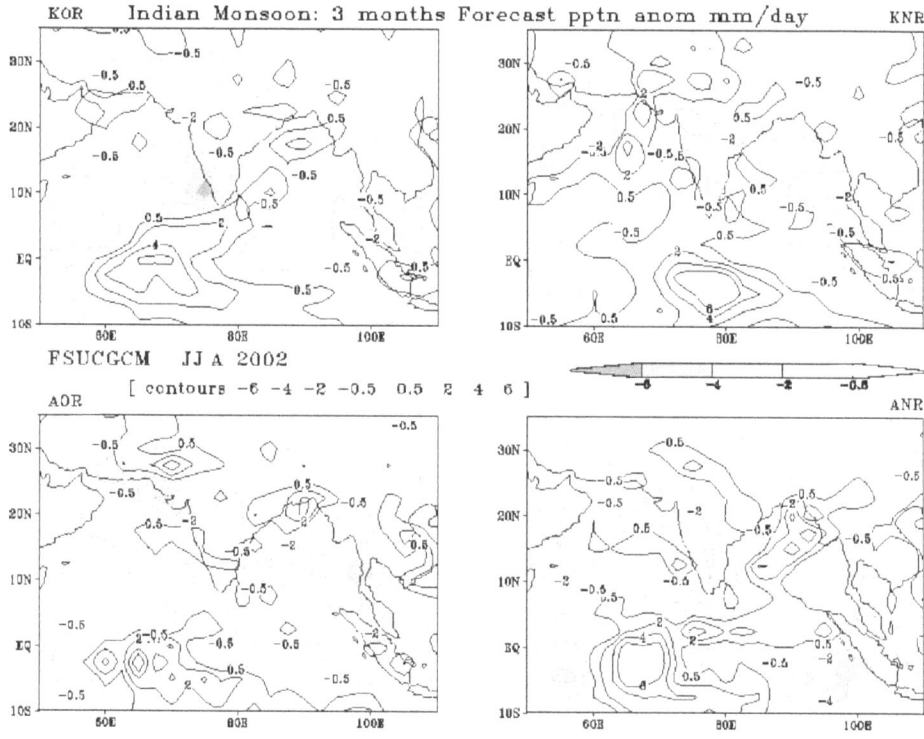

Figure 11

Three-month forecasts of distribution of anomalous rainfall for monsoon 2002 period (JJA) from four member coupled models.

coupled model. Figure 14 is similar to Figure 13 except for rainfall. For rainfall field the coupled model also behaves in a similar way also. However, the anomalous negative rain pattern was only seen in the month-1 and month-2 forecast composites. The month-3 forecast composite could capture the wind anomaly, but failed to capture the rain anomaly in advance.

A continuous 90-day seasonal runs valid for the 2002 JJA period (Figs. 10 and 12) showed the coupled model's ability to capture the large wind and precipitation seasonal anomalies well in advance. The composites of month-1, 2 and 3 forecasts valid for the JJA 2002 season (Figs. 13 and 14) also indicated the model's capability to capture the said circulation and rainfall anomalies reasonably well. Since most of the anomalies during the summer monsoon of 2002 occurred during the month of July (49% of all-India monsoon rainfall), it is of interest to see the month-1, 2 and 3 forecasts from the coupled model valid for July 2002 only. Here we show the ensemble mean forecasts from the four members of the FSU coupled model valid for July 2002 period (Figs. 15 and 16). The predicted anomalies of wind and precipitation for July 2002 look very similar to the composite JJA

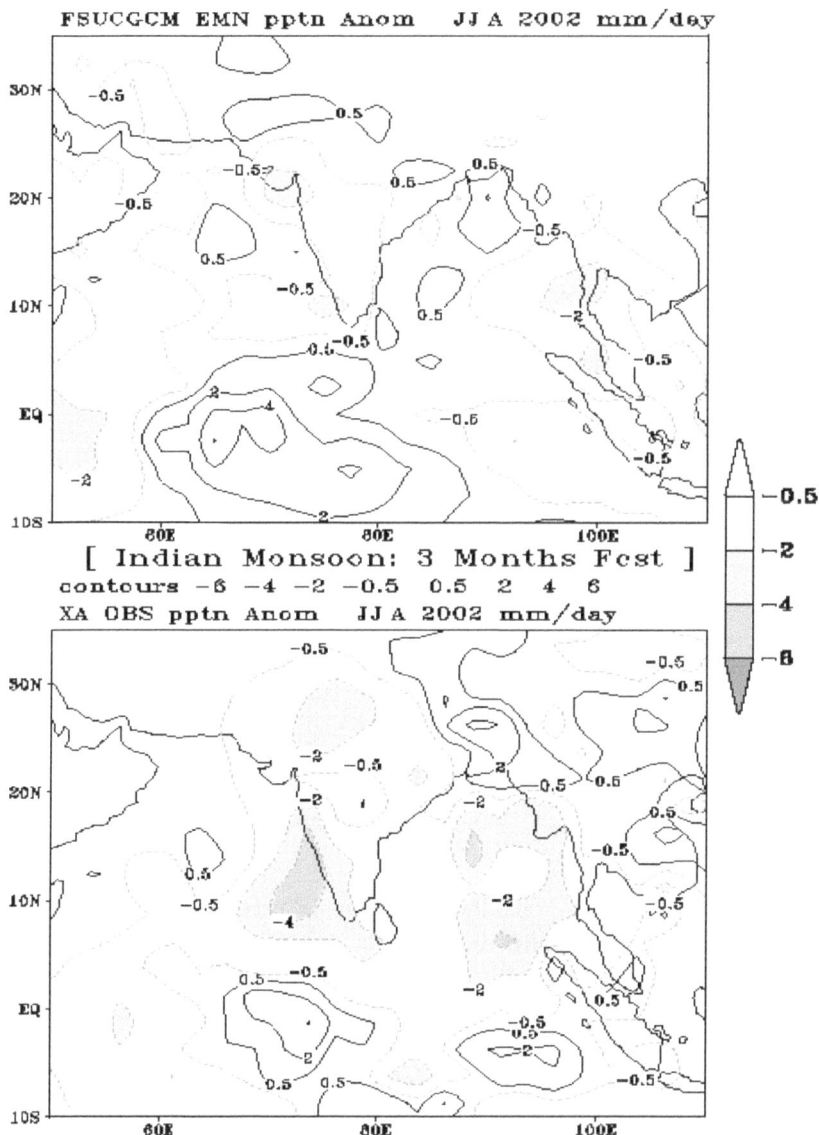

Figure 12
Three-month forecasts of distribution of anomalous rainfall for monsoon 2002 period (JJA) obtained from ensemble mean.

seasonal anomalies (Figs. 13 and 14). Since July had the largest observed anomaly as compared to June and August, it is obvious that the seasonal JJA anomaly plots will mostly depict the signatures from the July anomaly. All the month-1, 2 and 3 wind anomalies valid for July indicate the anomalous anticyclonic flow patterns over the Indian subcontinent. However, the Arabian Sea region

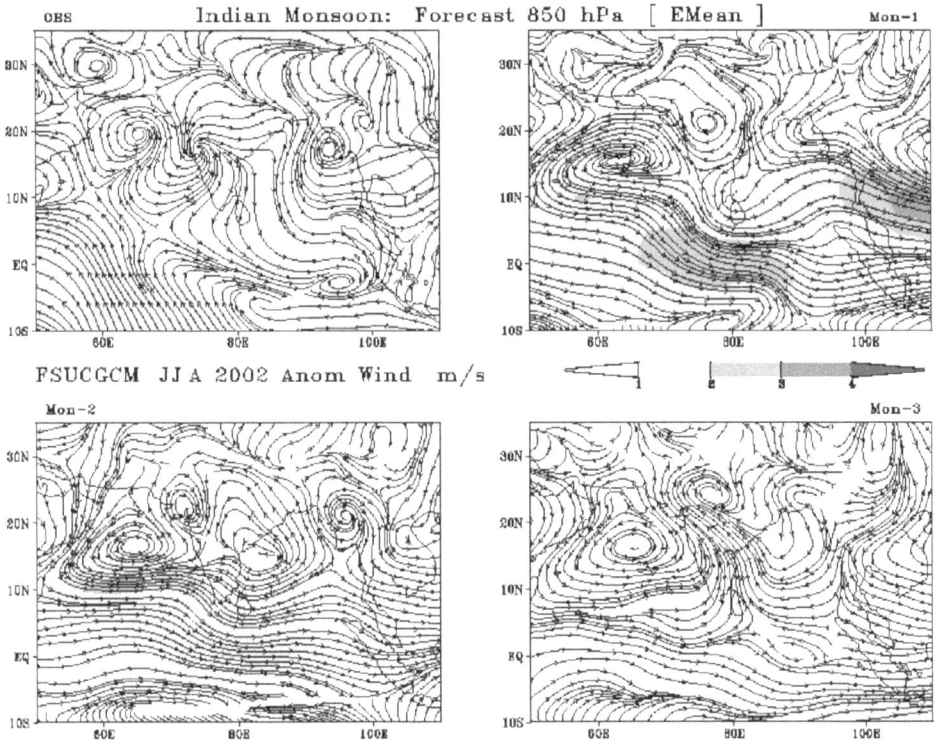

Figure 13
Observation and month-1, 2 and 3 forecasts of anomalous wind-flow pattern for the Indian region valid for
JJA 2002 monsoon period.

continued to show the anomalous cyclonic pattern that was seen in JJA seasonal
anomalies too. It may be noted that the best wind anomaly is seen in the month-1
forecast only. The forecast of rainfall anomalies for July 2002 (Fig. 16) captures
the negative observed anomalies well. However, the intensity of the rainfall
anomalies seen in observations was not captured by model. For month-2 and
month-3 forecasts, some area over the Indian subcontinent showed negative and
mild positive anomalies. Definitely with the increased length of forecast from
month-1 to month-3, the quality deteriorated. South of the equator the observed
positive anomaly in rainfall was predicted well by the model from month-1
through month-3 forecasts. It can be summarized here that the forecasts from the
seasonal run (90 days continuous run) valid for JJA 2002 have the best skill
compared to other combinations. This may be related either to the summer
monsoon predictability issues or due to the typical characteristics of this
particular coupled model. Using the coupled model, further diagnostics studies
are required to gain more insight to this anomalous 2002 summer monsoon and
the related predictability issues.

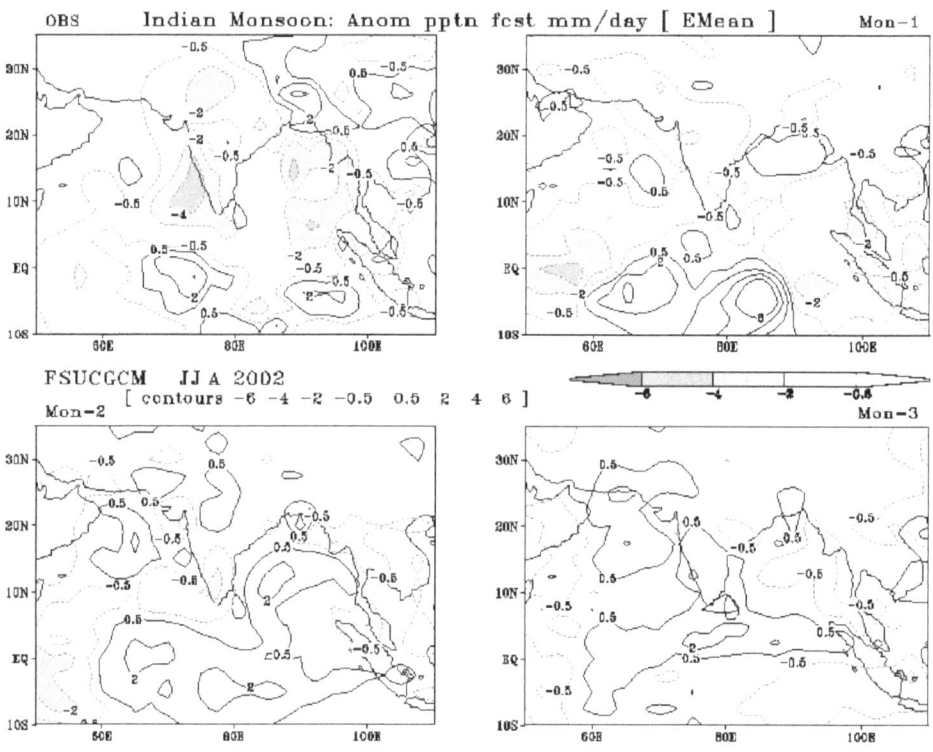

Figure 14
Observation and month-1, 2 and 3 forecasts of anomalous rainfall distribution pattern for the Indian region valid for JJA 2002 monsoon period.

5. Conclusions

The mean features of Indian summer monsoon and its variability, produced by versions of the 'Florida State University Coupled Ocean-Atmosphere General Circulation Model' (FSUCGCM) are investigated using seasonal hindcasts for the period 1987 to 2002. Lower-level wind flow patterns and rainfall associated with the summer monsoon season are examined from this fully coupled model seasonal integrations. By comparing with the observations, the mean monsoon condition simulated by this coupled model for the June, July and August period is seen to be reasonably realistic. The overall spatial low-level wind flow patterns and the precipitation distributions over the Indian continent and adjoining oceanic regions are comparable with the respective analyses. The model is able to produce the seasonal cycle of rainfall over the monsoon region. A reasonably realistic ITCZ rainfall pattern is reproduced in this coupled model. The model is also capable of simulating the interannual variability of SST anomaly over the equatorial Pacific region. The anomalous below normal large-scale precipitation and the associated

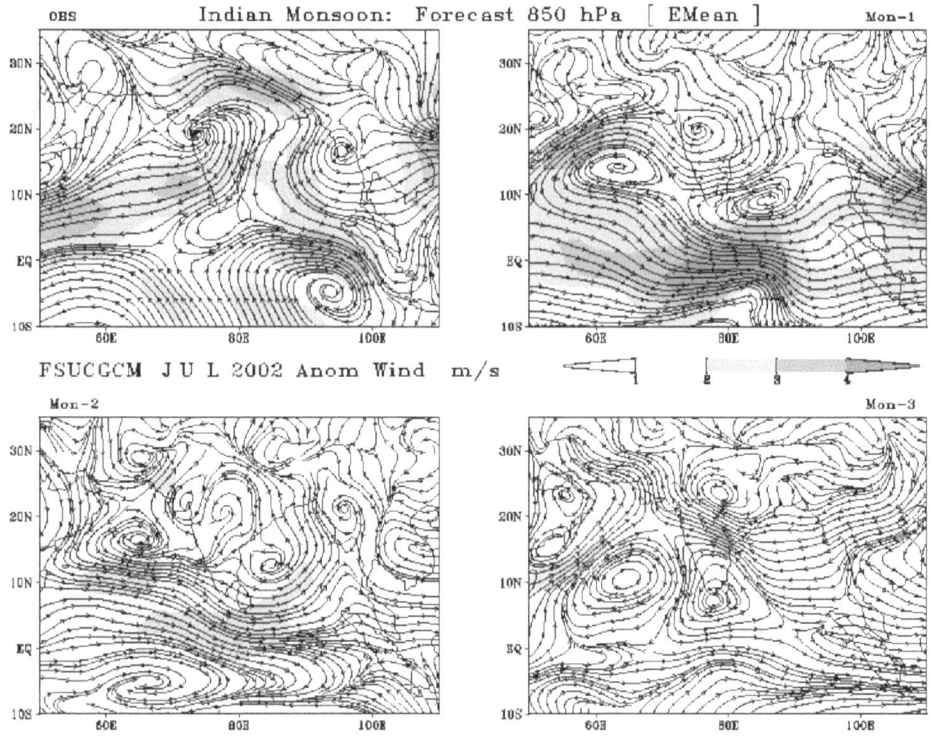

Figure 15

Observation and month-1, 2 and 3 forecasts of anomalous wind-flow pattern for the Indian region valid for July 2002 monsoon period.

anomalous low-level wind circulation pattern for the summer monsoon season of 2002 was predicted by the model three months in advance. For the Indian summer monsoon, the ensemble mean is able to reproduce the mean features better compared to individual member models. These four versions of the FSUCGCM are found to be suitable for use in the FSU multi-model superensemble forecasting system for the monsoon rainfall prediction over the Indian region.

6. Future Work

Seasonal prediction of realistic precipitation anomaly for the Indian monsoon generally has poor proficiency. This may be due to the internal chaotic dynamics associated with intraseasonal monsoon fluctuations and/or unpredictable land surface processes interactions. Improvements in land-surface parameterizations can possibly improve the skill of seasonal rainfall prediction further. From the model intercomparison studies we know that models can respond in different ways to even

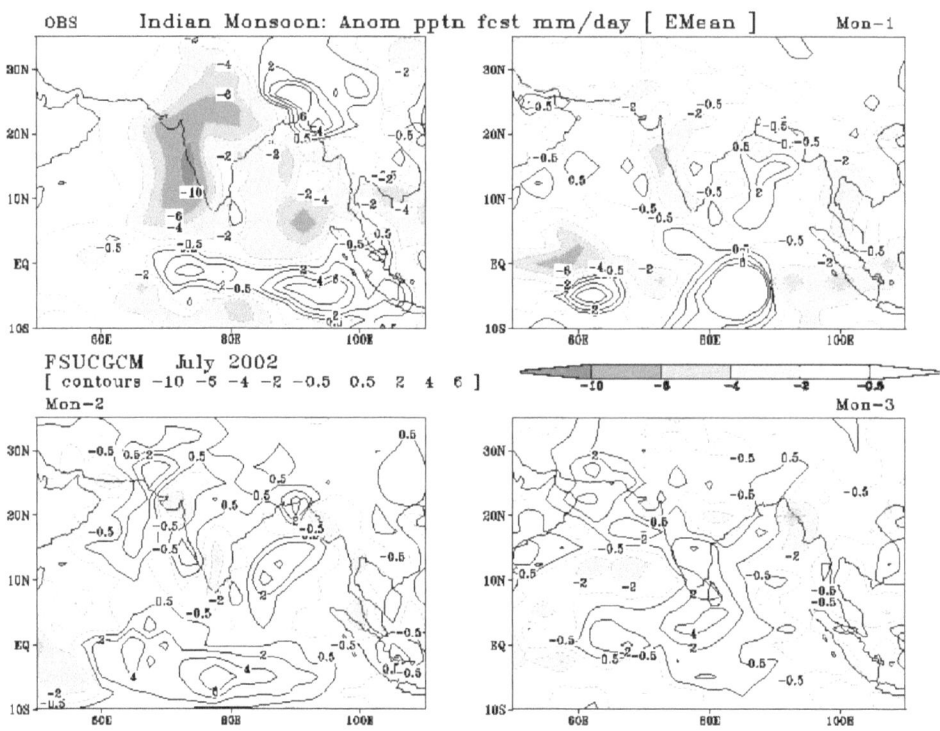

Figure 16
Observation and month-1, 2 and 3 forecasts of anomalous rainfall distribution pattern for the Indian region valid for July 2002 monsoon period.

prescribed identical SSTs and atmospheric conditions. In a coupled prediction system the incorporation of the ocean-assimilation (with ocean model) makes it more complicated to understand the evolution of the seasonal mean monsoon signal and noise. In a complex nonlinear monsoon system, the understanding of how the slowly varying boundary forcing and the internal dynamics evolve and interplay is still lacking. Proper understanding of tropospheric biennial oscillation (TBO) and its linkages to Indian monsoon precipitation variability in a modeling framework might give further insight to improve coupled model skills (MEEHL *et al.*, 2003).

On a regional scale, the high-frequency temporal variability can be captured only by higher resolution models. A higher horizontal resolution (T106) version of the FSUCGCM might better capture the spatial and temporal characteristics of the Indian monsoon. Particularly, simulation of the monsoon trough and associated rainfall is expected to be more realistic at T106 resolution (SPERBER *et al.*, 1994). The issue of what constitutes a reasonable lower bound on ensemble size was discussed in earlier studies (STERN and MIYAKODA, 1995; BRANKOVIC and PALMER, 1997). The increase in ensemble size might increase the confidence in being able to estimate reliably the impact of predicted SSTs on monsoon rainfall variability. We are working on designing experiments and real time runs with at least a ten-member ensemble for seasonal Indian summer monsoon rainfall forecasting. Using the FSU superensemble technique and the FSU coupled models, seasonal climate forecasting technique is being developed and tested for the Indian monsoon region for real time application.

Acknowledgements

The analyzed data used for model validation were taken from NCAR data archives. The research reported here was supported by the NSF grants ATM-0241517 and INT-0302172; NOAA grants NA16GP1365 and NA06GP0512; NASA grant NAG5-9662.

References

BACHIOCHI, D. and KRISHNAMURTI, T.N. (2000), *Enhanced Low-level Stratus in the FSU Coupled Ocean-atmosphere Model*, Mon. Wea. Rev. *128*, 3083–3103.

BRANKOVIC, C.M. and PALMER, T.N. (1997), *Atmospheric Seasonal Predictability and Estimates of Ensemble Size*, Mon. Wea. Rev. *125*, 859–874.

CHANG, C.B. (1979), *On the Influence of Solar Radiation and Diurnal Variation of Surface Temperatures on African Disturbances*, Rep. No. 79-3, Dept. of Meteorology, Florida State University, Tallahassee, Florida.

DEWITT, D.G. and SCHNEIDER, E.K. (1999), *On the Processes Determining the Annual Cycle of Equatorial SST: A CGCM Perspective,* Mon. Wea. Rev. *127(3)*, 381–395.

FASULLO, J. and WEBSTER, P.J. (2002), *Hydrological Signatures Relating the Asian Summer Monsoon and ENSO,* J. Climate *15*, 3082–3095.

GADGIL, S. and SAJANI, S. (1998), *Monsoon Precipitation in the AMIP Runs*, Clim. Dyn. *14*, 659–689.

GADGIL, S., SRINIVASAN, J., NANJUNDIAH, R.S., KRISHNA KUMAR, K., MUNOT, A.A., and KOLLI, R.K. (2003), *On Forecasting the Indian Summer Monsoon: The Intriguing Season of 2002*, Current Science *83*, 394–403.

GOSWAMI, B.N. and SHUKLA, J. (1991), *Predictability of a Coupled Ocean-atmosphere Model*, J. Climate *4*, 3–22.

GRELL, G.A. (1993), *Prognostic Evaluation of Assumptions Used by Cumulus Parameterizations*, Mon. Wea. Rev. *121*, 764–787.

KANG I.–S. and coauthors (2002), *Intercomparison of the Climatological Variations of Asian Summer Monsoon Precipitation Simulated by 10 GCMs*, Climate Dynamics *19* (5–6), 383–395.

KRISHNAMURTI, T.N., STEFANOVA, L., CHAKRABORTY, A., VIJAYA KUMAR T.S.V., COCKE, S., BACHIOCHI, D., and MACKEY, B. (2002), *Seasonal Forecasts of Precipitation Anomalies for North American and Asian Monsoons*, J. Met. Soc. Japan *80*(6), 1415–1426

KRISHNAMURTI, T.N., SURENDRAN, S., SHIN, D.W., TORRES, R.J.C., VIJAYA KUMAR, T.S.V., WILLIFORD, C.E., KUMMEROW, C., ADLER, R.F., SIMPSON, J., KAKAR, R., OLSON, W.S., and TURK, F.J., (2001), *Real-time Multianalysis-multimodel Superensemble Forecasts of Precipitation using TRMM and SSM/I Products*, Mon. Wea. Rev. *129*, 2861–2883.

KRISHNAMURTI, T.N., Kishtawal, C.M., ZHANG, Z., LAROW, T., BACHIOCHI, D., Williford, C.E., GADGIL, S., and SURENDRAN, S. (2000a), *Multi-model Superensemble Forecasts for Weather and Seasonal Climate*, J. Climate *13*, 4196–4216.

KRISHNAMURTI, T.N., KISHTAWAL, C.M., SHIN, D.W., and WILLIFORD, C.E. (2000b), *Improving Tropical Precipitation Forecasts from a Multianalysis Superensemble*, J. Climate *13*, 4217–4227.

KRISHNAMURTI, T.N., BACHIOCHI, D., LAROW, T., JHA, B., TEWARI, M., CHAKRABORTY, D.R., TORRES, R.J.C., and OOSTERHOF, D., (2000c), *Coupled Atmosphere-ocean Modelling of El-Niño of 1997–1998*, J. Climate *13*, 2428–2459.

KRISHNAMURTI, T.N., BEDI, H.S., and HARDIKER, V. (1998) *Introduction to Global Spectral Modeling* (Oxford University Press, New York, 1998) 253 pp.

KRISHNAMURTI, T.N., XUE, J., BEDI, H.S., INGLES, K., and OOSTERHOF, D., (1991), *Physical Initialization for Numerical Weather Prediction over the Tropics*, Tellus *43AB*, 53–81

KRISHNAMURTI, T.N. and BEDI, H.S. (1988), *Cumulus Parameterization and Rainfall Rates-3*, Mon. Wea. Rev. *116*, 583–599.

LACIS, A.A. and HANSEN, J.E. (1974), *A Parameterization for the Absorption of Solar Radiation in the Earth's Atmosphere*, J. Atmos. Sci. *31*, 118–133.

LAROW, T.E. and KRISHNAMURTI, T.N. (1998), *Initial Conditions and ENSO Prediction Using a Coupled Ocean-atmosphere Model*, Tellus *50A*, 76–94.

LATIF, M. (1987), *Tropical Ocean Circulation Experiments*, J. Phys. Oceanogr. *17*, 246–263.

MEEHL, G.A., BOER, G.J., COVEY, C., LATIF, M., and STOUFFER, R.J. (2000), *The Coupled Model Intercomparison Project*, Bull. Amer. Met. Soc. *81*(2), 313–318.

MEEHL, G.A., ARBLASTER, J.M., and LOSCHNIGG, J. (2003), *Coupled Ocean-atmosphere Dynamical Processes in the Tropical Indian and Pacific Oceans and the TBO*, J. Climate *16*, 2138–2158.

PALMER, T.N. and COAUTHORS (2004), *Development of a European Multi-model Ensemble System for Seasonal to Inter-annual Prediction (DEMETER)*, Bull. Am. Meteorol. Soc. 85(6), 853–872. Available online at http://www.ecmwf.int/research/demeter/data/bams_paper.pdf

SPERBER, K.R. and COAUTHORS (2001), *Dynamical Seasonal Predictability of the Asian Summer Monsoon*, Mon. Wea. Rev. *129*, 2226–2248.

SPERBER, K.R. and PALMER, T.N. (1996), *Interannual Tropical Rainfall Variability in a GCM Simulation Associated with the AMIP*, J. Climate *9(11)*, 2727–2750.

SPERBER, K.R., HAMEED, S., POTTER, G.L., and BOYLE, J.S. (1994), *Simulation of the Northern Summer Monsoon in the ECMWF Model: Sensitivity to Horizontal Resolution*, Mon. Wea. Rev. *122*, 2461–2481.

STEFANOVA, L. and KRISHNAMURTI, T.N. (2002), *Interpretation of Seasonal Climate Forecast Using Brier Skill Score, FSU Superensemble and AMIP-1 Dataset*, J. Climate *15*, 537–544.

STERN, W. and MIYAKODA, K. (1995), Feasibility *of Seasonal Forecasts Inferred from Multiple GCM Simulations*, J. Climate *8*, 1071–1085.

XIE, P. and ARKIN, P.A. (1996), *Analysis of Global Monthly Precipitation Using Gauge Observations, Satellite Estimates, and Numerical Model Prediction*, J. Climate *9*, 840–858.

XIOUHUA, Fu, BIN WANG, B., and LI, T. (2002), *Impacts of Air–sea Coupling on the Simulation of Mean Asian Summer Monsoon in the ECHAM4 Model*, Mon. Wea. Rev. *130*(12), 2889–2904.

(Received November 14, 2003, accepted March 5, 2004)

Pure appl. geophys. 162 (2005) 1455–1479
0033–4553/05/091455–25
DOI 10.1007/s00024-005-2679-6

© Birkhäuser Verlag, Basel, 2005

▎Pure and Applied Geophysics

Drought 2002 in Colorado: An Unprecedented Drought or a Routine Drought?

Roger A. Pielke, Sr.,[1] Nolan Doesken,[1] Odilia Bliss,[1] Tara Green,[1] Clara Chaffin,[1,2] Jose D. Salas,[3] Connie A. Woodhouse,[4] Jeffrey J. Lukas,[5] and Klaus Wolter[6]

Abstract — The 2002 drought in Colorado was reported by the media and by public figures, and even by a national drought-monitoring agency, as an exceptionally severe drought. In this paper we examine evidence for this claim. Our study shows that, while the impacts of water shortages were exceptional everywhere, the observed precipitation deficit was less than extreme over a good fraction of the state. A likely explanation of this discrepancy is the imbalance between water supply and water demand over time. For a given level of water supply, water shortages become intensified as water demands increase over time. The sobering conclusion is that Colorado is more vulnerable to drought today than under similar precipitation deficits in the past.

Key words: Drought, precipitation, Colorado, streamflow, snowpack, paleoclimatology.

1. Introduction

In reference to the 2002 drought, the Governor of Colorado stated in his 2003 State of the State address,

".... scientists tell us that this is perhaps the worst drought in 350 years."
(http://www.thedenverchannel.com/print/1913350/detail.html?use = print)

Clearly, such an assessment of drought severity depends on how drought is defined. Drought is characterized in a number of different ways, each with associated definitions of onset and recovery, duration, and related impacts. For example,

[1]Department of Atmospheric Science, Colorado State University, Fort Collins, 80523-1371, CO U.S.A
[2]currently at University of Idaho, Moscow, Idaho
[3]Civil Engineering Department, Colorado State University, Fort Collins, CO 80523-1372, U.S.A.
[4]Paleoclimatology Branch, National Climatic Data Center, NOAA, 325 Broadway, E/CC23 Boulder, CO 80305, U.S.A.
[5]Institute of Arctic and Alpine Research, University of Colorado, Campus Box 450 Boulder, CO 80309-0450, U.S.A
[6]NOAA-CIRES Climate Diagnostics Center, University of Colorado, Campus Box 216, Boulder, CO 80309-0216, U.S.A

meteorological drought could be measured by numbers of days below a specified precipitation threshold, or departure from a baseline average; an agricultural drought could be measured by soil moisture deficit and impacts on crops; and a hydrological drought could be measured by a period of precipitation deficit and impacts on water supply such as streamflow and surface and subsurface water storages. Spatial and temporal scales must also be considered in defining drought. The variety of ways to define drought makes a simple assessment of drought severity a difficult task.

In this paper, we explore the severity of the 2002 drought, defined by a variety of moisture-related variables including precipitation, snowpack, streamflow, reservoir storage, and tree growth. Although the 2002 drought is considered by some to be the third of a three-year drought, here we focus on 2002 as a single year event. The definition of the year varies somewhat according to variable measured, but in general, we consider it from fall 2001 though summer 2002. Its impact is gauged on the regional to statewide level.

Figure 1 shows the magnitude of the drought as determined by the U.S. Drought Monitor (http://drought.unl.edu/dm/), where the western third of the state is in the highest ("exceptional") category. In this display, drought has been defined based on the interpretation of available water deficit information by researchers at the National Drought Mitigation Center at the University of Nebraska at Lincoln, as well as input from a variety of experts in the field, including some of the co-authors of this paper. "Exceptional drought" refers to conditions found between once every fifty years or never before on record. This is one assessment of drought that our paper examines using a variety of analysis techniques.

2. An Evolution of the 2002 Drought in Colorado

The drought of 2002, with all of its devastating wildfires, profound water shortages and widespread crop losses, had its beginnings in the autumn of 1999. After a very wet spring in 1999 and a soggy August, precipitation patterns reversed and the fall of 1999 was very dry across most of Colorado. The winter of 1999–2000 followed with below average snow accumulation and much above average temperatures. The mountains of southwestern Colorado were particularly hard hit by a shortage of snow for winter recreation and summer water supply. With a very dry spring and early summer in 2000 over northeast Colorado and the South Platte watershed, drought conditions emerged quickly. In fact, the entire western U.S. was by then engulfed in a severe drought that resulted in the largest severe wildfire season in the last century for the western U.S. (http://www.nifc.gov/stats/wildlandfire-stats.html). A persistently hot summer made the situation worse, as transpiration rates were considerably higher than average over irrigated areas.

The 2001 Water Year was less extreme but still tended on the dry side. Colorado's northern and central mountains were the driest with respect to average. While spring

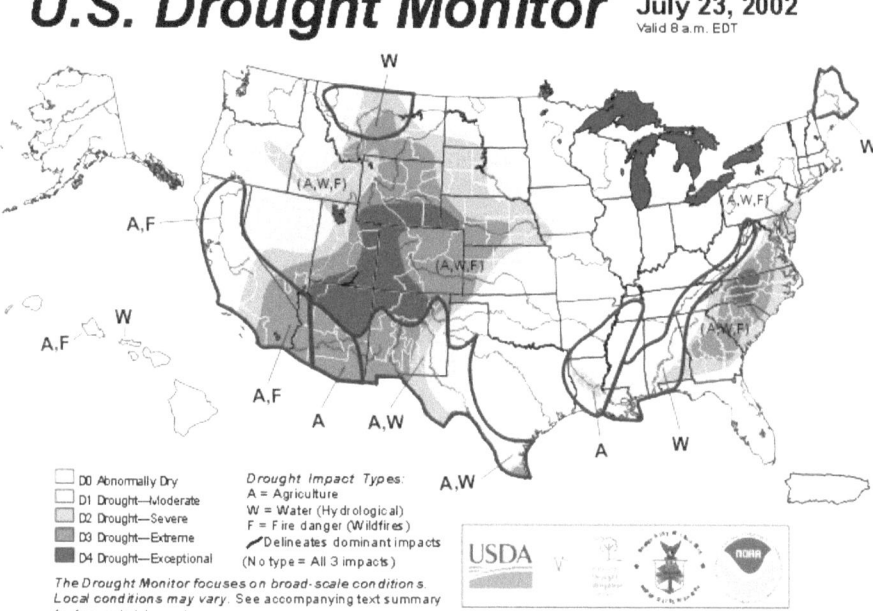

Figure 1

U.S. Drought Monitor for July 23, 2002 shows much of the state of Colorado in "exceptional" D4 drought (from National Drought Monitor, University of Nebraska – Lincoln, http://drought.unl.edu/dm).

and summer precipitation was relatively normal, hotter than average temperatures for the second summer in a row again resulted in high evaporation rates and continued depletion of soil moisture and surface water supplies. This set the stage for "The drought of 2002."

Beginning in September 2001, storm systems were few and precipitation was sparse across the Central Rockies. Much of western and southern Colorado received less than half the average September precipitation and temperatures were several degrees C above average across the entire state. Beneficial moisture fell from two storm systems that primarily affected the northeastern and east central counties of Colorado.

October weather patterns appeared more favorable as a variety of storm systems crossed the region. However, precipitation from passing storms was very light, and when the month was over precipitation totaled again less than half the average over the majority of the state. Some areas east of the mountains received no moisture at all. Temperatures were also mild ranging from about average near the Kansas border to over 2 degrees C above average over southwest Colorado.

Early November was unseasonably warm and dry. Most mountain slopes and peaks remained bare. Then, just in time for the Thanksgiving weekend, the snow began to fly. Dry powdery snow was widespread and quite deep in the mountains by the end of the month, although snow water content remained below average. In hindsight, the late November snow siege was really the only prolonged stormy period for the year, however, it was very helpful in starting the Colorado winter recreation season.

December brought many more opportunities for mountain snows, but most resulted in only a few centimeters here and there. The higher peaks and mountain ranges, particularly in northern Colorado, added some good snow, but the surrounding valleys stayed very dry. Temperatures, fortunately, were quite cold in the mountains and valleys, so there was little melting. Many areas of the state picked up less than half the December average and east of the mountains only a few millimeters of moisture was measured. Southeast Colorado fared a bit better due to a few storms coming up across Texas.

January 2002 brought seasonally cold temperatures to the state and above average snowfall for the Front Range urban corridor and the southeastern plains of Colorado. Unfortunately, January precipitation east of the mountains contributes very little to overall water supplies. In the mountains, January snows usually add significantly to the accumulating mountain snowpack. But in 2002, January precipitation in the mountains was much below average. Southwestern Colorado was the driest portion of the state with many stations in the San Juan, Animas and Dolores watersheds receiving less than 10% of the 30-year average.

February was also a disappointment. Despite cold temperatures and several storm opportunities, very little precipitation fell. North central counties did best with a few stations reporting near average snowfall and water content. But for most of Colorado, February was extremely dry with many stations reporting less than 25% of the long-term average. Because of the cold temperatures and frequent small snows, Colorado's huge winter recreation industry was able to limp along with surprisingly good snow conditions, but the snowpack water content by the end of February was only 80% of average at best in portions of northern Colorado, while in southern Colorado the snow water content was only about 40–50% of average.

March did not give many hints of the severe drought ahead. Widespread storms crossed the region at least every week, and temperatures were reluctant to begin the normal spring thaw. Unfortunately, none of the storms contributed the copious wet snows that Colorado spring snowstorms typically produce. Furthermore, the storms nearly skipped southeastern Colorado completely. Only northwestern Colorado ended up wetter than average for the month of March. Some parts of northern and central Colorado were near average. Most of Colorado however was very dry with nearly half the state less than 50% of the average.

By the end of March, the statewide snow water equivalent, as a percent of average, had dropped to 52% (Figs. 2 and 3). While not as bad as the winter of 1976–1977, these were still some very disappointing figures. Because of the heavy snows in late November, the seasonally cold temperatures and a relatively small precipitation deficit in the Front Range, and favorable publicity about good snow conditions for winter recreation, there was no strong public and government perception of a severe drought.

But then came April, and the reality of drought quickly hit home. The spring storms that sometimes dump heavy and widespread precipitation were non-existent in April. Almost no precipitation fell in eastern Colorado, and mountain precipitation was also meager. To make matters worse, April temperatures soared to record highs, especially in the mountains (Fig. 4), and mountain snow melted or evaporated at an alarming rate. Relative humidity on several afternoons fell to below 10%. Fire danger, which typically stays low to moderate through early June, was already high by mid-April, and the first severe forest fire of the season ignited 30 miles southwest of Denver on April 23rd (Snaking Fire). For the month as a whole, precipitation was less than 50% the average over three quarters of the state (Fig. 5). Temperatures ranged from about average near the Nebraska border to over 4 degrees C above average in the high valleys of the central mountains making this the warmest April on record for several mountain locations. Strong winds also occurred which enhanced evaporation losses beyond the seasonal average. Farmers trying to get crops planted had to apply early irrigation water resulting in premature depletion of the already limited water supplies.

May, while not quite as much warmer than average as April, was even drier. Only the northern Front Range area received significant moisture (Fig. 6). At a time of year when Colorado's rivers and streams are normally churning with snowmelt runoff, streamflow remained eerily placid. Irrigation water demand ramped up fast,

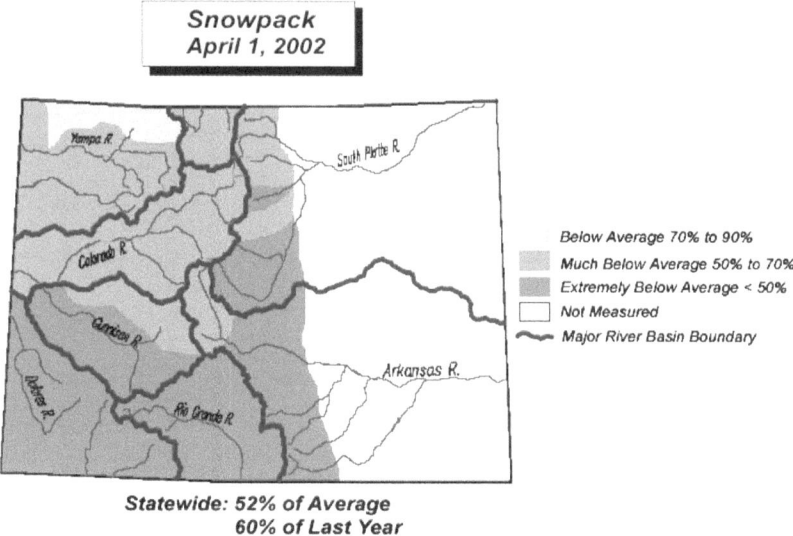

Figure 2
April 1, 2002 snowpack for state of Colorado (from the National Resources Conservation Service, NRCS, http://www.co.nrcs.usda.gov/snow/data/snmap402.html).

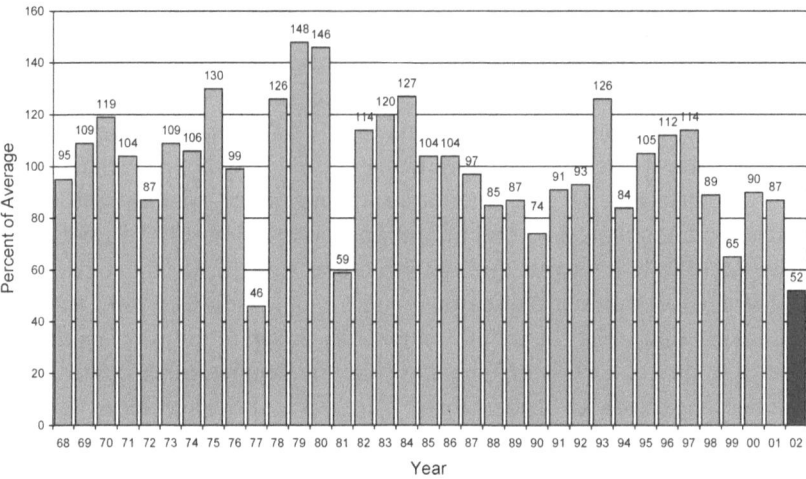

Figure 3

April 1 Snowpack percent of average for Colorado by year from 1968 through 2002 (from NRCS, Snow Survey Division).

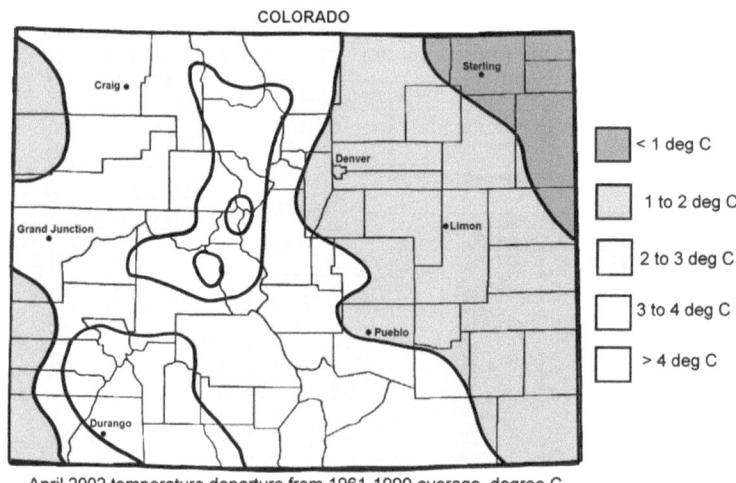

Figure 4

April 2002 temperature departures from the 1961–1990 average for the state of Colorado, USA (degrees C).

but it soon became obvious that supplies would not last through the growing season. Municipalities began to face the dire prospect that available water supplies might not provide for the typical summertime demand, so many areas began implementing

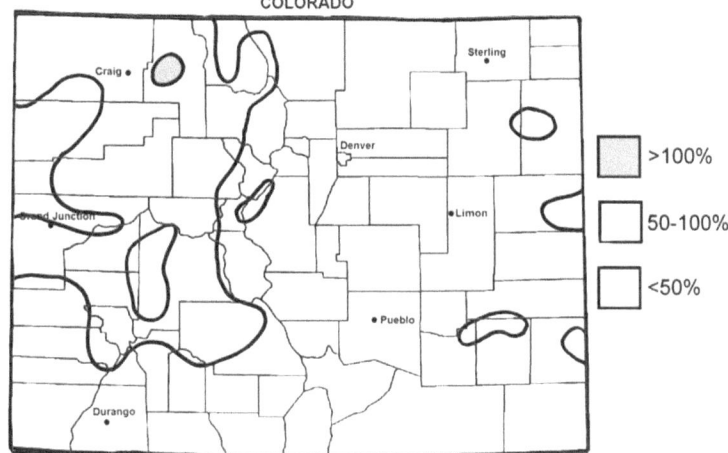

April 2002 precipitation as a percent of the 1961-1990 average.

Figure 5
April 2002 precipitation as a percent of the 1961–1990 average for the state of Colorado, USA.

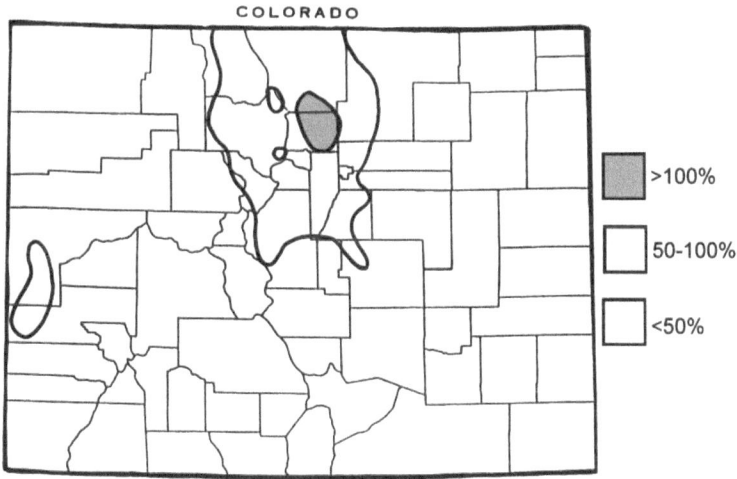

May 2002 precipitation as a percent of the 1961-1990 average.

Figure 6
May 2002 precipitation as a percent of the 1961–1990 average for the state of Colorado, USA.

strict water conservation regulations. More forest fires erupted and each new fire seemed to spread faster than the one before.

June arrived accompanied by relentless summer heat. Vegetation that normally grows lush and tall during the spring barely greened up. By June, relative humidity often dropped to less than 10%, and bans on outside burning were enforced over much of the state. Temperatures routinely climbed to the 30–40° C range at lower elevations east and west of the mountains. Dry air allowed nighttime temperatures to dip to comfortable levels most every night. Little or no precipitation fell for the entire month of June over western Colorado (Fig. 7). East of the mountains, a few thunderstorms occurred and some locales enjoyed respectable rainfall amounts. Parts of Cheyenne County, for example, reported more than 100 mm of rain in June. But with persistent high temperatures, frequent strong winds, and low humidity, the rain scarcely greened the native vegetation. Winter wheat crop conditions continued their rapid deterioration, and ranchers quickly sold or moved all or parts of their herds in response to the poor range conditions and high cost of feed. The most severe fires of the season erupted in June including the Hayman fire southwest of Denver, which quickly grew to be the largest documented forest fire in Colorado (557 km^2) since records have been kept.

July brought a few changes. While precipitation was again below average statewide, and temperatures were above average for the fourth consecutive month, some increase in humidity was observed later in the month. Initially, wildfire smoke could be seen almost every day, but eventually, as humidity rose, fires spread more slowly, and some were successfully extinguished. July is normally the most lightning prolific month of the year, but in 2002 thunderstorms were few. This helped the fire situation by reducing the number of natural ignitions. There were some focused locations with showers and thunderstorms during July. A few small, localized areas, mostly in or near the mountains, ended up with near

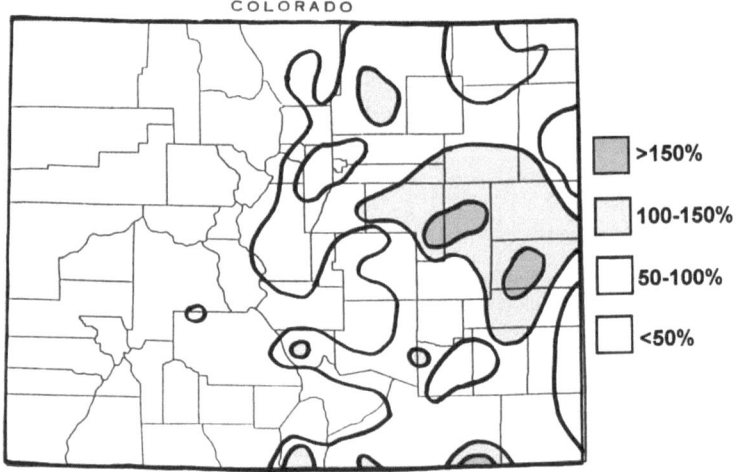

June 2002 precipitation as a percent of the 1961-1990 average.

Figure 7
June 2002 precipitation as a percent of the 1961–1990 average for the state of Colorado, USA.

average rainfall for the month. But most areas remained dry. The eastern plains were parched with most stations reporting less than 30% of their average July precipitation. Even where irrigation water held out, crops withered under the stress of heat and low humidity. Many irrigation water supplies came to an end, and crop failure ensued. By late July, Colorado was in a very serious drought. Furthermore, drought conditions were not limited just to Colorado but extended over much of the Great Plains and Rocky Mountain States (Fig. 1). In hindsight, the drought pattern that evolved though July 2002 started 3–4 years earlier but intensified in the year 2002.

August arrived with some optimism. The first several days of the month were not quite as hot, and subtropical moisture helped to fuel more afternoon showers and thunderstorms. But the monsoon moisture surge was brief and soon ended. By the 10th of August heat and low humidity returned accompanied by another round of fast-spreading fire activity. Crop and range conditions continued to deteriorate as did streamflows and water levels in the state's largest reservoirs. By mid-August, media reports likened this to the great Dust Bowl of the 1930s. Temperatures during the day occasionally reached over 38° C temperatures in Front Range cities. As the month neared its end, a subtle change in weather patterns brought a round of spring-like thunderstorms loaded with hail and high winds to portions of eastern Colorado. The hail did little damage, however, since so few crops were still growing in late August. For the state as a whole, August precipitation was still below average, but unlike previous months there were some large areas of eastern Colorado that received heavy rains.

Humid and stormy weather continued into September. For the first time since August 2001, the majority of Colorado received above average rainfall. Temperatures were still warmer than average, but with the cooler air of fall, frequent showers and a few soaking rains, grasses actually began to green up a bit. Quite a few stations accumulated at least double the average monthly rainfall. Even the bone-dry areas of southwest Colorado got some much appreciated moisture with some areas reporting over 100 mm of moisture for the month. With cooler weather imminent, and the growing season drawing to a close, the worst of the 2002 drought was at last behind us.

Fig. 8 shows precipitation for the entire 2002 Water Year as a percent of the 1961–1990 average. For the first time since such records have been kept, the entire state was below average and the majority of the state was less than 70% of average. The driest areas of the state below 50% were Weld County, an area surrounding Colorado Springs, Pueblo and Rocky Ford, a section near Durango, and portion of the San Juan Mountains and east to Del Norte and Center. These areas generally covered the sites in Table 1 where 2002 Water Year was the driest year on record.

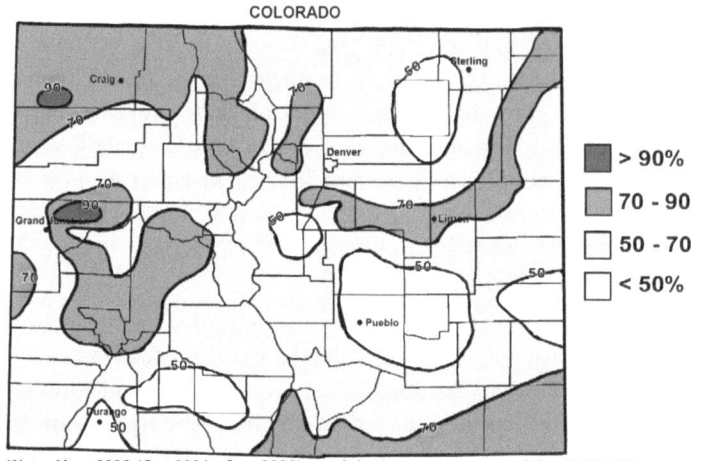

Water Year 2002 (Oct 2001 - Sep 2002) precipitation as a percent of the 1961-1990 averages.

Figure 8
Water Year 2002 (October 1, 2001 – September 30, 2002) precipitation as a percent of the 1961–1990 average for Colorado, USA.

3. *Quantitative Analysis*

There is no question that the state of Colorado suffered a serious drought in 2002. The impacts to municipal water supplies, agriculture, recreation and streamflows were exceptionally severe. This section of the paper investigates the question as to whether or not the severity of water deficits were out of proportion to the actual precipitation deficit.

A. *September 1, 2001 to August 31, 2002 Precipitation*

The first evaluation concerns the observed precipitation in eight experimental Colorado climate divisions (Fig. 9) for the core period of the Colorado drought. These climate divisions are based on the similarity of historical precipitation anomaly patterns, and allow for a more representative assessment of Colorado climate anomalies than conventional NCDC climate divisions (WOLTER, 2003). Table 1 presents accumulated precipitation data for 12-month (Sep. 2001 – Aug. 2002), 13-month (Sep. 2001 – Sep. 2002) and 12-month 2002 Water Year periods (Oct. 2001 – Sep. 2002). The magnitude of the standard deviations below the average for each station for each time period are also shown. Two observing sites in each of the 8 regions were analyzed. Stations were ranked for the period-of-record and 1941–2002 time period (the longest period of record in common for all stations). While the time period available varies (the higher altitudes have shorter records), the data can place the 2002 drought in perspective.

Table 1

Precipitation Accumulation Analysis for 12-month (Sep. 2001–Aug. 2002), 13 month (Sep. 2001–Sep. 2002) and 2002 Water Year (Oct. 2001–Sep. 2002) compared to Period-of-Record (POR) and 1941–2002 period. In parentheses are the magnitudes of the standard deviations below the average for the time period, based on the available data for each station.

Climatic Stations	Region	Period of Record (POR)	Sep. 2001 – Aug. 2002 (12 months)			Sep. 2001 – Sep. 2002 (13 months)			2002 Water Year (Oct. 2001 – Sep. 2002)		
			POR Rank	1941–2002 Rank (SD)	mm	POR Rank	1941–2002 Rank (SD)	mm	POR Rank	1941–2002 Rank (SD)	mm
Grand Lake 1NW	1	1940–2002	1	1 (1.87)	319	2	2 (1.42)	407	1	1 (1.71)	327
Taylor Park	1	1941–2002	1	1 (2.02)	265	2	2 (1.65)	324	3	3 (1.51)	303
Grand Junction WSO	2	1892–2002	8	5 (1.39)	141	31	20 (0.63)	205	43	27 (0.34)	201
Meeker	2	1891–2002	7	5 (1.78)	263	8	6 (1.64)	303	7	7 (1.58)	276
Montrose No. 2	3	1896–2002	3	3 (1.54)	148	15	8 (1.02)	203	29	1 (0.80)	193
Mesa Verde NP	3	1923–2002	1	1 (2.30)	189	1	1 (1.97)	250	3	3 (1.79)	246
Del Norte 2E	4	1920–2002	1	1 (2.55)	81	3	3 (1.96)	132	3	3 (1.80)	119
Center 4 SSW	4	1891–2002	1	1 (2.69)	62	1	1 (2.20)	97	4	4 (1.92)	94
Colorado Springs WSO	5	1892–2002	1	1 (2.13)	165	1	1 (2.01)	198	2	1 (1.97)	172
Pueblo WSO	5	1891–2002	1	1 (2.65)	96	1	1 (2.64)	113	1	1 (2.56)	101
Rocky Ford 2SE	6	1892–2002	1	1 (2.33)	92	1	1 (2.36)	108	1	1 (2.31)	238
Cheyenne Wells	6	1897–2002	4	2 (1.67)	235	10	5 (1.39)	291	9	5 (1.48)	252
Akron 4E	7	1905–2002	1	1 (2.02)	239	1	1 (1.83)	277	1	1 (1.88)	95
Leroy 7WSW	7	1891–2002	4	2 (1.91)	269	4	2 (1.82)	294	4	2 (2.05)	243
Kassler	8	1899–2002	8	4 (1.49)	319	9	5 (1.44)	351	6	4 (1.48)	309
Fort Collins	8	1890–2002	3	2 (1.96)	200	3	2 (1.84)	237	5	4 (1.66)	214
Stations Ranked Driest			9	9		6	6		4	6	

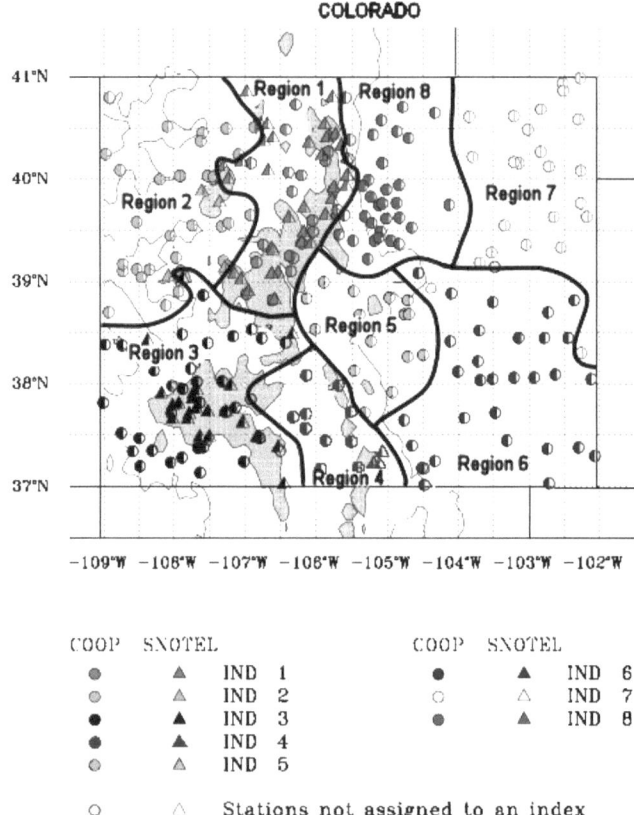

Figure 9
Climate divisions for the state of Colorado (from Klaus Wolter, NOAA-CIRES Climate Diagnostic Center).

For 9 of the 16 sites, the 12-month time period of September 1, 2001 to August 31, 2002 was the driest for the period-of-record and the 1941–2002 record. Table 1 also presents the 13-month time period with September 2002 added. This month was obviously relatively wet, as only 6 sites were the driest during the period-of-record and the 1941–2002 period. Using the 2002 Water Year (October 1, 2001 to September 30, 2002), 4 of the 16 sites were the driest of record (or 5 of the 16 sites for the 1941–2002 ranking). Clearly, the time period examined (even shifted by one month or adding a month) provides a different perspective on the severity of the precipitation deficit in Colorado. Grand Junction, for example, was 1.39 standard deviations below the mean for the September 2001 to August 2002 time period, but this was reduced to 0.34 standard deviations below the mean for the Water Year and 0.63 for the September 2001 to September 2002 time period.

If we assume the precipitation for 12-month and 13-month time periods follows a normal probability distribution and adopt a probability of 1% or less as being exceptional, for the period September 2001 to August 2002, 5 of the 16 stations were in this category (Del Norte, Center, Mesa Verde, Pueblo, and Rocky Ford). The assumption of a statistical normal distribution is appropriate for time scales longer than 12 months (MCKEE et al., 1993). We apply this assumption for our 12-month and 13-month data sets since the data should be at least close to this distribution. For the Water Year and the period September 2001 to September 2002, 2 sites (Pueblo and Rocky Ford) were in this class out of the 16 sites. Using a 5% or less probability, for the period September 2001 to August 2002, 13 sites were in this category. For the Water Year, 10 were in this class, while for the 13-month time period, 11 were in this class. To place these probabilities in context, a 5% probability means that there is a 1 in 20 chance of precipitation being below the observed value for that time period, while a 1% probability indicates a 1 in 100 chance. This analysis suggests that most of these precipitation observing sites had a serious drought but the deficit, even in the core period of the drought, was not exceptional. Of course, in making this conclusion, we are accepting these station as being representative of their respective climate divisions.

In all three time periods evaluated, over the common interval of record and using the actual observed data rather than the standard deviations to determine whether the drought was exceptional, four stations consistently ranked the driest (Colorado Springs, Pueblo, Rocky Ford, and Akron), suggesting that this drought, as a single year ranks as the most severe precipitation deficit on record for some of south central and southeastern Colorado and a small portion of the South Platte Basin in northeastern Colorado. For these four locations, the deviations from the mean for the September 2001 to August 2002 time period in terms of the standard deviation were 2.13, 2.65, 2.33 and 2.02 below the average, respectively. To place these values in context, assuming a normal probability distribution, a value of 2.65 would have a probability of occurrence of 0.4% in any one year. Even for the Water Year and for the September 2001 to September 2002 time period, the precipitation was more than two standard deviations below average except for Colorado Springs (-1.97 for the Water Year) and Akron (-1.88 for the Water Year and -1.83 for the 13 months).

In the historical perspective, the Dust Bowl years of the 1930s were more severe over parts of the Eastern Plains of Colorado, while the northern Front Range of Colorado was drier in the mid-1950s. Based on the instrumental record, however, observed statewide precipitation anomalies in 2001–2002 were among the most severe of the last century. Ironically, the most populated region of the state, the northern Front Range (region 8 in Fig. 9) had the least severe precipitation deficit, but was affected by harsher drought conditions over the upstream mountains (region 1). The use of a statewide average, of course, masks the actual variations with Colorado which is why we choose to use climate divisions within the state.

B. Cumulative Precipitation Plots

The 2002 Water Year data presented in Table 1 can be presented on a month-by-month basis, along with cumulative series of the 30-year average, the driest year, and the wettest year (Fig. 10). These plots show the absence of wet months for the period October 1, 2001 to September 30, 2002. The cumulative precipitation values for the 2003 Water Year illustrate the recovery for much of the state, while the southwestern part of the state, in particular, remained in a precipitation deficit, albeit not as severe as the previous year.

C. Snowpack

Figure 11 presents the snowpack for the major water basins in Colorado during the first half of 2002, while Figure 12 presents the cumulative plot for the years 1999–2000, 2000–2001, 2001–2002, 2002–2003, as well as the average. The deficit for 2001–2002 is clearly evident, including the early melt of the snowpack. Thus, the 1 April 2002 statewide average that was the second lowest after 1977 (Fig. 3), fell to the lowest level on record (since 1968) by 1 May 2002. By comparison, the 1977 snow drought was embedded in a string of near-average to above-average years (Fig. 3), thus minimizing its impact on the state in terms of reservoir management. It is noteworthy that the last four snow seasons depicted here share an early melt-out date along with below-normal snowpack for most of the season. This increases the length of time that snowmelt water is exposed to evaporative losses in open reservoirs. Thus not only was the snow accumulation particularly below average in 2002, the early loss of snow prevented it from being used later in the spring season. Moreover, the wet years that occurred before this drought promoted vegetation growth, which increased the transpiration demand for water. Such an increase of transpiration would reduce the amount of water available for river runoff.

Figure 13 illustrates the regional temperature departures for the 2002 Water Year. For the month of April 2002, these warmer than average temperatures explain the early snow melt in the mountains, as well as a more rapid than usual physical evaporation and transpiration of what little soil moisture existed in the soil throughout the state.

▶

Figure 10

Cumulative precipitation totals for 2002 Water Year, 2003 Water Year, 30-year average, and maximum and minimum years for regions (a) Grand Lake, (b) Meeker, (c) Mesa Verde National Park, (d) Center, (e) Pueblo, (f) Rocky Ford, (g) Akron and (h) Fort Collins, Colorado. One inch = 25.4 mm.

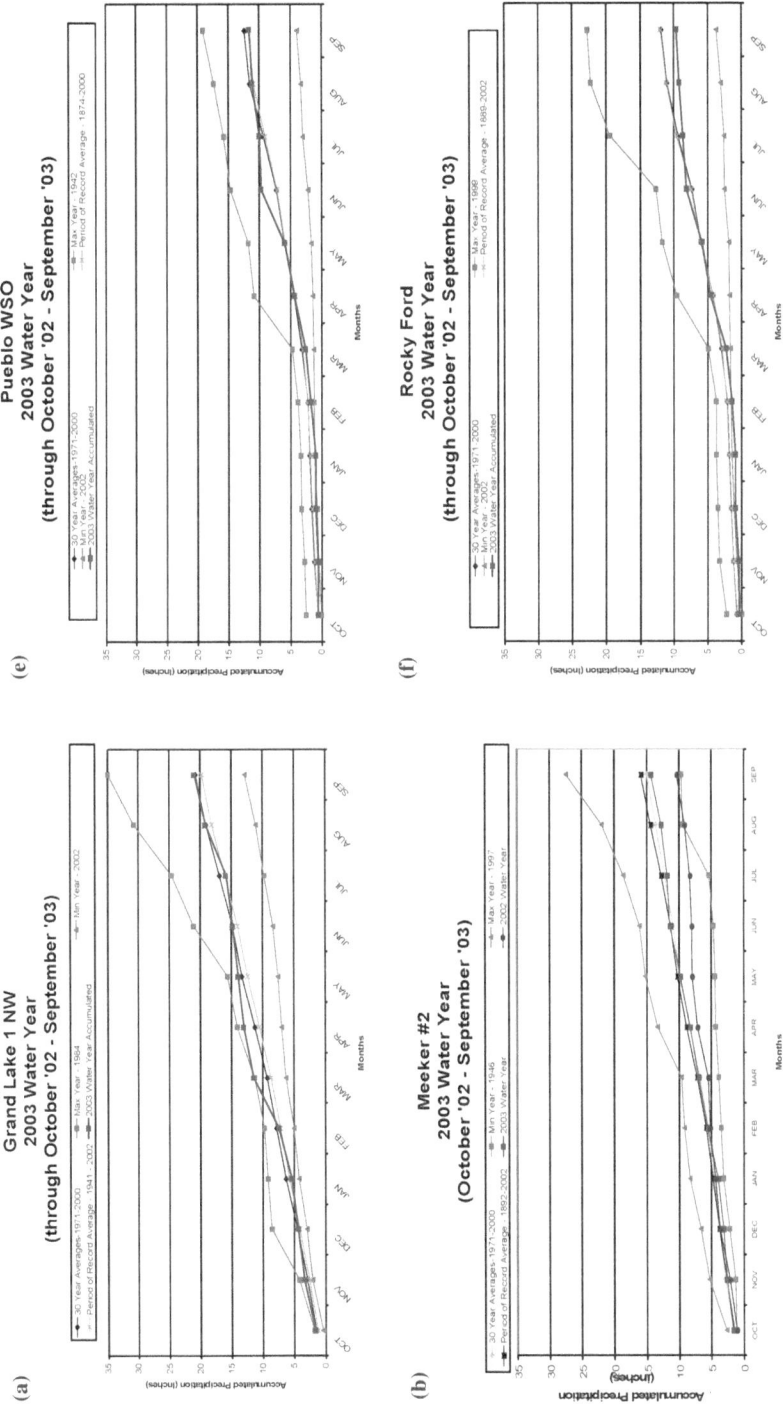

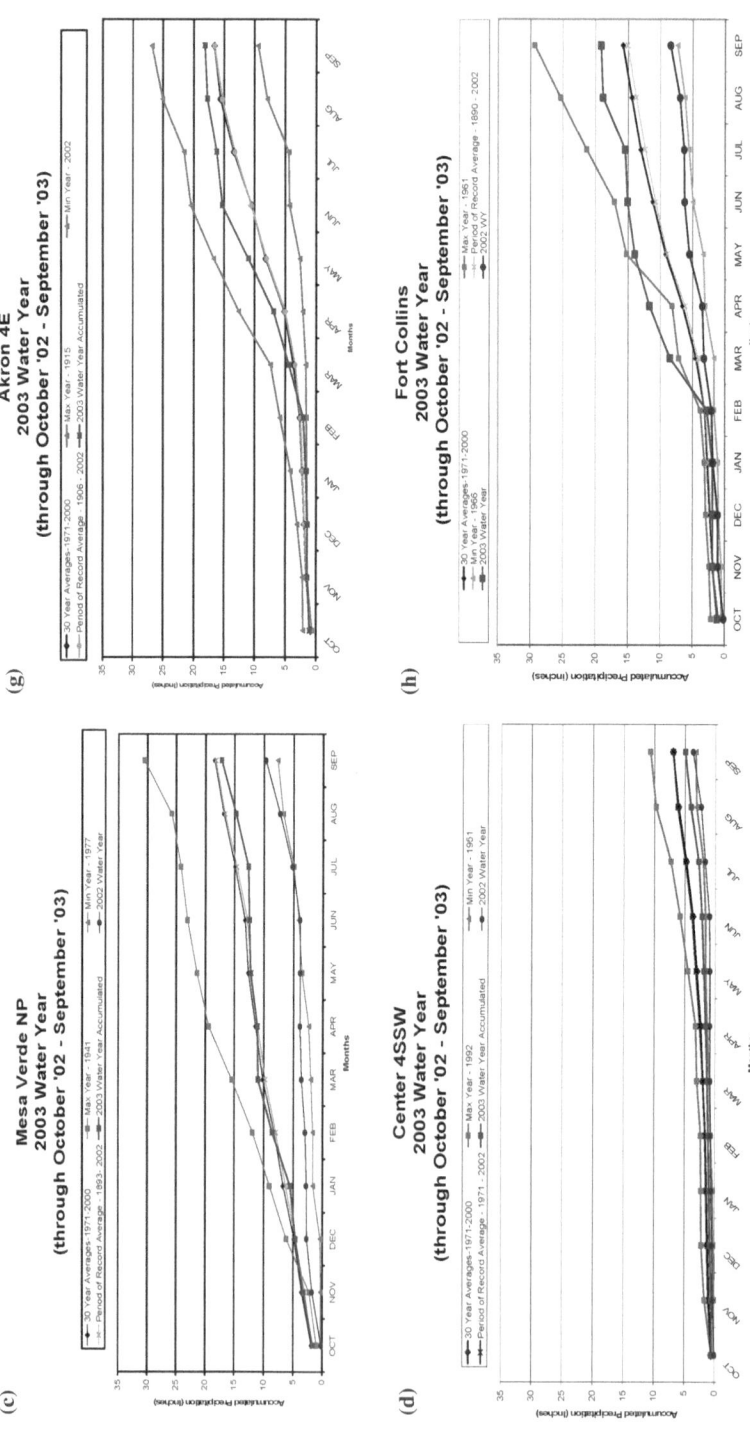

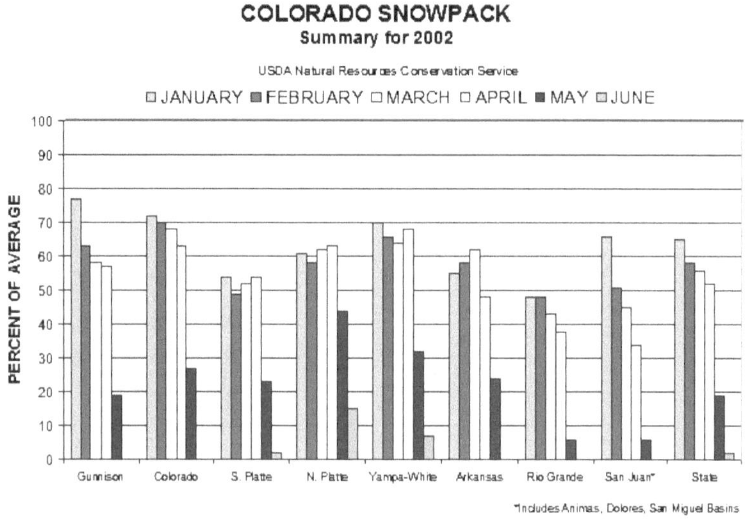

COLORADO SNOWPACK
Summary for 2002
USDA Natural Resources Conservation Service
JANUARY FEBRUARY MARCH APRIL MAY JUNE
PERCENT OF AVERAGE
100
90
80
70
60
50
40
30
20
10
0
Gunnison
Colorado
S. Platte
N. Platte
Yampa-White
Arkansas
Rio Grande
San Juan*
State
*Includes Animas, Dolores, San Miguel Basins

Figure 11
Colorado basin snowpack as a percent of average for months January to June 2002 (from USDA, NRCS).

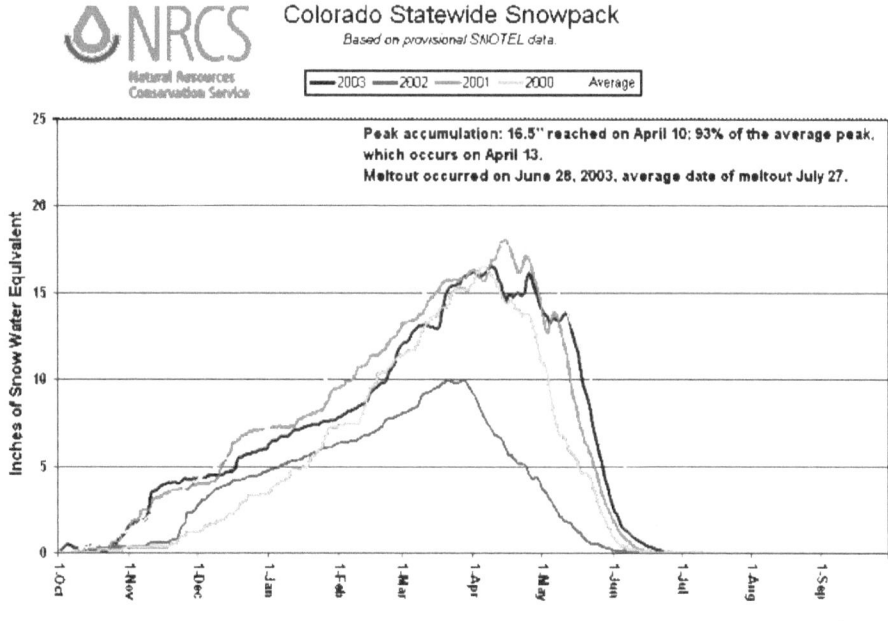

NRCS
Natural Resources Conservation Service
Colorado Statewide Snowpack
Based on provisional SNOTEL data.
2003
2002
2001
2000
Average
Peak accumulation: 16.5" reached on April 10; 93% of the average peak, which occurs on April 13.
Meltout occurred on June 28, 2003, average date of meltout July 27.
Inches of Snow Water Equivalent
25
20
15
10
5
0
1-Oct
1-Nov
1-Dec
1-Jan
1-Feb
1-Mar
1-Apr
1-May
1-Jun
1-Jul
1-Aug
1-Sep

Figure 12
Cumulative statewide snowpack for Colorado for years 1999–2000, 2000–2001, 2001–2002, 2002–2003, and average (from USDA, NRCS). One inch = 25.4 mm.

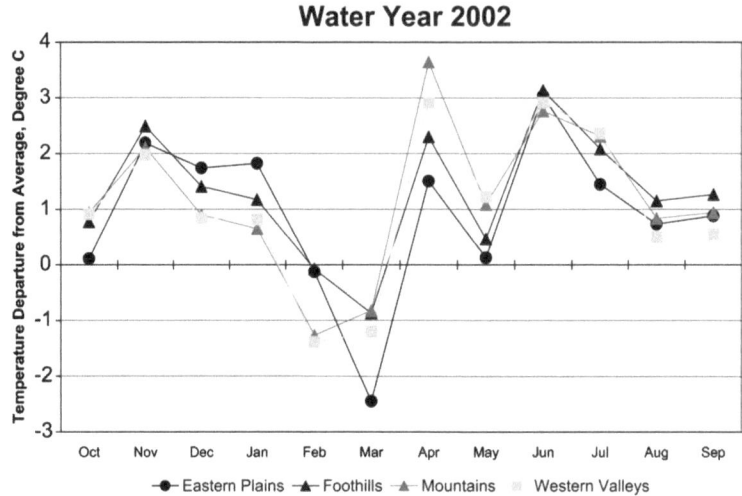

Figure 13
Temperature departures from average for the 2002 Water Year for the Eastern Plains, Foothills, Mountains and Western Valleys. Notice the warm temperatures for April 2002 in the Mountain region, greater than 3°C above average.

D. Streamflow

From the water resources perspective, streamflow is a key hydrological process that summarizes various atmospheric, land surface, and subsurface components of the hydrologic cycle. It is particularly useful for water resources managers because it reflects the water that may be available at a given diversion point of a stream or may be entering lakes and reservoirs. Because water supply hydraulic structures are generally designed to meet projected water demands, drought analysis typically involves the relationship of both water supply and water demand. Thus, when streamflow in a given time period becomes smaller than the demand, a deficit occurs and a sequence of continuous deficits may become a hydrological drought. To illustrate this issue for Colorado, this section presents a detailed example of the severity of the deficit in the Water Year 2001–2002 taking as example the streamflow data of the Poudre River.

The time series of naturalized annual flows of the 2002 Water Year (Oct. 2001 - Sept. 2002) for the Poudre River in northeastern Colorado at the Mouth of the Canyon station for the period 1884–2002 is shown in Figure 14. It indicates that the Water Year 2002 had the lowest flow in the historical record. Streamflow records for the major river basins in the state showed similar decreasing flows throughout the period 1999–2002 with the Water Year 2002 being the smallest or near the smallest on record. Considering that the average annual flow for the entire record is about 299,011 acre-ft., the time series in Figure 14 shows a wet period of about 40 years in

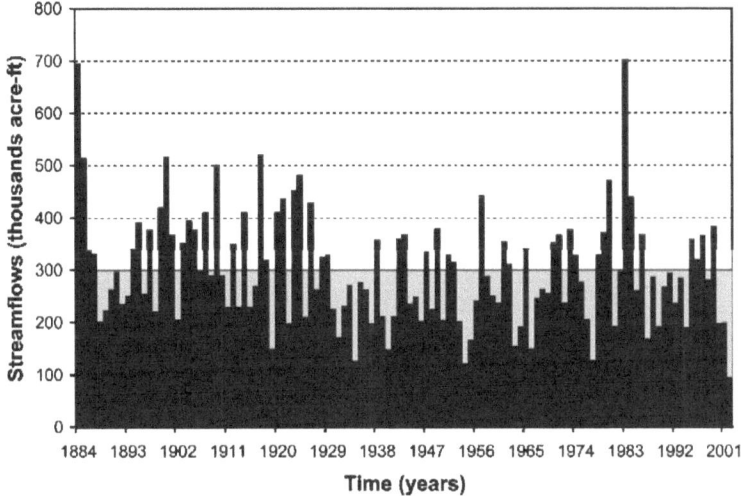

Figure 14
Annual flow records of the Poudre River for the period 1884–2002. The figure shows some extreme drought events such as those in the 1930s, 1950s, and the drought of the 2000s. Note that the 2002 flow is the smallest value in the entire record (after SALAS *et al.*, 2005). One acre-ft = 1233.5 cubic meters.

the first part of the record and a drier period in the rest of the record. In addition, the time series indicates that various drought episodes have occurred in the Poudre River throughout the historical record, such as those of the 1930s, the 1950s, as well as the 3-year drought in the period 2000–2002. Because the standard deviation of the annual flows is 106,512 acre-ft, the 2002 streamflow (95,000 acre-ft) is 1.9 standard deviations below the mean. However, the annual streamflows are somewhat skewed (0.98), so perhaps a better picture of the drought severity in that year may be obtained from the transformed annual flow data. In such cases, the 2002 transformed annual flow is 40.2, which is 2.8 standard deviations below the mean (of the transformed flows). Either case illustrates the severity of the drought in that year. Also Figures 15 (a) and (b) show respectively in the original and transformed flow domains, the plots of the monthly streamflows during the Water Year 2002 compared to the long-term mean monthly flows, the mean monthly flows minus one monthly standard deviation, and the mean monthly flows minus two monthly standard deviations. Clearly, both Figures 15 (a) and (b) show that the monthly flows for the year 2002 especially for the months of May, June, and July are significantly low.

To characterize further the severity of the 2002 drought, we use the concept of return period (mean recurrence interval). For our study the 119 years of historical annual streamflow data were statistically analyzed and the first-order autoregressive (AR-1) model fitted to the transformed flows. The model was tested based on various fitting techniques and comparing certain statistics obtained from the historical and

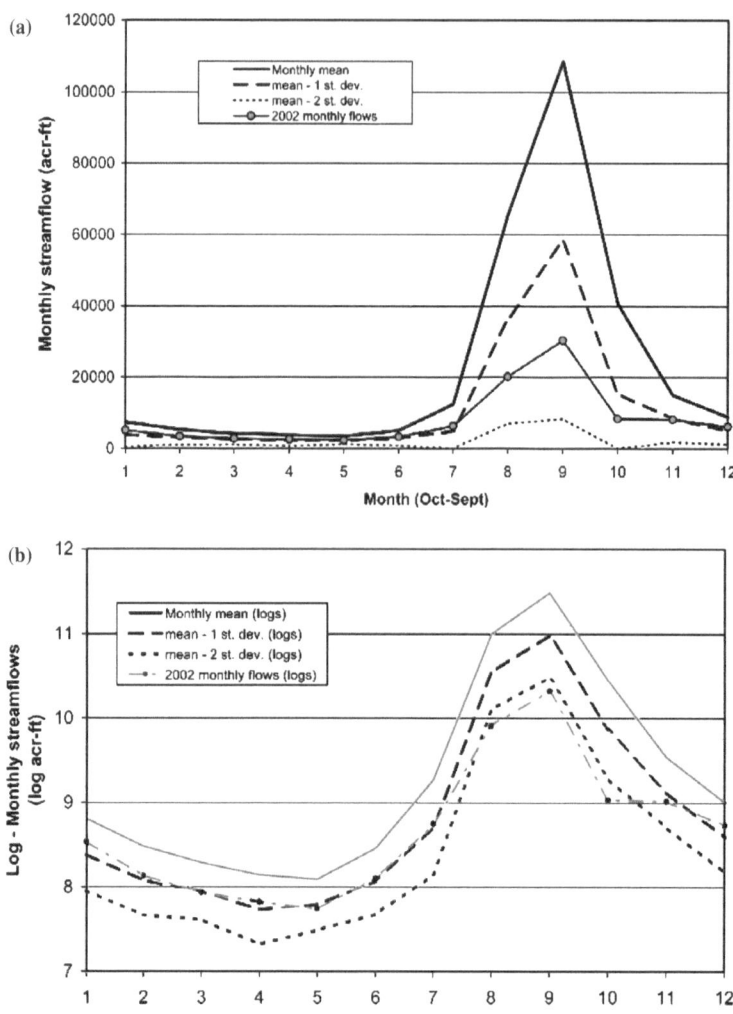

Figure 15
Comparison of the 2002 monthly flows of the Poudre River versus the monthly mean, monthly mean minus one standard deviation, and monthly mean minus two standard deviations in the (a) original flow domain and (b) the log-transformed flow domain. Note the significant low flow conditions especially for the 3-month period May, June, and July. One acre-ft = 1233.5 cubic meters.

generated samples (e.g., SALAS, 1993). The AR(1) model was then used for simulating a 200,000-year sample from which drought severity was determined. For ease of reference, we use the following notation. The deficit threshold D_0 is defined as a fraction of the threshold water demand x_0, i.e. $D_0 = \lambda x_0$, so that for a single year drought $0 \leq \lambda \leq 1$, and λ is called the deficit coefficient. The water demand threshold used for the drought analysis of the Poudre River is the long-term mean, i.e.,

$x_0 = 299,011$ acre-ft and the deficit threshold D_0 is the drought in the year 2002 that reached a deficit of 204,011 acre-ft, i.e., $\lambda = 0.682$.

Given that we are concerned with characterizing the severity of a single year drought event, it may be tempting to estimate the return period of the 2002 Water Year deficit (204,011 acre-ft) by the usual frequency analysis of the historical deficits. This analysis will give an estimate of about 120 years of return period. However, because the data are autocorrelated such an approach would not provide an accurate estimate of the return period. Thus, we estimated the return period of 1-year droughts exceeding a specified level of deficit based on the concept of mean interarrival time. The estimated return periods for various levels of deficit are shown in Figure 16 for both the generated and the historical data. Only a few points are shown for the historical data because of the lack of enough drought events from which to calculate the return periods. Nevertheless Figure 16 shows a close agreement between the generated and historical results. Thus, from Figure 16 the estimated return period of the 2002 drought with deficit of 204,011 acre-ft is *436 years!* In conclusion the analysis of this river (and other rivers in the state appear to have such extreme deficiencies of flow), shows that the impact of this drought was exceptional with respect to water supplies.

E. Reservoir Storage

The snowpack deficit, and resultant reduced river flow, of course, would be expected to result in less reservoir storage. As discussed in Section C above, the

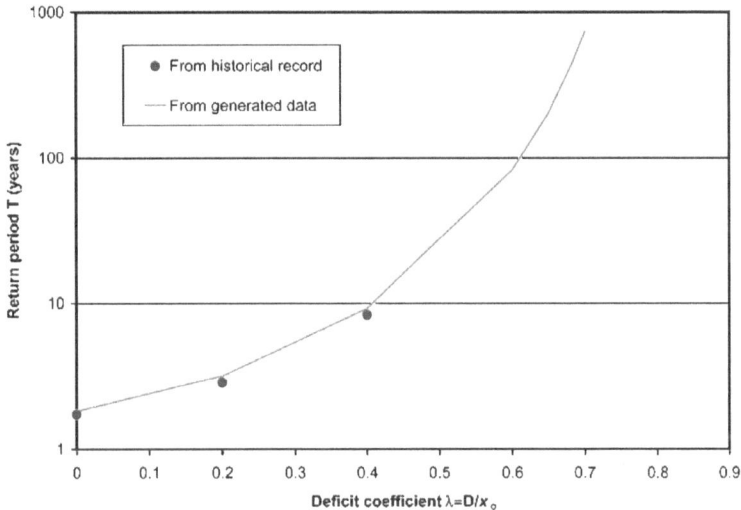

Figure 16

Comparison of the return period (T) of single-year deficit obtained from the generated flow record versus the T obtained from the historical record for various values of the deficit coefficient λ.

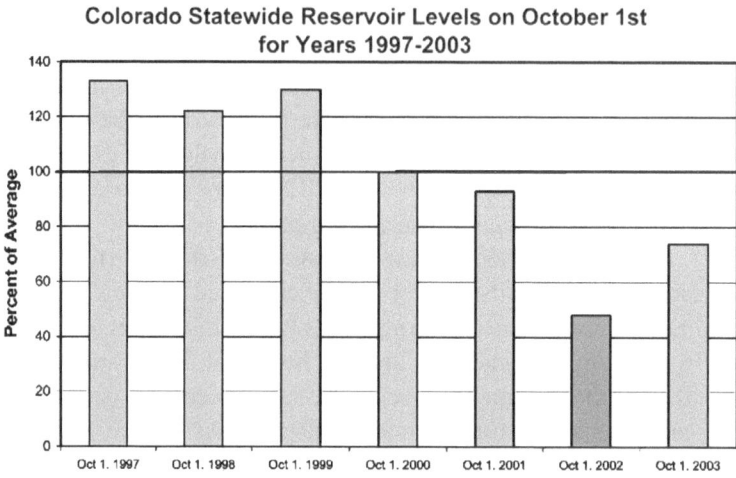

Figure 17

Colorado statewide reservoir storage levels as a percent of average for the end of the growing season (from the NRCS).

warmer than average maximum temperatures shortened the melt season, as well as resulted in greater evaporation losses. The reservoir storage percentage at the end of the growing season in 2002, shown in Figure 17, shows the lower than average volume for the state. As the warm, dry growing season continued in 2002, the large depletion of water storage as a result of above average irrigation and municipal water demand resulted in the implementation of water restrictions, as discussed in Section 2.

F. Paleo-historical Perspective

Instrumental records of precipitation, snowpack, and reservoir storage provide a temporal context for assessing the 2002 drought that ranges from several decades to slightly more than a century, depending on the record. This time frame can be extended with paleoclimatic proxy data, which document a potentially broader range of natural climate variability. In much of the western U.S., lower elevation coniferous trees have proved to be excellent proxies for hydroclimatic variability, as tree growth at lower elevation generally responds to variations in available moisture. In Colorado, variations in tree-ring widths of these conifers are closely correlated to seasonal precipitation (spring in particular) as well as metrics that integrate climate conditions prior to and/or concurrent with the growing season, such as total water year flow and winter snowpack. Thus, variations in tree-ring widths are a proxy for past moisture variability and can be used to reconstruct past climate. In western Colorado, tree-ring data have been useful for high-quality reconstructions of Water Year streamflow, April 1 snow water equivalent (WOODHOUSE, 2003), the Standard

Precipitation Index (SPI), and seasonal precipitation, indicating tree sensitivity to drought measured in a number of ways.

Twelve tree-ring chronologies from sites in the basins of Gunnison River and the main stem of the Colorado River were resampled in June, 2003 to update them to 2002. Each chronology is comprised of dated and measured series from about 20–30 trees (two samples per tree) which have had growth trends and high order autocorrelation, both related primarily to biological factors, removed (FRITTS, 1976; COOK and KAIRIUKSTIS, 1990). The start dates of the tree-ring chronologies range from A.D. 1135 to A.D. 1440, allowing an assessment of the 2002 ring width within the time frame of last five to eight centuries. When evaluated for the 20th century (1900–2002), 2002 is the narrowest ring in just three of 12 chronologies, but when the 12 chronologies are averaged together, 2002 is the narrowest ring, followed by 1902, 1977, and 1954. When the 12 chronologies are averaged for the full common chronology period, 1440–2002, 2002 is the third narrowest out of 563 years. In individual chronologies, 2002 ranks in the narrowest 6th percentile to the narrowest half a percentile, so it is clearly a very low growth year even in the context of 500–800 years. The two narrowest rings in the 12-chronology average are 1685 and 1851. Thus, 2002 is the narrowest tree-ring width in the Gunnison/Upper Colorado region of western Colorado in 150 years, reflecting drought severity as defined by tree growth. Preliminary results suggest 2002 tree growth was at least as suppressed in most areas of the Front Range as well.

4. Conclusion

The evaluation of the severity of the 2001–2002 drought varies according to variable measured and spatial scale considered. Although this paper presents only a subset of the data that recorded this drought in Colorado, some conclusions can be drawn.

From a precipitation perspective, the 2001–2002 drought in Colorado was almost certainly not a statewide record; however, it was the driest for the available period September 1, 2001 to August 30, 2002 record for 9 of 16 representative sites. However, shifting this period by just one month, or adding one month, eliminated the majority of the observing sites from the driest on record. Nonetheless, for some parts of the state (the southern Front Range in particular), it was the most severe single drought year in the instrumental record (back to the late 19th century).

For statewide snowpack, low seasonal snowpack and warm April temperatures resulted in early snowmelt and losses due to evaporation in 2002, but 1977 was a year of comparable severity in the 35-year record available (as defined by the April 1 snowpack), followed by 2002. By May 1, however, the 2002 snowpack was reduced to below that of 1977. These dry and warm conditions are reflected in streamflows and reservoir storage, which integrates conditions from previous years. Streamflow records for the major river basins in the state showed similar decreasing flows throughout the period 1999–2002 with the Water Year 2001–2002 being the smallest or near the smallest on record. As an example, the analysis of the 119 years of streamflow records for the Poudre River showed that the severity of the water deficit in the Poudre in the Water Year 2002 was of the order of 400+ years return period which confirms the severity of this aspect of the drought. 2002 shows to be a year where a particularly severe water shortage had developed.

The tree-ring record, providing a longer-term context for evaluating 2002, shows tree growth for a large portion of western Colorado to be the lowest in 150 years. This small growth increment, as with snowpack, streamflow, and reservoir storage, reflects the cumulative effect of both moisture deficits and warm temperatures.

Although precipitation deficits were not exceptional in all areas of the state, evaporation losses, hot temperatures, and higher than average municipal and irrigation demand, resulted in a drought event that severely impacted many economic sectors in Colorado, and provided a "wake-up call" for many Colorado water management agencies.

The magnification of the impacts, therefore, with respect to the actual precipitation deficit indicates Colorado society is now more vulnerable to short-term drought than in the past. This sobering message is the one the policy makers need to digest and react to.

Acknowledgements

This work was supported by the Colorado Climate Center under the Colorado Agricultural Experiment Station, the Drought Analysis and Management Laboratory of Colorado State University, as well as the Western Water Assessment (WWA) project, funded through NOAA-Office of Global Programs, and the Denver Water Board.

REFERENCES

COOK, E.R., and KAIRIUKSTIS, L.A., *Methods of Dendrochonology: Applications in the Environmental Sciences* (Kluwer Academic Publishers, 1990).

FRITTS, H.C., *Tree Rings and Climate* (Academic Press 1976).

MCKEE, T.B., DOESKEN, N.J., and KLEIST, J. (1993) *The Relationship of Drought Frequency and Duration of Time Scales*, Eighth Conf. Appl. Climatol., 17–22 January 1993, Anaheim, California.

SALAS, J.D., *Analysis and modeling of hydrologic time series*. In *Handbook of Hydrology* (ed. Maidment, D.R.) (McGraw Hill 1993), 72 pp.

SALAS, J.D., FU, C., CANCELLIERE, A., DUSTIN, D., BODE, D., PINEDA, A., and VINCENT, E. (2005), *Characterizing the Severity and Risk of Droughts in the Poudre River, Colorado*, ASCE Jour. of Water Resources Planning and Management, accepted for publication.

WOLTER, K. (2003), *Climate Projections: Assessing Water Year (WY) 2002 Forecasts and Developing WY 2003 Forecasts*. CSU Drought Conf. Proceed., Fort Collins, CO, 9 pp + 8 figures (available from the author upon request).

WOODHOUSE, C.A. (2003), *A 431-Year Reconstruction of Western Colorado Snowpack*, J. Climate *16*, 1551–1561.

(Received November 12, 2003, accepted March 5, 2004)

 To access this journal online:
http://www.birkhauser.ch

Pure appl. geophys. 162 (2005) 1481–1510
0033–4553/05/091481–30
DOI 10.1007/s00024-005-2680-0

© Birkhäuser Verlag, Basel, 2005

❙ Pure and Applied Geophysics

Intraseasonal Variability of the Summer Monsoon over the North Indian Ocean as Revealed by the BOBMEX and ARMEX Field Programs

P. Sanjeeva Rao,[1] and D.R. Sikka[2]

Abstract—During the summer monsoon season over India a range of intraseasonal modulations of the monsoon rains occur due to genesis of weather disturbances over the Bay of Bengal (BOB) and the east Arabian Sea. The amplitudes of the fluctuations in the surface state of the ocean (sea-surface temperature and salinity) and atmosphere are quite large due to these monsoonal modulations on the intraseasonal scale as shown by the data collected during the field programs under Bay of Bengal Monsoon Experiment (BOBMEX) and Arabian Sea Monsoon Experiments (ARMEX). The focus of BOBMEX was to understand the role of ocean-atmospheric processes in organizing convection over the BOB on intra-seasonal scale. ARMEX-I was aimed at understanding the coupled processes in the development of deep convection off the West Coast of India. ARMEX-II was focused on the formation of the mini-warm pool across the southeast Arabian Sea in April-May and its role in the abrupt onset of the monsoon along the Southwest Coast of India and its further progress along the West Coast of India. The paper attempts to integrate the results of the observational studies and brings out an important finding that atmospheric instability is prominently responsible for convective organization whereas the upper ocean parameters regulate the episodes of the intraseasonal oscillations.

Key words: BOBMEX, ARMEX, intraseasonal oscillation, SST, ocean-atmosphere coupled system.

Introduction

The intraseasonal oscillation (ISO) of the monsoon in the form of active-break cycle of the monsoon and the synoptic scale monsoon disturbances in modulating the rainfall over the Indian subcontinent have been prominently emphasized since 1950. An international field program in the form of the International Indian Ocean Expedition (IIOE) was carried out during 1963 to 1965, which brought out several new features of the summer monsoon meteorology. It was followed by the Monsoon Experiments in 1970 (ISMEX-73, MONEX-77 and MONEX-79) which well established the mean structural components of the monsoon. The major transient

---

[1]Earth System Science Division, Department of Science and Technology, New Delhi, 110 016, India
E-mail: psrao@alpha.nic.in
[2]40 Mausam Vihar, New Delhi, 110 051, India

disturbances and their usual formation areas became well recognized. During the peak phase of the 'Asian Summer Monsoon' there are one or two episodes, lasting for about a week or so each, when the grand organization of the monsoon trough across 70 to 140°E with 3 or 4 synoptic scale disturbances embedded in it are also observed. The monsoon trough after an active monsoon spell shifts toward the foothills of Himalayas to herald a sequence of weak or break monsoon conditions (RAMAMOORTHY, 1969; GADGIL and JOSEPH, 2003). The role of strong monsoon onset and the Somali current on the Arabian Sea cooling during the summer monsoon and fresh water discharge by *in situ* rainfall and river runoff in the northern BOB have been recognized (RAO, 1980; VARKEY *et al.*, 1996). The role of two important low-frequency ISOs on the quasi-periodic scales of 10–20 days (KRISH-NAMURTI and ARDUNUY, 1980) and 30–50 days (SIKKA and GADGIL, 1980; YASUNARI, 1980; KRISHNAMURTI and SUBRAMANIAM, 1982 and several others since then) was recognized. The 10–20 day scale oscillation moves westward from the west Pacific along 10–20°N towards the BOB (CHATTERJEE and GOSWAMI, 2003). The 30–50 day scale oscillation propagates northward from the near-equatorial Indian Ocean toward northeast Arabian Sea and the north BOB and remained an important focus of low-frequency monsoon variability on ISO scale. The ISO has been linked with the active-break cycle of the monsoon. There is considerable interannual variability in the intensities and phases of the ISO modes (SINGH *et al.*, 1992). The structure of the ISO is very similar to the structure of the interannual variability of the monsoon (GOSWAMI and AJAY MOHAN, 2001).

2. Indian Climate Research Programme and the Associated Field Research Programs: BOBMEX and ARMEX

During 1990s the Indian monsoon researchers launched field programs, 'Monsoon Trough Boundary Layer Experiment' (MONBLEX) and the 'Land Surface Processes Experiment' (LASPEX). MOTBLEX was carried out in 1990 with focus on the study of boundary layer processes along the NW/SE-oriented monsoon trough and the interaction between the moist convection dominated the southeastern end of the trough in the north BOB and adjoining land areas and the dry convective dominated the northwestern end of the trough over Rajasthan. LASPEX was carried out during 1997–1998 to understand the seasonal and intraseasonal processes in the land-vegetation-atmosphere system along western India. The Indian oceanographic community had also launched a long-term program to monitor the oceanic climate over the Indian Seas by deployment of met-ocean buoys. In view of the availability of special observational systems for atmosphere and ocean to study the weather and climate of India, the joint ocean-atmosphere community in India proposed launching of the Indian Climate Research Programme (ICRP) (DST, 1995). ICRP included carrying out special field research programs on the coupled ocean-atmospheric processes considered important for the study of monsoon. Under this program field research programs viz. Bay of Bengal Monsoon Experiment (BOBMEX) and the Arabian Sea Monsoon Experiment (ARMEX) have been carried out.

2.1. Broad Research Objectives of the BOBMEX and ARMEX

2.1.1. BOBMEX: BOBMEX scientific objectives were focused on the study of the coupling of the ISO in the atmosphere and upper ocean thermohaline structure and their roles in regulating the monsoon. The region of intensive observation was between 12°N to 20°N in the BOB. BHAT et al. (2001) described in detail the BOBMEX implementation and initial scientific results.

2.1.2. ARMEX: ARMEX was implemented in two phases (SANJEEVA RAO, 2005). ARMEX-I was focused on the study of offshore trough, embedded mesoscale vortex and heavy rainfall events along the West Coast. The region covered was the east Arabian Sea off the West Coast of India (Fig. 1). ARMEX-II was focused on the development and collapse of pre-monsoon mini warm pool over the SE Arabian Sea (Lakshadweep Sea) and onset of the monsoon. The region of intensive observations was the SE Arabian Sea and the offshore region of the West Coast of India (Fig. 2).

3. Scientific Results

3.1. BOBMEX

Two research ships, ORV Sagarkanya (SK) and INS Sagardhwani (SD) were deployed with all meteorological and upper ocean observational systems. The quasi-stationary location of SK was near 18°N 87°E, which is close to the normal formation of monsoon lows and depressions. The SD in its quasi-stationary location was positioned near 13°N 87°E, which is along the latitude of the strong lower-tropospheric wind regime over the central Bay. Time series observations of ocean-atmosphere parameters were recorded at two deep met-ocean buoys functional near 18.5°N 87.5°E and 12.2°N 90.7°E. Weather radar at Chennai, Visakhapatnam, Paradeep and Kolkata along the East Coast of India as well as INSAT cloud imagery and outgoing longwave radiation (OLR) provided information about the organization of convective episodes over the BOB. The period of quasi-stationary observations were: for SK—27 July to 06 August 1999 and 13–24 August 1999; and for SD—17–22 July 1999, 30 July to 05 August 1999, 12–16 August 1999 and 25–28 August 1999. Of these periods 30 July to 05 August 1999 and 13–16 August 1999 were common to both the research vessels.

BHAT et al. (2001) have described some details about the instrumentation, calibration procedures followed and the major results about the modulation of organized convection during the intensive observational phase (25 July–25 August 1999) of the BOBMEX. Several other studies, have reported on different aspects of

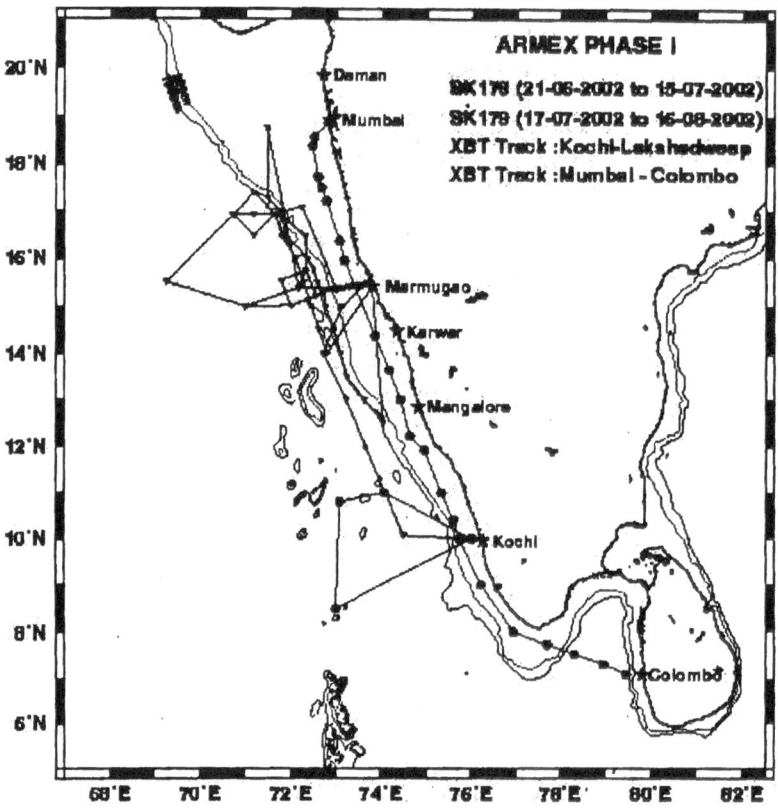

Figure 1
Region of intensive observations and platforms deployed during the ARMEX-I (2002).

the BOBMEX. We shall only highlight the major results with regard to the ISO of organized convection in response to ISO of the atmospheric circulation and the ISO of the thermohaline circulation of the upper 100 m of the north and central BOB.

3.1.1. ISO of organized convection - Atmospheric features: After a rather quiet or weak period of monsoon activity in the first fortnight of July 1999, the monsoon showed signs of revival during 17–21 July and the monsoon trough was re-established in the north BOB by 23 July. During the period 17–22 July, lower tropospheric winds showed strengthening over SD location and the monsoon cloud band passed north of it during 21–24 July. The wind flow at 850 hPa level on 28 July, 03 August and 07 August showed that the monsoon trough was dipping into the north BOB (data not shown). Three cyclonic circulations lying over West Bengal on 28 July, 3 August and 7 August as remnants of three different low pressure systems which had formed over north Bay during 27 July to 6 August. The last cyclonic circulation disappeared on 12 August and the monsoon trough shifted close to the

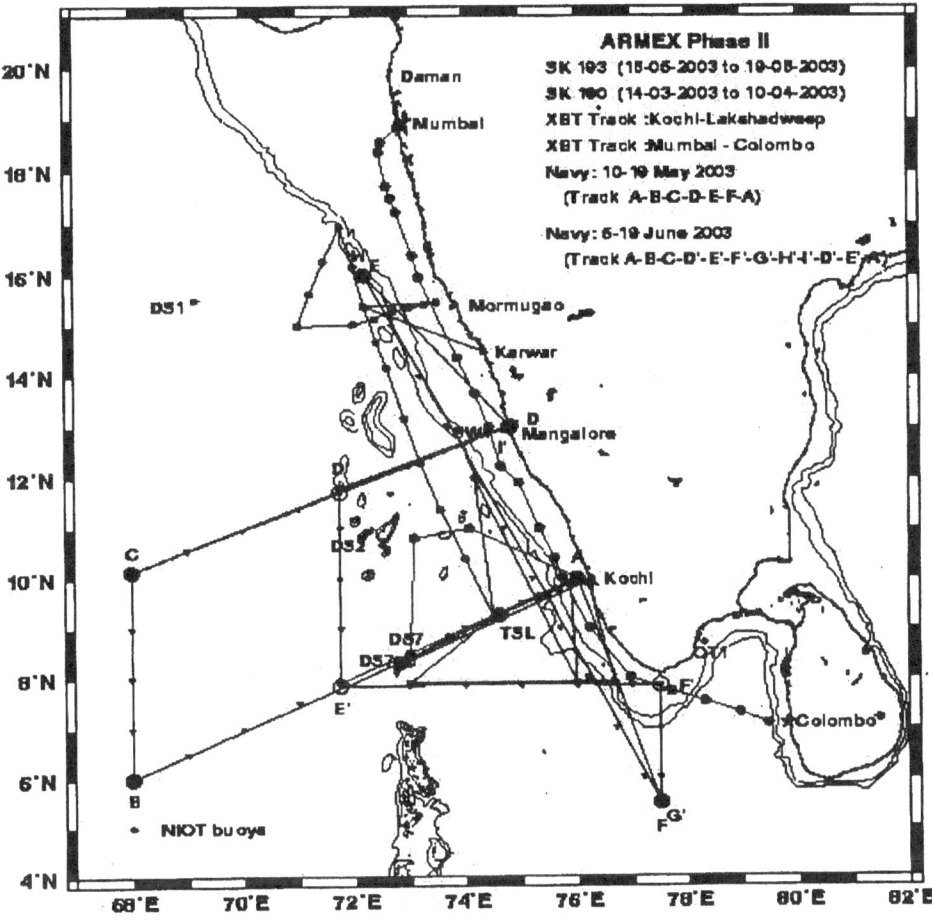

Figure 2
Region of intensive observations and platforms deployed during the ARMEX-II (2003).

foothills of Himalayas. Figure 3 (after KALSI, 2003) shows averaged OLR values below 180 Wm^{-2} at the block 12.5 – 15.0°N and 86.2 – 88.7°E during 16 July to 31 August 1999. This period also witnessed fall in the OLR (below 180 Wm^{-2}) over the adjoining north BOB (block 16.2 – 21.2°N and 87.5–90.0°E) almost simultaneously but with a lag of about one to two days in different sequences (Fig. 3). The period 21–26 July over the southern block followed by a short period of weakening of convection (OLR > 180 Wm^{-2}) during 27–30 July at the southern and 28–30 July at the northern block. The period 31 July to 01 August also witnessed low values of OLR in the northern block and the southern blocks. A brief spell of higher than 180 Wm^{-2} OLR values was observed but the OLR values remained below 220 Wm^{-2}, typical of deep convection. Thus the period of quasi-stationary position of SK between 31 July to 06 August could be reckoned as the core of active

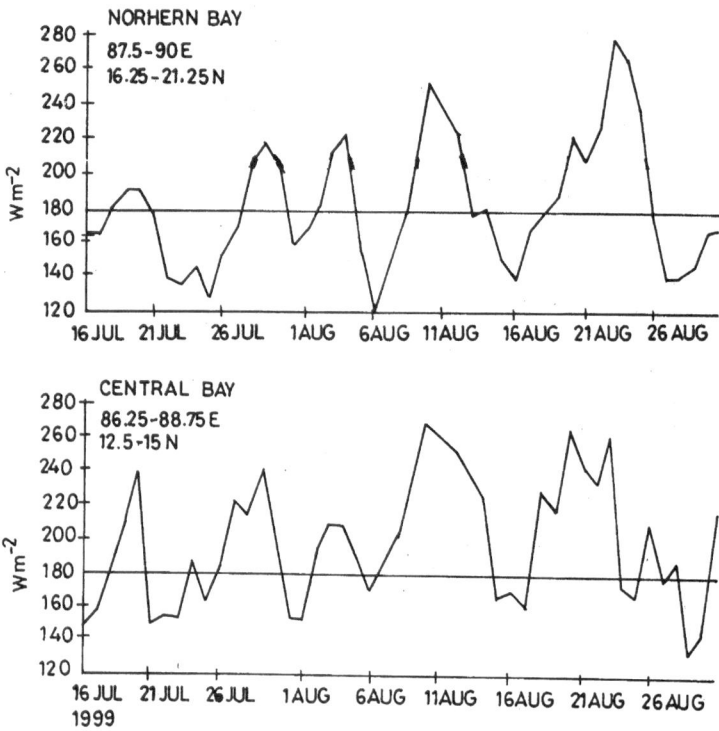

Figure 3

Fluctuations in OLR during active and weak spells of convection in the northern and central Bay of Bengal during the BOBMEX-1999 after KALSI (2003).

convection. Thereafter the monsoon weakened in the north and the central BOB from 9 to 25 August, except that a cloud cluster scale mesoscale convection existed in the vicinity of SK location between 14–17 August in association with an upper air cyclonic circulation at 850–700 hPa levels. During this period too the OLR values had dropped to below 180 Wm^{-2}. The monsoon convection over the BOB showed revival from the central block between 23–25 August, which shifted in a northwesterly direction and occupied the north BOB between 26–31 August. By that time SK had moved out of the area as its intensive phase of the planned field observational program was over. There was a brief interruption in the weak monsoon conditions over the north Bay as the cloud cluster scale disturbance over the area between 14–17 August, produced heavy rainfall on SK location (> 20 cm) and also along coastal Orissa and coastal West Bengal during this period. Thereafter the monsoon remained rather weak over the north BOB upto 24 August. The northward moving convection from the central Bay interacted with the meso-scale organization of cloud cluster in the northern Bay during 14–17 August. Two episodes of the northward migrating ISO occurred over the BOB during the BOBMEX field phase — one during 21–25

July and the other during 23–25 August 1999, which were responsible for modulating the monsoon activity on the low frequency ISO scale (30–40 days).

Table 1 shows the contrasting features of the atmospheric conditions during the active convective period (26 July to 09 August, 1999) and rather suppressed convective episode (10–24 August, 1999) of the ISO cycle over the north BOB. Three spells of rainfall exceeding 3 cm daily occurred during 31 July to 1 August, 5–6 August and 15–16 August, 1999 over SK location. All the features show that strong winds at the surface, strong wind shear between lower and upper troposphere, cyclonic vorticity and convergence at the top of the atmospheric boundary layer, upward motion at 500 hPa level and strong moisture convergence support active convection with low OLR and higher rainfall in the presence of apparent heat source and moisture sink during the active phase of ISO. The reverse occurs during the suppressed phase of the ISO. This is similar to what has been reported by MOHANTY and DAS (1986). Again the relative vorticity associated from the horizontal shear in the zonal component of the wind (not shown) at top of the atmospheric boundary layer at 900 hPa (representative of the Indian monsoon) remained mostly positive and above the normal value over the region 20–28°N along 80°E during the active convection phase and negative and below the normal between 12–26°N. This was similar to what happened in active and weak monsoon spells of July 1972 (SIKKA and GADGIL, 1978). There are opinions with regard to the initiation of northward moving equatorial zone. One of them emphasizes the role of planetary scale wave number one Madden-Julian oscillation (MJO) kicking up of convection as it approaches the Indian longitudes. The other emphasized on the competition between the regionally oscillating heat sources over the equatorial and north BOB. We have not examined the role of MJO as the emphasis during BOBMEX was to document the regional aspects. Probing of the links with the MJO would require further study.

BHAT *et al.* (2002) have emphasized the sharp decrease in mid-troposphere moisture content during the suppressed phase of BOB convection during BOBMEX. Such a feature has been known from the results of earlier studies, too. It was also observed that the lower tropospheric winds were stronger at about 2.5° latitude south of the center of low pressure systems. There was a clear reduction of moisture in the layer 850–500 hPa during the suppressed convective episode as the layer is stabilized in such an episode in comparison to the active phase of convection. Sea-level pressure was below normal (about 4 hPa) and the pressure gradient between the south and north BOB (8–20°N) was quite high (about 12–14 hPa) during the active phase of convection. Thus, strengthening of winds, fall in pressure over the northern BOB and higher pressure gradients preceded and existed simultaneously with the establishment of the active phase. This would suggest that it was the atmospheric dynamics (atmospheric instability), associated with the trough phase of the regional ISO cycle, which was responsible for the initiation of the active convective phase. Deep convection was then sustained by overlapping formation of three synoptic scale monsoon low-pressure systems in about a two-week period (28–30 July, 2–4 and 6–8

Table 1

Contrasting atmospheric features over the northern Bay of Bengal (17.5°N – 20.5°N and 87.5–90.0°E) during active convective episode (26 July to 09 August, 1999) and suppressed convective episode (10 to 24 August, 1999) of the ISO during BOBMEX.

Sl. No.	Features	Units	Active Convective Episode	Suppressed Convective Episode
1	Surface wind strength (at SK position)	ms^{-1}	12	06
2	Wind shear between 900 and 200 hPa	ms^{-1}	-40	-30
3	Sea-level pressure	hPa	1002	1008
4	Surface relative humidity	%	88	82
5	Number of low pressure systems		3	1 (mesoscale cluster)
6	Average OLR	Wm^{-2}	175	215
7	Rainfall	% of normal	+30	-20
8	Relative vorticity at 900 hPa	$\times 10^{-5} s^{-1}$	+2 (cyclonic)	-2 (anti-cyclonic)
9	Convergence/divergence at 900 hPa	$\times 10^{-5} s^{-1}$	-2 (convergence)	+1 (divergence)
10	Vertical velocity at 500 hPa	$hPa\ hr^{-1}$	-2.5 (upward)	+1.0 (downward)
11	Moist static energy at surface (500 hPa)	JKg^{-1}	355 (340)	360 (335)
12	Apparent heat source between 850–500 hPa	$Deg\ d^{-1}$	+7 (heating)	-4 (cooling)
13	Apparent moisture sink	$Deg\ d^{-1}$	+8 (heating)	-1 (cooling)
14	Lapse rate between 900–400 hPa	–	Highly moist column with lapse rate close to moist adiabatic	Reduction in moisture between 750–400 hPa with lapse rate > moist adiabatic.

August, 1999). Moist static stability in the lower troposphere increased during the active phase, but recovered within a day or two of the intervening period between the movement of one low-pressure system and the formation of the following overlapping low-pressure system over the northern BOB. Recovery of the moist static instability therefore makes the environment ready for the formation and growth of the following low-pressure system. There were short periods during the active spell in which the cloudiness had temporarily decreased and gaps in deep convection were observed, followed again by the organization of deep convection. The preferred period of the regional ISO (30–50 days) happened to be 33 days (21 July to 23 August, 1999) during BOBMEX. It may be mentioned that the rainfall over whole of India was nearly 10% below normal averaged for the active episode but this deficiency became over 40% in the suppressed episode. Thus inspite of formation of three overlapping monsoon disturbances during the active phase, the monsoon rainfall over India in its entirety did not exceed the normal as these disturbances weakened soon after crossing the coast and did not penetrate west of 80°E.

3.1.2. ISO in the Ocean-Atmosphere interface parameters: BOBMEX data have been utilized for the study of fluctuations during the active and suppressed convective phases of the ISO in the ocean-atmosphere interface parameters like SST and evaporative sensible heat and radiative fluxes from BHAT et al. (2001), BHAT (2002) and GHANEKAR et al. (2003). These fluctuations occur on the diurnal, synoptic (2 to 5 days) and ISO scales (active–suppressed phases of organized convection). Salient features of these fluctuations are summarized below:

(a) SST: By the time the BOBMEX observational program began over northern BOB, the SST in the region had already dropped to about 28.5°C — typical of the peak monsoon season. It fluctuated in a narrow range between 28.3°C to 28.8°C during the active phase under the synoptic scale variability of active monsoon convection. With the establishment of suppressed convective phase, SST over the region rose to 29°C, between 10–12 August, but abruptly fell to the lowest value of 26.8°C on 16–17 August under the influence of an intense mesoscale cloud cluster, which directly lay over the SK location and produced heavy rainfall. The fall in SST was due to lack of insolation, evaporative cooling and convective downdrafts that accompanied the heavy rainfall over SK location. The SST, thereafter, steadily rose between 17–24 August from 26.8°C to 29.3°C. Thus the rise in SST in this phase was about 2.5°C. But for the strong surface waters cooling during the 15–17 August episode, the SST in northern BOB would have continued to rise from the low value of about 28°C in the peak active phase during 1–5 August to near 29.5°C of the peak suppressed phase. The drop of SST on the low-frequency ISO scale over the northern BOB is about 2°C, which is larger than the intra-annual variability of SST in the July–August period for the same region. This significant variability of the SST on

ISO scale obviously contributes to regulate the ISO of the monsoon flow. The atmospheric instability promotes growth of monsoon disturbances and the associated lack of insolation due to deep convection, cool downdrafts accompanied by rainfall, evaporative flux and sensible heat flux loss promotes cooling of the surface waters. The ocean supplies evaporative and sensible heat fluxes to the atmosphere to support convective formation. With the cooling of the northern BOB during active monsoon phase, the atmosphere is restored to a relative stability. The overlapping formation of monsoon disturbances was arrested with the weakening of the lower tropospheric winds, rise in sea-level pressure and reduction in pressure gradients between near-equatorial region and the northern BOB. Since there were simultaneous observations of ocean-atmosphere parameters over SK and SD locations during two short periods, we summarize in Table 2, fluctuations which occurred in some key parameters on these two locations.

SENGUPTA and RAVICHANDRAN (2003) studied the ocean-atmosphere interface fluxes on the met-ocean buoys in northern BOB for the monsoon season of 1998 and found ISO in SST to be between 1.5 to 2.0°C at 18°N 88°E and about 1.0°C at 13°N 87°E. VINAYCHANDRAN et al. (1989), SARMA et al. (1997) and SEETARAMAYYA et al. (2001) using other data sets in the case of active and suppressed convection episodes, had shown drop (rise) of SST, decrease (increase) of solar radiation and increase (decrease) in latent heat flux and sensible heat flux with active (suppressed) convective episodes. BOBMEX data of SK during the core period of active convection and cloud cluster scale convective episode (14–16 August 1999) have also shown that the northern BOB lost net energy to support organized convection and as a result cooled. Figure 4 shows the day-to-day fluctuations in the heat budget of the ocean at the surface over SK location. During the active phase of 4–8 August the latent heat and the total heat loss from the ocean are quite high. During the weak phase fluxes are rather low. The waters over the central BOB over SD location continued to remain under a small amount of net surface heating as the skies were less cloudy and the stronger winds had moved somewhat to the north, close to the

Table 2

Fluctuations in some key ocean-atmosphere interface parameters over SK and SD locations during simultaneous observations periods.

Period	Parameter/Variable	Unit	SK location	SD location
30 July – 05 August, 1999	SST	°C	28.5	28.5
	Evaporative flux	Wm^{-2}	200	150
	Sensible heat flux	Wm^{-2}	30	10
	Net surface heating	Wm^{-2}	−150	20
13–16 August, 1999	SST	°C	27.5	28.7
	Evaporative flux	Wm^{-2}	250	150
	Sensible heat flux	Wm^{-2}	50	10
	Net surface heating	Wm^{-2}	−200	50

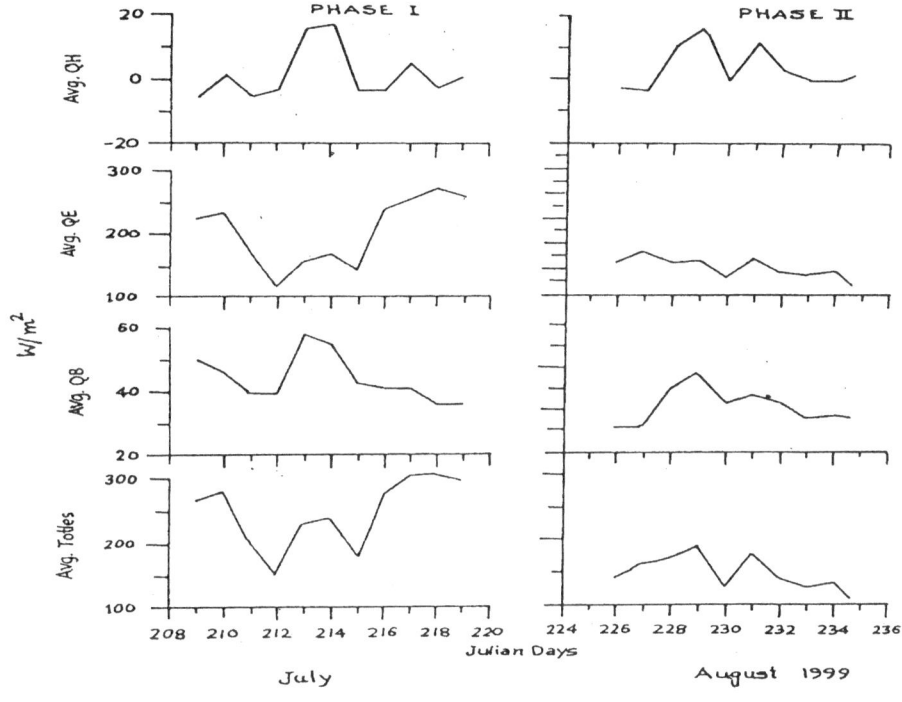

Figure 4

Daily variations in heat budget components at the ocean surface over SK location (sensible heat flux (Q_H), latent heat fluxes (Q_E), effective back radiation (Q_B) and total heat loss during two stationary phases during 27 July to 8 August, 1999 active convection and 14–24 August, 1999 suppressed convection.

center of the monsoon disturbed area. The SSTs reverted to the conditions of the suppressed convection within 2–3 days of the decrease in deep convection. Incidentally both the simultaneous observation periods for SK and SD fell into active convective episodes over the SK location — the first one under the overlapping active spell and the second one under influence of mesoscale cloud cluster. The fluctuation of the SST over SD location remained within a narrow range of 28.5 to 29.0°C throughout the BOBMEX period. We have no direct simultaneous measurements in the south and the north of the BOB when the convection fluctuated between two broad convective episodes over the northern BOB. However, since latitudinal extent of the organized monsoon cloud band is between 5–7 degrees covering central to northern BOB, the southern BOB would remain relatively cloud-free during that period. SSTs in this period would therefore increase over southern Bay as the winds are also weak. Suppressed convection over the northern Bay results in warming up of the northern Bay and almost simultaneous active convection over the southern Bay cools its SST. There is a high probability that the SST over the northern BOB drops to 27.5–28.0°C after its active convection phase as it rises over the southern Bay where the skies are simultaneously clear. When the peak phase of the suppressed convection occurs over the northern Bay, SST would rise to 29.5°C and it would drop to about 28.0–28.5°C over the southern Bay under the revival process of the monsoon, which begins in the southern Bay (SIKKA and GADGIL, 1980; YASUNARI, 1980). Thus increase of SST gradients between northern and southern Bay, with northern Bay being slightly warmer (> 29°C) than south Bay (~28°C), would promote northward movement of the monsoon cloud band and hence revival of monsoon over the northern Bay.

The above heuristic argument leads us to conclude that the SST changes over the northern and southern Bays take place under the ISO cycle of the peak monsoon (July and August), as observed at the beginning of the advance phase of the monsoon in early June when the SST was warmer in the northern Bay and somewhat cooler in the southern Bay. We believe that the ISO in the north-south gradients in the SST between south and north of the BOB, as well as hydrological feedbacks to the atmosphere along the land-ocean boundary adjoining the northern BOB are responsible for the northward migration of the monsoon cloud band from the near-equatorial BOB. Thus the relative strengths of the two heat sources — one in the near-equatorial warm waters of the Bay when the active phase of the monsoon lies over the northern BOB play crucial role in modulation of the monsoon convection.

3.1.3 Intraseasonal oscillation in the thermohaline circulation over the Bay of Bengal. Thermohaline structure of the upper 100 m of the BOB basin undergoes complex changes on the annual cycle (VARKEY et al., 1996). The basin was poorly observed until the IIOE and since 1964 oceanographic surveys of the basin were carried out periodically by Indian scientists. The basin is not only influenced by the semi-annual reversals of surface winds during the winter and summer monsoons, but also visited by tropical cyclones in the pre- and post-monsoon seasons. The basin is also land-locked on three sides. It is dominated by upwelling along the East Coast of India and transient oceanic eddies of cyclonic and anticyclonic nature do exist. A large part of the basin is also affected by deep convection from May to November. The heavy rainfall over central and northern BOB during the summer monsoon and the river runoff from the Ganga, Brahmaputra, Meghana, Mahanadi rivers on its north, and Irrawadi on its east and Godavari and Krishna rivers to its west are responsible for a tremendous amount of fresh water discharges into the coastal regions of northern, eastern and western BOB. Heavy rainfall over the BOB as shown by the 'Global Precipitation Climatology Project (GPCP)' is well known. Fresh water influx would considerably modify salinity structure in the uppermost 50 m layer of central and northern Bay during the monsoon season. The near-surface Indian monsoon current enters from the southwest along 8–10°N across the Arabian Sea and as such the salinity in the southern BOB is rather high (33–34 PSU). The water balance (evaporation-precipitation) is considerably negative for the northern BOB (VARKEY *et al.*, 1996) as it is the seat of deep convection within the Indian monsoon regime. SANILKUMAR *et al.* (1994) had discussed some features of the thermohaline structure of the northern Bay using MONTBLEX data. During the BOBMEX period SK and SD took intensive CTD observations and we use the data during their stationary time series observation periods to describe ISO of the near-surface thermohaline structures. Table 3 provides the data for SK and SD locations. The following aspects are worth mentioning:

(i)   There is small change in the average values of the SST both at SK and SD locations among different periods. The daily range of SST at SK during 22 July to 6 August was small (~0.5°C). But in the period 13 to 23 August this range increased to about 1°C due to processes already explained.

(ii) Sea-surface salinity (SSS) fluctuations over SD locations in all observational phases are within narrow range (< 0.5 PSU). However, there was considerable change in the average SSS during different observational phases at SK location. The salinity dropped by two units from the first period to the second period. The first major drop occurred from the average value of 32.6 PSU on 29 July to 28.7 PSU (drop by 4 PSU) after 4 days on 03 August. This occurred under the excessive river water discharge and *in situ* rainfall during the active convective episode when a depression and a low pressure area crossed the coast and gave heavy rainfall in the coastal belt. Surface salinity increased to 30 PSU on 6 August as the rainfall had temporarily ceased after the low pressure area of 03 August had shifted inland and weakened. It dropped to a very low value of 27.5 PSU between 13–18 August (drop of 2.5 PSU) due to heavy *in situ* rains and partly advection of another freshwater plume from the coastal area after the inland rainfall associated with depression between 7–9 August. Thus the ISO of the surface salinity over SK location is rather large (2.5–3.0 PSU) as this region is affected by advection of fresh water plumes from the coastal belt as well as heavy *in situ* rainfall (VINAYCHANDRAN and MURTY, 2002). For lack of more data it is rather difficult to say with certainty as to the relative contribution of the two sources of fresh water, to the fall in SSS. However, it is safe to guess that the role of *in situ* daily rainfall of more than 5 cm could be almost equal to that of the river runoff plumes.

(iii) Mixed layer depth (MLD) varies between 60–75 m in different observational phases over SD location. However, MLD is nearly half (~37 m) over SK location. Shallowness of the MLD in the northern BOB, whose upper part is covered by fresh water would be responsible for larger amplitudes in SST changes over SK location and rather rapid warming of the SST after the skies clear over the northern Bay.

(iv) A barrier layer is present almost each day over the SK location at about 15–25 m depth which stratifies the density structure in the mixed layer. The thickness of the barrier layer is 20–25 m. Minor inversions in mixed layer

Table 3

Thermohaline structure of the upper 50 m layer of the northern Bay of Bengal during BOBMEX (range of variations given in brackets)

Location	Period	SST (°C)	SSS (PSU)	Mixed layer depth (m)	Bottom of the barrier layer(m)	Barrier layer thickness (m)
17.5°N 87.0°E SK	27 July to 06 August, 1999	28.5 (28.3–28.6)	30.4 (32.9–28.7)	37 (35–40)	19	20
	13—23 August, 1999	28.7 (28.0–28.9)	28.3 (27.5–29.9)	38 (35–42)	19	25
13.1°N SD	17–22 July, 1999	28.7 (28.6–28.8)	33.1 (33.0–33.3)	63 (57–67)	57 (55–60)	27 (20–30)
	30 July to 05 August, 1999	28.6 (28.4–28.7)	33.4 (33.3–33.4)	58 (55–65)	51 (45–55)	23 (20–25)
	12–16 August, 1999	28.5 (28.2–28.6)	33.4 (33.3–33.7)	75 (73–80)	47 (35–50)	22 (20–25)
	25–28 August, 1999	28.5 (28.2–28.7)	33.5 (33.5–33.6)	75 (70–82)	44	27 (20–40)

temperature with strength of about 0.2–0.5°C also exist near the barrier layer in SK location. Over SD location weak barrier layer exists on most of the days but at a depth of about 45–55 m. Shallow mixed layer with barrier layer and associated temperature inversions would help in more rapid heating of the sea surface over the northern BOB compared to that over the central BOB. This could also contribute to the higher amplitudes in SST over the northern BOB which provides greater stability of the upper most 20 m and inhibits mixing in ISO scale in comparison to that over the central BOB.

3.2 ARMEX-I: Off-shore Trough and Embedded Vortices

Heavy rainfall episodes occur along preferred zones along the West Coast in association with a north-south oriented off-shore trough near the surface level in which sometime cyclonic vortices on mesoscale (100–200 km horizontal extent) are embedded (GEORGE, 1956). The prevailing surface winds over the central and east Arabian Sea are from the west during June to August, but on some occasions the winds at some coastal stations along the West Coast become southerly and even southeasterly. This would indicate presence of off-shore trough with possibility of presence of a mesoscale vortex within some portion of the trough. Using MONEX 1979 data MUKHERJEE and SHYAMALA (1984) had studied the presence of a mesoscale vortex off Mumbai. GROSSMAN and DURRAN (1980) investigated the deep convection off the Konkan Coast during the MONEX (Arabian Sea) field phase and suggested the important role of frictional convergence and topography for the triggering of heavy rainfall events along preferred regions of the coast. FRANCIES and Gadgil (2002) have investigated the heavy rainfall episodes which occur along the West Coast during summer monsoon season.

Monsoon onset had occurred along the Kerala Coast on 30 May, 2002. After a sluggish start the monsoon strengthened along the east Arabian Sea by mid-June and strong to vigorous monsoon conditions prevailed across Goa – Konkan-south Gujarat coast in the second fortnight of June. An episode of very heavy rainfall was recorded along north Konkan-south Gujarat Coast during 26–28 June, 2002 during which period 24-hr point rainfall ranged between 20–60 cm. This heavy rain spell was brought about by the formation of an off-shore vortex (SHYAMALA, 2003). A monsoon low pressure system was moving from the east approached Gujarat-Maharastra region on 25 June. Associated with this large-scale system, the off-shore trough along Konkan-Gujarat coast intensified and a vortex-like structure in satellite photographs was observed on 27 June (Fig. 5). Several researchers have made attempts to assimilate ARMEX observations to simulate this off-shore vortex and the associated deep convection by using mesoscale (MM5) model with different horizontal resolutions, ranging between 60 to 5 km (BHASKAR RAO and PRASAD, 2004; DAS et al., 2003). The intensity of point rainfall in the model-simulated vortex did not exceed 10–20 cm in different runs, except the very high resolution (5 km)

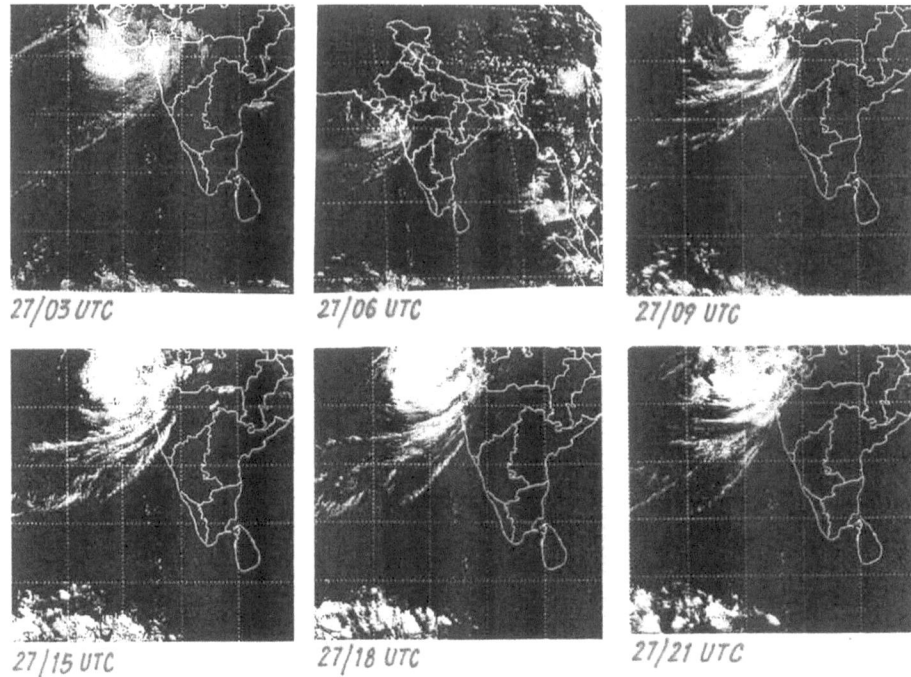

Figure 5
Satellite photographs of the cloudiness associated with mesoscale off-shore vortex of north Konkan—south Gujarat Coast on 27 June, 2002. (Source: India Meteorological Department).

nonhydrostatic model. These studies have brought out that for simulating the mesoscale off-shore trough/vortex, it is necessary that intensive observations are provided in the sensitive area for the formation of the system such that mesoscale models could handle its prediction in a more realistic manner.

Another aspect brought out by the radiosonde observations taken onboard SK (cruises SK-178 and SK-179), which was located just about 200 km southwest of the centre of the vortex, was the presence of temperature inversions near 750–600 hPa level and abrupt reduction in moisture in the lower-middle troposphere (BHAT, 2003). These inversions apparently resulted from the subsidence caused on the western flank of the mesoscale deep convection associated with the vortex. The monsoon flow passed on to a weak phase on 30 June and remained extremely weak during the entire month of July such that the monthly rainfall on All India scale for July 2002 was below the normal by 49%. This was an unusual drought-monsoon season for India in which failure of monsoon for month of July alone resulted in all India drought. SIKKA (2003) and GADGIL et al. (2003) have investigated the drought of 2002 from two different perspectives. SIKKA (2003) showed that the drought was linked to an evolving warm El Nino episode in central Pacific. GADGIL et al. (2003) examined the role of the warm eastern equatorial Indian Ocean during April-August 2002 as causative factor for the incidence of drought. A special feature of the tropospheric thermal structure during July 2002 was the prevalence of temperature inversions extending from the central Arabian Sea to northwest India across Gujarat and Rajasthan and then their spread towards central India (across Gwaliar, Nagpur and Raipur) as well as over the Peninsular India up to 15°N. Tropospheric inversion layers were present at various levels between 850 hPa to 600 hPa with depth of about 50 to 100 hPa and magnitude of temperature rise being 1–3°C within the inversion layer. The widespread presence of such a structure day after day on extensive scale over India did not allow the monsoon convection to develop fully. As seen from Table 4 the inversions over western India were more frequent than over east Arabian Sea. The frequency of inversions over eastern Indian stations like Kolkata, Bhubaneswar, Patna and Ranchi were much less than over western and central India. Persistent inversions over the northwest and central India was due to large-scale subsidence in the lower-mid troposphere. The Arabian Sea branch of the monsoon had almost collapsed during most of July 2002 with sea-level pressure anomaly varying between 3–5 hPa. Even though the BOB monsoon remained moderately strong during July, large-scale convection could not penetrate west of

Table 4

Frequency of tropospheric inversions in different layers between 850 and 650 hPa levels over the eastern Arabian Sea (over Sagar Kanya) and Indian Stations during July 2002.

Location / Station	Number of days with inversions 00UTC Profile	12UTC Profile
Sagar Kanya position 15–17°N 69–73°E (26 June to 14 July)	19	19
Sagar Kanya Position 15–17°N 71–74°E (17–26 July)	0	1
15.5°N72.2°E (Stationary position) (27–31 July)	3	5
July month as a whole	18	24
Goa	5	12
Mumbai	21	17
Aurangabad	17	9
Ahmedabad	26	22
Jodhpur	22	22
New Delhi	12	12
Lucknow	5	7
Patna	3	6
Ranchi	1	4
Kolkatta	2	3
Bhubaneswar	1	1
Bhopal	7	11
Nagpur	7	12
Gwalior	6	16
Raipur	11	16
Vishakapatnam	10	8
Hyderabad	7	12

85°E under the presence of large-scale subsidence prevailing over western, central and northern India.

In all, there were three clear cases of off-shore trough along the West Coast with embedded vortices during July to August 2002 viz. (i) 15–16 June off Goa, (ii) 26–28 June off north-Konkan-south Gujarat Coast, and (iii) 7–10 August off Karnataka-Goa Coast. The trough was active during 3–4 August along the northern Karnataka coast when SK reported a good spell of rainfall, highly moist lower to middle troposphere, fall in SST (28.3 to 27.9°C) and a fall of about 1.0 PSU in SSS. The activity moved northward and lay near Goa-Konkan Coast between 6–10 August 2002. Due to cool downdrafts associated with deep convection surface air temperature fell from 27°C on 6 August, to 25°C on 8 August over SK. The winds had also strengthened to over 10 ms^{-1}, the surface pressure fell by about 2 hPa and surface relative humidity increased to over 90 percent. Very heavy rainfall (over 15 cm per day) occurred along the Konkan Coast on 8–9 August under the deep convection organized by this mesoscale system. In summary ARMEX-I indeed confirmed the role of off-shore trough with embedded vortex for triggering spells of deep convection along the West Coast of India. The existence of offshore trough may be linked to the large-scale flow with modulation brought about by ocean-atmosphere fluxes, coastal geometry and orographic features. Detailed studies on these aspects are needed. Mesoscale models could be able to handle the prediction of this phenomenon provided additional data from intensive observational systems, specially designed for the purpose are put in place in every monsoon season.

Figure 6 shows day-to-day variations in OLR, SST and wind speed over the location of SK during 25 June to 4 August, 2002. Notice the high OLR value of over 250 Wm^{-2} on the subsiding flank of the off-shore vortex during 26–28 June, 2002. It is an observational fact that on the flanks of rising limb of an intense vortex, compensatory subsidence is often enhanced. The SST over SK location fell by about 0.3°C as a result of strong monsoon winds in spite of the presence of subsidence. OLR values over the east Arabian Sea and most of Indian land area, during most of July were between 220–240 Wm^{-2}, indicative of shallow cloudiness under weak monsoon conditions with subsidence in the lower-mid troposphere. SST over SK location did not show much fluctuation in the first fortnight of July 2002. However, it fell gradually from 28.7°C on 17 July to 28.0°C on 4 August. During this period wind speed increased from 8 ms^{-1} to 12 ms^{-1} and the OLR reached the marginal value of 200 Wm^{-2} for deep convection during 31 July to 1 August as the monsoon strengthened. The period 3–4 August 2002 had witnessed intensification of the off-shore trough and hence convective activity.

3.3 ARMEX-II: Build-up of the SE Arabian Sea Warm Pool and Onset of Monsoon 2003

3.3.1. Evolution of the warm pool: Under the annual cycle, the equatorial Indian Ocean begins to warm up on both sides of the equator by mid-February. The

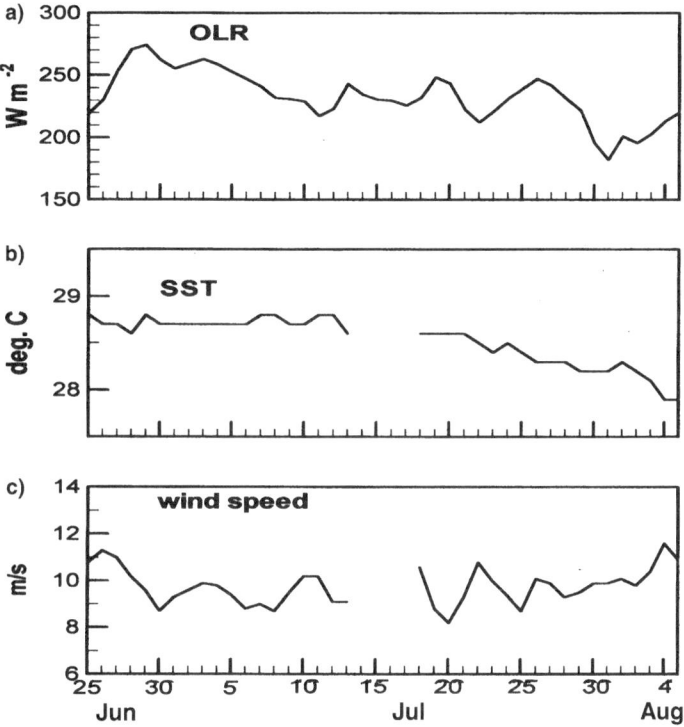

Figure 6
Day-to-day variations in OLR, SST and wind over the location of SK during 25 June – 4 August, 2002 during ARMEX-I.

warming region shifts northward across the equator by mid-March as the Sun approaches the equator and by mid-April the warmest waters of the north Indian Ocean lie around 5–10°N from 45–100°E. Annual cycle of the SST over the Indian Ocean is such that the northern region of the Arabian Sea cools by 3–4°C from November to February. This cooling is observed in the central Arabian Sea (10–15°N) by 1.5–2.5°C and into the south Arabian Sea (5–10°N) by 1.0–1.5°C between November to February. The SST warming between February to May in the north, central and south Arabian Sea is 3.5°C, 2.5°C and 1.5°C, respectively. The summer monsoon cooling from May to August is most marked in the central Arabian Sea except for the cooling associated with the coastal upwelling regions of the Somali Coast (4–5°C) and off southern parts of the West Coast of India (2–3°C). The warming of the north Indian Ocean from February to May is primarily due to solar heating in the absence of large-scale cloudiness and covers the entire longitudinal domain from 45–100°E. However, the peak values of SST in April-May are observed across the SE Arabian Sea — Lakshadweep Sea area (5–10°N, 70–73°E). The SE Arabian Sea is therefore a dominant tropical warm waters region in April-May within which a mini warm pool over the Lakshadweep area, is situated which was first noticed by SEETARAMAYYA and MASTER (1984) and further studied by several workers (JOSEPH, 1990; VINAYCHANDRAN and SHETYE, 1991; SHENOI *et al.*, 1999; RAO and SIVAKUMAR, 1999, 2000). SHANKAR and SHETYE (1997) examined the dynamics of the Lakshadweep SST high (mini-warm pool) and linked the preferential warming of the area to the intrusion of low salinity waters of the BOB into SE Arabian Sea, as a result of the westward moving Rossby waves from the BOB generated by downwelling coastal Kelvin waves between December to March. The cessation of the surface summer monsoon current in the near-equatorial SE Arabian Sea in November and the rainfall in the NE monsoon season along the Kerala Coast might also contribute to the formation of low salinity waters over the SE Arabian Sea. Thus, the fresh water lens due to the above combined effects and consequently the presence of barrier layer over the SE Arabian Sea during winter makes the thermohaline structure of near-surface waters quite stable. As a result the solar heating from mid-February to March is concentrated in the upper 20 m layer of the waters. During April-May the SST in the warm pool reaches 30–31°C each year. As the Northern Hemisphere near-equatorial trough (NHET) is established near 5°N toward late April, formation and westward movement of cloud clusters occur along 70–75°E in the near equatorial belt south of Lakshadweep Sea which modulate SST on synoptic scale by about 0.5°C in different spells. In several studies collapse of the warm pool has been linked with the process of onset of the monsoon over the Lakshadweep Sea and Kerala Coast. The monsoon onset is brought about in many years (50% on the average) by the formation of an intense vortex (depression/ tropical storm) over the SE Arabian Sea towards the end of May/early June (SEETARAMAYYA and MASTER, 1984; JOSEPH, 1990; RAO and SIVAKUMAR, 1999).

The warm pool collapses under the impact of the burst of the monsoon winds within about a week to fortnight of the invasion of the strong winds over the Lakshadweep Sea area. In view of the importance of the formation of the warm pool and its possible role in monsoon onset, ARMEX-II was mounted in the spring of 2003 to understand the processes involved in the buildup of the warm pool, its association with the onset of summer monsoon and its final collapse. Since the core of the mini 'warm pool' region is rather small in horizontal extent, a cluster of met-ocean buoys were operated within the area covered by 8.3–15.55°N to 69.1–74.3°E. Besides, SK surveyed the area in two cruises (SK-190 and SK-193) during 14 March to 10 April 2003 (at the beginning of the warm pool formation) and (ii) during 15 May to 19 June 2003 (at the peak phase of the warm pool and its collapse after the onset of monsoon). During the SK-190 cruise the ship was located at a stationary position near 09.13°N 74.30°E from 23 March to 2 April 2003 to take time-series observations. Again during the SK-193 cruise time-series observations were repeated near the same stationary position from 23 May to 7 June, 2002. At the end of the second stationary position the ship moved in a NE direction and sailed off-Karnataka Coast to end the experiment at Goa on 19 June. Two vessels of the Naval Hydrographic Office (INS Matunga and INS Investigator) also took part during the second phase taking observations in a north-south track off the Karnataka-Kerala Coast (Fig. 2). In order to understand the processes involved with the buildup and collapse of the warm pool, a fortnightly XBT survey was also carried out from May 2002 to August 2003 between Cochin (Kerala Coast) and Minicoy - Karavati (Lakshadweep Sea) region. Besides bimonthly XBT surveys off the West Coast from Mumbai to Colombo were carried out from October 2002 to August 2003. These XBT surveys monitored the evolution of thermal structure in the critical period of November to March when the signature of low-salinity waters appear and peak over the SE Arabian Sea. We now describe the evolution and peaking of the warm pool as well as its collapse in 2003 with the onset and advance of the monsoon with a variety of data collected during the ARMEX-II field phase.

(a) Build up of the Warm pool and SST fluctuations during March-April 2003

SK occupied the first time series location on 23 March 2003. By that time SST's in the SE Arabian Sea had already increased from 29.5 to 30.5°C during 14–22 March 2003. The time series observations between 23 March to 7 April indicated that daily average SST fluctuated between 30.6 to 29.9°C within narrow diurnal range of about 0.8°C (Table 5). The slight fall in SST on 28–29 March resulted from strengthening of the wind speed accompanied with sea level pressure fall by 2–3 hPa, cloudiness increased and air temperature also dropped under cool downdrafts from clouds. Thus high input of solar radiation, evaporation, sensible heat exchange and OLR during most of the 16-day period maintained the SST to about 30°C in the warm pool region. Met-ocean buoy located at 8.31°N 72.65°E (DS 7) had stopped functioning by the beginning of April but another buoy located in very close neighborhood at 8.31°N 72.66°E (DS 7A) continued to function till 18 May, 2003. There was an abrupt fall in surface salinity from about 34.5 PSU to 33.5 PSU in 3 days between 6 to 9 April and the salinity remained rather low until 25 April, 2003. However, SST continued to rise gradually during this period and it reached the value of 31.5°C on 19 April, 2003. Another met-ocean buoy located at 10.62°N 72.51°E (DS2) observed the following SST values which confirm the peaking of SST by 19 April and persistence of high SST over the area until 7 May 2003.

Table 5

Daily variations in SST in the SE Arabian Sea

Date	Daily average SST(°C)	Diurnal range of SST (°C)
23 March 2003	30.1	29.8–30.5
27 March 2003	30.2	30.0–30.8
29 March 2003	29.8	29.6–30.2
01 April 2003	30.8	29.8–30.6
05 April 2003	30.0	29.9–30.1
Average for the period during 23 March to 5 April 2002 over SK	30.3	–

(b) Warm pool during May – June 2003

During the second phase of SK cruise (May-June 2003) it was stationary between 23 May to 06 June, 2003 at the same position of the earlier cruise. During this period, average daily SSTs of 30.6° to 30.7°C were observed. Over DS2 location also daily average SST for 25 May to 1 June, 2003 and 2–7 June, 2003 was 31.2°C and 31.1°C,

Period	Average daily SST (°C)	Maximum value of SST (°C)
25–29 March 2003	31.2	32.1
16–18 April 2003	31.7	32.2
27 April – 7 May 2003	31.8	33.3

respectively, indicating a fall of about 0.6°C due to increase in cloudiness from the high value of 31.8°C observed during 25 May to 7 June, 2003. Table 6 shows further details about the SST buildup over the three met-ocean buoys (DS7, DS7A and DS2). A fairly constant SST status was maintained over the Lakshadweep Sea between 23 May and 7 June, both at SK and adjoining DS2 location, as no major synoptic oscillation occurred on the daily scale. Daily average SST at SK location had further increased from around 30.0–30.6°C from 23 March to 5 April period to

Table 6

Fluctuations in daily SST over the cluster of buoys in the SE Arabian Sea during pre-monsoon season of 2003. (Diurnal range of SST is given in bracket).

Position of the Buoy 08.31°N 72.65°E		Position of the Buoy 08.31°N 72.66°E		Position of the Buoy 10.62°N 72.51°E	
Date	SST (°C)	Date	SST (°C)	Date	SST (°C)
05 March	29.2 (0.3)	05 March	29.1 (0.4)	05 March	29.5 (0.6)
10 March	29.7 (0.6)	10 March	29.6 (0.9)	10 March	30.2 (0.8)
15 March	29.8 (0.5)	15 March	29.9 (0.6)	15 March	30.0 (0.8)
20 March	29.7 (0.3)	20 March	29.8 (0.6)	20 March	30.1 (0.6)
25 March	30.2 (0.3)	25 March	30.3 (0.8)	25 March	30.8 (1.3)
30 March	30.4 (0.7)	30 March	30.4 (0.4)	30 March	30.4 (0.5)
05 April	30.2 (0.3)	05 April	30.1 (0.3)	05 April	30.5 (0.5)
10 April	30.2 (0.6)			10 April	30.9 (0.7)
15 April	30.8 (0.2)			15 April	31.4 (1.2)
20 April	31.0 (0.3)			20 April	31.1 (1.2)
25 April	31.1 (0.2)			25 April	30.9 (0.3)
30 April	30.7 (0.3)			30 April	31.5 (1.0)
5 May	31.2 (0.4)			5 May	31.6 (0.7)
10 May	30.7 (0.3)			10 May	31.1 (0.2)
15 May	30.5 (0.3)			15 May	30.5 (0.2)
				20 May	30.8 (0.7)
				25 May	31.2 (1.0)
30 May	30.5 (0.3)			30 May	31.3 (0.6)
5 June	30.4 (0.4)			5 June	31.0 (0.2)
10 June	30.3 (0.2)			10 June	30.4 (0.4)

23 May to 7 June period in just about one month. It had also remained even higher (31.2–31.8°C) over the adjoining buoy DS2. The area remained more or less cloud-free and the winds were rather weak except for two short spells around 28–29 March and 2–3 April when the region was affected by low pressure/cloudiness and SST fell by about 0.5°C. The warming phase over the buoy locations showed a minor oscillation of 15-days with amplitude of about 0.5°C. In spite of high SST values of greater than 30°C over a 31-days period, no deep convection developed over the region. SSTs remained quite de-coupled from convection as the lower troposphere was dominated by subsidence under the influence of the Arabian Sea high. The broad region of the high SST had shifted to the central Arabian Sea during the first week of June as noticed from NCEP SST analysis (not shown). The Lakshadweep high SST region lay to the SE of the big warm pool over the central Arabian Sea. Weekly rainfall figures for the meteorological subdivisions of Lakshadweep, Kerala and coastal Karnataka remained considerably below normal.

Table 7 shows observed pentad rainfall over 8 stations located along Lakshad-weep, Kerala Coast and Karnataka Coast. The monsoon rainfall considerably intensified over Lakshadweep-Kerala-Karnataka coastal belt only during 13–20 June, 2003 when the strong monsoon winds (over 10 m/sec) had invaded the Lakshadweep Sea area. It was only after the invasion of strong winds that SSTs fell bellow 29°C between 15–20 June when the sky remained cloudy under deep convective organization, with high turbulent exchanges of heat (latent and sensible) and cold downdrafts occurring over the region. This was the major episodic fall in SST (by about 2°C) within about 5 days. This characterizes the impact of monsoon on the SE Arabian Sea, which is known as the SE Arabian Sea cooling as the mini-warm pool collapses under the impact of the onset of monsoon over Kerala first reported by RAO (1980). During July and August 2003, SST over the area fluctuated generally between 28.3°C to 28.8°C over the SE Arabian Sea, depending up on the strength of the surface monsoon winds over the area.

(c) Surface layer temperature inversions during winter and heating of the Lakshadweep Sea in the pre-monsoon season

Table 7

Observed pentad rainfall (mm) at some stations during 1–20 June, 2003.

Region	Station	1–5	6–10	11–15	16–20
Lakshadweep Islands	Amni Devi	0.0	71.7	161.6	83.0
	Minicoy	7.7	17.4	117.1	43.5
Kerala Coast	Trivandrum	0.0	0.6	36.0	91.0
	Kochi	3.7	94.5	54.3	131.4
	Cannanore	0.6	72.4	134.2	327.7
Karnataka Coast	Mangalore	0.0	79.8	77.7	401.3
	Hanovar	0.0	11.2	176.8	410.2
	Karwar	0.0	26.3	202.3	555.5

As already mentioned two XBT lines were organized under the ARMEX to document the heating from the preceding post-monsoon of 2002 season to the succeeding pre-monsoon season of 2003. The following inferences could be drawn from the analysis of this XBT data set.

(i)   SST near Mumbai coast (18°N 72°E) in October 2002 was about 29.3°C. This was dropped to 28.5°C in December and to 28.4°C in February. It rose to 30.7°C by the end of May.

(ii) In the central portion off the West Coast near to 74.5°E, SST in winter is about 28.9°C and it is 30.2°C toward end of May / early June.

(iii) In the southern portion near 7.5°N 78.5°E, the winter SST is 27.9°C and the pre-monsoon SST in May is 28.5°C.

(iv) No surface layer temperature inversion was observed along the track, except for the southern location (7.5°N 78.5°E) for the launch on 21 December, 2002, where an inversion was observed between 40 m to 60 m depths with water temperatures being 28.0°C and 28.7°C at the two depths, respectively.

The frequency of XBT along the Mumbai-Colombo line being nearly bimonthly interval, the data from this XBT line could only reveal that the winter SST is the lowest (~28.4°C) and the winter cooling decreased toward the south and some upwelling prevailed in the Gulf of Mannar area. The Lakshadweep Sea was intensively monitored by the XBT line between Kochi-Minicoy-Karvathi track (Fig. 2) with about a fortnightly frequency. Hence the fluctuation of heating the surface, the thermal characteristic of uppermost 100 m depth of waters and the surface salinity were analyzed. The SST data showed that after the monsoon season cooling, SST on the average increased to about 29.5°C (range 29.3–29.9°C) in mid-October to end of November period over the area (post-monsoon SST maximum). Surface salinity, on the average in November 2002 near 10°N 76°E was 34.8 PSU (range 33.6 to 35.5 PSU). Over the rest of the track surface salinity on the average was 35.8 PSU (range of 35.1 to 36.9 PSU). Thus higher salinity surface waters lay in the vicinity of Lakshadweep Islands. On the average surface salinity in the winter period in the coastal waters off Cochin (8–10°N 75–78°E) remained near 34.4 PSU (range 34.2. to 34.6 PSU). Over the rest of the Lakshadweep Sea the wintertime average salinity was 35.25 PSU (range 34.1 to 36.7 PSU). Thus surface salinity in the coastal waters off the Kerala coast was lower by about 1 PSU than over the rest of the Lakshadweep Sea. But the spatio-temporal range of sea-surface salinity fluctuations over the Lakshadweep Sea was much higher (~2.5 PSU) to that of the waters off Kerala Coast. This would indicate either fresh water influx over the Kerala Coastal waters due to rains or runoff of the NE monsoon over Kerala Coast or penetration of the BOB low salinity waters. We have also noted the surface layer inversions at varying depths between 10–70 m and with varying temperature increase from 0.7°C to 1.1°C (mostly between 0.4–0.8°C) during November to February with maximum frequency of their occurrence between mid-January to mid-February, 2003 (Table 8). These inversions appeared near the coastal waters off Kerala (9–10°N 75–76°E) in November-December and penetrated westward over the Lakhsadweep area in January-February and disappeared in March-April. The inversions appeared when the SST dropped from about 29.5°C in November to 28.5°C in January-February and disappeared when the SST rose to 30.0°C and above after mid-March. Surface layer temperature inversions in the Lakhsadweep Sea during winter were first noted by THADATHIL and GHOSH (1992) with the bathythermograph data collected over 1132 stations during the ORV Gaveshani and ORV Sagarkanya cruises over the area for a 10-years period (1976–1986). Temperature difference in the inversion layer lay between 0.2–1.2°C and the depths of the inversions varied between 10–80 m with the maximum frequency in the range of 30-50 m. They examined several reasons for the existence of the inversions. According to THADATHIL and GOSH (1992) the most probable mechanism for their formation was the thermohaline forced advection of colder and lesser saline waters of the BOB origin over the warmer and more saline Arabian Sea waters along the West Coast of India. SHETYE et al. (1991) suggested a northward flowing coastal current along the West Coast. This northward flowing coastal current may bring low salinity waters of the BOB origin into coastal waters off the central and southern parts of the West Coast of India. SHANKAR et al. (2004) have also examined the XBT profiles of October 2002 to April 2003 over the Lakshadweep Sea and found that the inversions begin to appear in November and disappear in April and are the signatures of the low salinity BOB waters laying over the higher salinity waters of the Arabian Sea origin within the surface layer (10–80 m). Also, the inversions are sustained on the seasonal scale by sharp haline stratification of the warm layer (barrier layer). DURAND et al. (2004) simulated these inversions in a dynamical-numerical model with 0.5° horizontal resolution and 10 m vertical resolution in the upper 120 m. We agree that chief causative factor of the existence of surface layer inversions may be the haline stratification as suggested by other researchers. However we suggest that there may be also contribution from the wintertime cooling of the SST. The surface waters cool preferentially in the December-February period. The warmer waters left within the mixed layer and close

Table 8

Surface layer inversions over Lakshadweep Sea during November 2002 to March 2003

Month	Frequency	Percentage of profiles with inversions	Range of Inversion Strength (°C)
November 2002	2 out of 21 profiles	10	0.23–0.35
December 2002	6 out of 11 profiles	55	0.2–0.78
January 2003	16 out of 18 profiles	89	0.20–0.98
February 2003	18 out of 20 profiles	90	0.33–1.34
March 2003	9 out of 24 profiles	37	0.23–1.09

to the top of the thermocline after October-November warming could be trapped and remain about 0.5°C warmer than the SST. The warming of the SST during the beginning of the pre-monsoon season which would reach to over 30°C, would destroy the below surface inversion layer as the barrier layer collapsed in the dry pre-monsoon season. As a result the entire mixed layer warms up by mid-March without any inversion layer after the collapse of the barrier layer.

(d) Onset of Monsoon in 2003 Season

SSTs in the SE Arabian Sea - Lakshadweep Sea had reached the peak values of 31–32°C in April-May and were near this range until 6 June, 2003. Yet in spite of the warm pool having been fully developed until mid-May, the monsoon onset did not occur until the end of the first week of June. Monsoon struck the Kerala Coast on 8 June and moved northward on 15 June. SST over the area remained fairly high during 8 to 14 June, 2003. Monsoonal cooling occurred during the period 15–20 June when the rains had advanced to the Karnataka-Goa Coast (12–15°N) and the surface winds had substantially increased over the SE Arabian Sea. Monsoon was late by over one week and not accompanied with the formation of an intense vortex in 2003 even though the SST over the SE Arabian Sea had remained above 30.5°C. This would suggest that it is not necessary that high SSTs over the Lakshadweep Sea would promote early or more vigorous monsoon onset process. The ocean is somewhat decoupled with atmosphere over the SE Arabian Sea in the pre-monsoon season. Monsoon onset or the formation of vortex occurs as a result of atmospheric instability as the lower tropospheric winds strengthen and cyclonic wind shear builds up to the north of the maximum wind. The Sea state provides only a favorable environment for the atmospheric instability to grow. We also suggest that the northward advance of the monsoon along the West Coast of India is facilitated by the buildup of the SST gradients between the SE Arabian Sea and NE Arabian Sea. By the end of May or early June SSTs begin to decrease somewhat from their peak pre-monsoon warm stage over the SE Arabian Sea whereas the SSTs continue to rise over the NE Arabian Sea (as the monsoon has yet to reach this area) and peak there by mid-June. Thus a gradient of about 1.0°C is set up between the SE Arabian Sea and NE Arabian Sea by mid-June. Setting of such a N/S gradient would promote convective organization to shift towards the warmer waters and hence promotes the northward march of the monsoon convective zone along the West Coast of India. This aspect would, however, need more investigation to confirm it.

4. Concluding Remarks

Several aspects of the ISO in the regionally coupled ocean-atmosphere system have been examined with the help of data collected during the BOBMEX, ARMEX-I and

ARMEX-II experiments. The integrated data analyses have lead us to the following conclusions:

(i)   Seasonal warming of the BOB and the Arabian Sea provide a favourable environment for the organization of convection, prior to the onset of monsoon.

(ii)   Strengthening of the low-level jet (LLJ) triggers the formation of monsoonal disturbed episodes on ISO scale which are responsible for decrease in insolation and increase in rainfall, followed by significant fall in salinity of surface waters due to local rainfall and river-water discharges. ISO play important role in monsoon dynamics as well as considerably modulate thermohaline circulation over the northern BOB.

(iii) Ocean-atmospheric exchanges under the warm ocean surface fuel convective episodes resulting in cooling of the SST suggesting ocean's role in regulating the monsoon and in turn being impacted by it.

(iv)   During ISO related suppressed convective episodes over the northern BOB, SST rises substantially and the north-south gradients in SST recover which in turn create a favorable environment for the propagation of convection from the southern Bay to northern BOB in the mid-monsoon season, just like that it occurs in the onset and advance of the monsoon in June.

(v)   The wind induced changes in the lower-troposphere play a crucial role in triggering large-scale convective episodes. Changes in the vertical stability of the atmosphere during different episodes of the monsoon on the ISO scale do occur and need to be further illustrated. The ocean surface responds to these modulations and regulates the monsoon processes, which have higher fluctuations in parameters like rainfall, atmospheric moisture content in the lower mid-troposphere, SST and salinity compared to the intra-annual fluctuations in these parameters.

(vi)   Incidence of strengthening of the off-shore trough during the active phase of advancing monsoon and during the mid-season revival of the monsoon, sometime accompanied with mesoscale cyclonic vortex, have been noted. Mesoscale models are capable of simulating these isolated deep convective systems, provided critical data inputs are made available over the region of their formation. Modeling aspects need to be further examined.

(vii) Pre-monsoon warming of the SST over the SE Arabian Sea and Lakshadweep Sea region occurs under strong insolation and low surface wind conditions due to the subsidence occurring in association with lower-tropospheric high pressure cell over the Arabian Sea. SSTs in March climb to above 30°C and remain so until mid-May. In spite of high SSTs in this area organized convection remains decoupled from the SSTs due to subsidence in the lower troposphere. The monsoon onset towards end of May-early June over the SE Arabian Sea occurs due to rush of strong LLJ. This sets up dynamic instability in the lower troposphere and organizes convection on synoptic scale. This aspect needs to be looked further through dynamical modeling approach. The

SST gradient set up between the SE Arabian Sea and NE Arabian Sea by mid-May promotes northward march of organized convection and monsoon rains along the West Coast of India.

(viii) Presence of surface layer inversions over the SE Arabian Sea during November to March between 10–80 m depth results from a combination of factors like (a) the haline stratification due to presence of low salinity waters of the BOB in southern and central parts of the West Coast in Winter, (b) rainfall over Kerala and SE Arabian Sea in November-December under low surface wind conditions, and (c) trapping of the warm waters of the mixed layer of October-November period as the winter cooling of surface waters may provide a shielding effect under low wind conditions.

Acknowledgement

The authors acknowledge several research groups involved in the BOBMEX and ARMEX programs who shared their processed data with the authors (G.S.Bhat, V.V.Gopalakrishna, S.S.C. Shenoi, S.R. Kalsi, S.P. Ghanekar, B.N. Goswami, U.C. Mohanty. D. Sengupta, CK Rajan and others). Thanks are also due to the Director, National Institute of Ocean Technology, Chennai; Head, National Centre for Medium Range Weather Forecasting, New Delhi and Director General of Meteorology, India Meteorological Department, New Delhi for providing various data products used in this study. We are grateful to Prof. VS Ramamurthy, Secretary, DST and Dr. B.D. Acharya, Head, Earth System Science Division, Department of Science and Technology for the encouragement to undertake this study. We also thank the anonymous reviewers for their critical comments on the earlier manuscript.

REFERENCES

BHASKAR RAO, D.V. and PRASAD, D.H. (2005), Numerical Simulation of a Heavy Rainfall Event during *RMEX-I with and without Data Assimilation,* Mausam, 56, 121–130.

BHAT, G.S., GADGIL, S., HARISH KUMAR, P. V., KALSI, S. R., MADHUSOODANAN, MURTY, V. S. N., PRASADA RAO, C. V. K., RAMESH BABU, V., RAO, L. V. G., RAO, R. R., RAVICHANDRAN, M., REDDY, K. G., SANJEEVA RAO, P., SENGUPTA, D., SIKKA, D. R., SWAIN, J., and VINAYACHANDRAN, P. N. (2001), *BOBMEX - The Bay of Bengal Monsoon Experiment,* Bull. Am. Meteorol. Soc. *82,* 2217–2243.

BHAT, G. S. (2002), *Near-surface Variations and Surface Fluxes over the Northern Bay of Bengal during the 1999 Indian Summer Monsoon,* J. Geophy. Res. *107,* 4336.

BHAT, G. S; CHAKRABORTY, A., NANJUNDIAH, R. S., and SRINIVASAN, J. (2002), *Vertical Thermal Structure of the atmosphere during Active and Weak Phases of Convention over the Northern Bay of Bengal: Observations and Model Results,* Current Science *81,* 296–302

BHAT, G. S. (2003) *Convective Inhibition Energy of the Inversion and the Suppressed Rainfall Observed over the Arabian Sea during July 2002,* Presented at ARMEX Workshop, 22–23 Dec. 2003, Chennai, India.

CHATTERJEE, P. and GOSWAMI, B.N. (2003), *Structure, Genesis and Scale Selection of the Tropical Quasi Bi-weekly Mode*, Q. J. R. Metrol. Soc. *128*.

DAS, S, DASGUPTA, M., MOHANTY, U. C., and PRASHANTI, K. (2003), *Assimilation of ARMEX Observations in a Mesoscale Model and its Impact on Weather Forecast*, Paper presented at ARMEX Workshop, 22–23 Dec. 2003, Chennai, India.

DST (1995), *Indian Climate Research Programme-Science Plan*, Department of Science and Technology, New Delhi, India.

DURRAND, F., SHETYE, S. R., VIALARD, J., SHANKER, D., SHENOI, S. S. C., ETHE, C., and MADEC, G. (2004), *Impact of Temperature Inversions on SST Evolution during the Pre-summer Monsoon Season*, Geophys. Res. Lett.

FRANCIES, P. A. and GADGIL, S. (2002), *Intense Rainfall Events over the West Coast of India*, CAOS Report, Centre for Atmospheric and Oceanic Sciences, IISc, Bangalore, India.

GADGIL, S., VINAYCHANDRAN, P. N., and FRANCIS, P. A. (2003), *Droughts of Indian Summer Monsoon: Role of Cloud Cover over the Indian Ocean*, Current Science *85*, 1713–1719.

GADGIL, S. and JOSEPH, P.V. (2003), *On Breaks of the Indian Monsoon*, Proc. Ind. Acad. Sci. (E & P. Sci.) *112*, 529–556.

GEORGE, P.A. (1956), *Effect of Offshore Vortex on Rainfall along the West Coast of India*, Ind. J. Met. Geophys *7*, 235–240.

GHANEKAR, S.P., MUJUMDAR, V.R., SEETARAMAYYA, P., and BHIDE, U.V. (2003), *Ocean-atmosphere Interaction and Synoptic Weather Conditions with the two Contrasting Phases of Monsoon during BOBMEX-1999*, Proc. Ind. Acad. Sci. (E & P Sci.) *112*, 283–293.

GOSWAMI, B. N. and AJAY MOHAN, R. S. (2001), *Intra-seasonal Oscillation and Interannual Variability of the Asian Summer Monsoon*, J. Climate *14*, 1180–1198.

GROSSMAN, R.L. and DURRAN, D.R. (1980), *Interaction of Low-level Flow with the Western Ghat Mountains and Off-shore Convection in the Summer Monsoon*, Mon. Wea. Rev. *112*, 652–672.

JOSEPH, P.V. (1990), *Warm Pool over the Indian Ocean and Monsoon Onset*, Tropical Ocean-Atmos. News Lett *53*, 1–5.

KALSI, S. R. (2003), *Synoptic Weather Conditions during BOBMEX*, Proc. Ind. Acad. Sci. (E & P. Sci), *112*, 239–253.

KRISHNAMURTI, T.N. and ARDUNUY, P. (1980), *The 10–20 Day Westward Propagating Mode and Breaks in the Summer Monsoon*, Tellus. 32, 15–25.

KRISHNAMURTI, T. N. and SUBRAMANIAM, D. N. (1982), The 30–50 Day Mode at 850 mb during MONEX, J. Atmos. Sci. *300*, 1290–1306.

MOHANTY, U.C. and DAS, S. (1986), *On the Structure of the Atmosphere during Suppressed and Active Convention over the Bay of Bengal*, Proc. Ind. Acad. Sci. *52*, 625–640.

MUKHERJEE, A.K. and SHYAMALA, B. (1984), *The Dynamics of an Off-shore Vortex in the East Arabian Sea and the Associated Rainfall*, Mausam *35*, 233.

RAMAMOORTHY, K. (1969), *Monsoon of India—Some Aspects of the Break in Indian southwest Monsoon during July and August*, Forecasting Manual *IV*, 18.3, Ind. Met Dept. Pune, India.

RAO, R.R. (1980), *Cooling and Deepening of the Mixed Layer in the Central Arabian Sea during Monsoon-77: Observations and Simulations*, Deep Sea Res. *33*, 1413–1424.

RAO, R.R. and SIVAKUMAR, R. (1999), *On the Possible Mechanism and Evolution of a Mini-warm Pool during the Pre-summer Monsoon Season and the Onset Vortex in the Southeastern Arabian Sea*, Q. J. R. Meteorol. Soc. *125*, 787–809.

RAO, R.R. and SIVAKUMAR, R. (2000), *Seasonal Variability of Near-surface Thermal Structure and Heat Budget of the Mixed Layer of the Tropical Indian Ocean from a New Global Ocean Temperature Climatology*, J. Geophys. Res. *105*, 995–1015.

SANILKUMAR, K.V., MOHANKUMAR, N., Joseph, M.X., and Rao, R.R. (1994), *Genesis of Meteorological Disturbances and Thermohaline Variability of the Upper Layer in the Head Bay of Bengal during MONTBLEX-90*, Deep Sea Res. *41*, 1569–1581.

SANJEEVA RAO, P. (2005), *Arabian Sea Monsoon Experiment (ARMEX): An Overview*, Mausam, 56, 1–6.

SARMA, Y. V. B., SEETARAMAYYA, P., MURTY, V. S. N., and RAO, D. P. (1997), *Influence of the Monsoon Trough on Air-sea Interactions in the Head Bay of Bengal during the South West Monsoon of 1990 (MONTBLEX-90)*, Boundary Layer Meteorol, *821*, 517–526.

SEETARAMAYYA, P. and MASTER, A. (1984), *Observed Air-sea Interface Conditions and a Monsoon Depression during MONEX-79*, Arch. Met. Geophys. Biocl. Ser. *A. 33*,61–67.

SEETARAMAYYA, P., NAGAR, S.G., and MULLAN, A.H. (2001), Response of the Northern Bay of Bengal (Head Bay) to Monsoon Depression during MONTBLEX-90, The Global Atmos. Ocean Sys. *7*, 325–345.

SENGUPTA, D. and RAVICHANDRAN, M. (2003), *Oscillations of Bay of Bengal Sea Surface Temperature during 1998 Summer monsoon*, Geophysics Res. Lett.

SHANKAR, D. and SHETYE, S. R. (1997), *On the Dynamics of the Lakshadweep High and Low in the South eastern Arabian Sea*, J. Geophys. Res. *102* (C-6), 12551–12562.

SHANKAR, D., GOPALKRISHNA, V.V., SHENOI, S.S.C., SHETYE, S.R., RAJAN, C.K. ZACHARIA, J., ARALIDIDAD, N., and MICHAEL, G.S. (2004), *Observational Evidence for Westward Propagation of Temperature Inversions in the Southeastern Arabian Sea*, Geophys. Res. Lett.

SHENOI, S.S.C., SHANKAR, D., and SHETYE, S.R. (1999), *On the Sea-surface Temperature High in the Lakshadweep Sea before the Onset of the Southwest Monsoon*, J. Geophys. Res. *104* (C-7), 15703–15712.

SHETYE, S.R., GOUVERIA, A.D., SHENOI, S.S.C., MICHAEL, G.S., SUNDER, D., ALMEIDA, A.M., and SANTANAM, K. (1991), *The Coastal Current off Western India during the Northeast Monsoon*, Deep Sea Res. *38*, 1517–1529.

SHYMALA, B. (2003), *Observational Studies of Off-shore Vortices during ARMEX-2002*, Presented at the ARMEX Workshop, 22–23 Dec. 2003, Chennai, India.

SINGH, S.V., KRIPLANI, R.H., and SIKKA, D.R. (1992), *Interannual Variability of the Madden-Julian Oscillation in the Indian Summer Monsoon Rainfall*, J. Climate *5*, 973–978.

SIKKA D.R. and GADGIL, S. (1978), *Large-scale Rainfall over India and its Relationship with Lower and Upper Tropospheric Vorticity*, Ind. J. Met. Hydro. Geophys. *29*, 219–231.

SIKKA, D.R. and GADGIL, S.(1980), *On the Maximum Cloud Zone and the ITCZ over the Indian Longitudes during the Southwest Monsoon*, Mon. Wea. Rev. *108*, 1840–1853.

SIKKA, D.R. (2003), *Monsoon Monitoring and Forecasting Drought of 2002*, Proc. Ind. Nat. Sci. Acad. 69A.

THADTHIL, P. and GHOSH, A. (1992), *Surface Layer Temperature Inversion with the Athil Arabian Sea during Winter*. J. Oceanogr. *48*, 293–304.

VARKEY, M.J., MURTY, V.S.N., and SURYANARAYANA, A. (1996), *Physical Oceanography of the Bay of Bengal and Andaman Sea*, Ann. Rev. Oceanography and Marine Biol. *34*, 1–70.

VINAYCHANDRAN, P.N., SADHURAM, Y., and BABU, V.R. (1989), *Latent and Sensible Heat Fluxes under Active and Weak Phases of the Summer Monsoon of 1986*, Proc. Ind. Acad. Sci. (E & P. Sci.), *98*, 213–222.

VINAYCHANDRAN, P.N. and SHETYE, S.R. (1991), *The Warm Pool in the Indian Ocean*, Proc. Ind. Acad. Sci. (E & P Sci.), *100*, 165–175.

VINAYCHANDRAN, P.N. and MURTY, V.S.N. (2002), *Observations of Barrier Layer Formation in the Bay of Bengal during Summer Monsoon*. J. Geophys. Res.107.

YASUNARI, T. (1980), *A quasi-stationary Appearance of 30-40 Day Period in the Cloudiness Fluctuations during the Summer Monsoon over India*, J. Meteorol. Soc., Japan. *59*, 336–354.

(Received March 23, 2004; accepted July 7, 2004)

To access this journal online:
http://www.birkhauser.ch

Pure appl. geophys. 162 (2005) 1511–1541
0033–4553/05/091511–31
DOI 10.1007/s00024-005-2681-z

Pure and Applied Geophysics

A Study on Climatological Features of the Asian Summer Monsoon: Dynamics, Energetics and Variability

U.C. MOHANTY, P.V.S. RAJU, and R. BHATLA

Abstract—A continuing goal in the diagnostic studies of the atmospheric general circulation is to estimate various quantities that cannot be directly observed. Evaluation of all the dynamical terms in the budget equations for kinetic energy, vorticity, heat and moisture provide estimates of kinetic energy and vorticity generation, diabatic heating and source/sinks of moisture. All these are important forcing factors to the climate system. In this paper, diagnostic aspects of the dynamics and energetics of the Asian summer monsoon and its spatial variability in terms of contrasting features of surplus and deficient summer monsoon seasons over India are studied with reanalysis data sets. The daily reanalysis data sets from the National Centre for Environmental Prediction/National Centre for Atmospheric Research (NCEP/NCAR) are used for a fifty-two year (1948–1999) period to investigate the large-scale budget of kinetic energy, vorticity, heat and moisture. The primary objectives of the study are to comprehend the climate diagnostics of the Asian summer monsoon and the role of equatorial convection of the summer monsoon activity over India.It is observed that the entrance/exit regions of the Tropical Easterly Jet (TEJ) are characterized by the production/destruction of the kinetic energy, which is essential to maintain outflow/inflow prevailing at the respective location of the TEJ. Both zonal and meridional components contribute to the production of kinetic energy over the monsoon domain, though the significant contribution to the adiabatic generation of kinetic energy originates from the meridional component over the Bay of Bengal in the upper level and over the Somali Coast in the low level. The results indicate that the entire Indian peninsula including the Bay of Bengal is quite unstable during the summer monsoon associated with the production of vorticity within the domain itself and maintain the circulation. The summer monsoon evinces strong convergence of heat and moisture over the monsoon domain. Also, considerable heat energy is generated through the action of the adiabatic process. The combined effect of these processes leads to the formation of a strong diabatic heat source in the region to maintain the monsoon circulation. The interesting aspect noted in this study is that the large-scale budgets of heat and moisture indicate excess magnitudes over the Arabian Sea and the western equatorial Indian Ocean during surplus monsoon. On the other hand, the east equatorial Indian Ocean and the Bay of Bengal region show stronger activity during deficient monsoon. This is reflected in various budget terms considered in this study.

Key words: Reanalysis, summer monsoon, kinetic energy, vorticity, heat and moisture.

1. Introduction

The Indian summer monsoon is characterized with rainfall regimes, onset/withdrawal phases, break and active conditions and synoptic disturbances. The basic

Centre for Atmospheric Sciences, Indian Institute of Technology-Delhi, Hauz Khas, New Delhi 110 016, India.

forcing of the Asian summer monsoon is provided by the annual cycle of solar radiation interacting with different heat capacities of the tropical ocean and land areas (LI and YANAI, 1996) and their respective geographical arrangements. The tropical ocean regions, particularly, the Arabian Sea, Bay of Bengal and Indian Ocean act as main reservoirs of heat and moisture in supplying the necessary energy to the establishment and maintenance of the large-scale monsoon circulation and associated monsoon activity over the Indian subcontinent. There is complex feedback between the flow field and the heating, especially through the interaction between moist convection and large-scale flow (MOHANTY *et al.*, 1983), which is poorly understood. The orientation of the orographic barriers over the Indian subcontinent also modifies the circulation considerably, consequently the summer monsoon system becomes a very complex array of weather phenomena. The Tibetan plateau and the Himalayan Mountains play an important role in the observed monsoon, in the form of total barrier to low-level meridional winds. The rainfall is caused mainly as the low-level flow meets the mountain barrier and increases the strength of the heating and confines it to the region south of the Tibetan plateau. At the same time, it has been established that the sensible heating, resulting from the absorption of solar radiation by the elevated Tibetan plateau, provides a heat source which strengthens the upper level anticyclone.

The prominence of diagnostic studies is well recognized as an important component of the Global Atmospheric Research Program in elucidating the dynamics of tropics (PEARCE, 1979; KUNG and TANAKA, 1983; KUNG and SMITH, 1974). Although, several works were reported on atmospheric diagnostics, some studies (MOHANTY *et al.*, 1982a,b; PEARCE and MOHANTY, 1984) elucidated onset and maintenance aspects of the Asian summer monsoon. Presumably, this is due to the paucity of data of the tropics. Nevertheless, the advent of multifarious operational centers around the world having global analysis of meteorological fields, facilitated the understanding of the dynamics of the Asian summer monsoon and its variability. Significant contributions to the diagnostic studies of the atmospheric energetics have been made by LORENTZ (1955), OORT (1964) and NEWELL (1970). Several studies (KRUEGER and WINSTON, 1975; KANAMITSU and KRISHNAMURTI, 1978; RAMESH *et al.*, 1996 and others) have been carried out to analyze the contrasting circulation features and energetics of normal and deficient monsoon seasons. Further, a detailed comparison of the evolution of certain parameters such as outgoing long-wave radiation, sea-surface temperature, stream-function anomalies, divergent circulation and precipitation patterns, etc. was made by KRISHNAMURTI *et al.* (1989, 1990) to elucidate some of the differences between deficit (1987) and surplus (1988) monsoon seasons over India. The thermodynamic characteristics of the Asian summer monsoon are studied by RAMESH *et al.* (1999) and RAO *et al.* (2003) with a global analysis-forecast system. They found that the model forecast failed to simulate analyzed atmospheric variability in terms of mean circulation, which is indicated by underestimation of various terms of heat and moisture budgets with an increase in the forecast period. Despite revealing basic information and elucidating the complexity of the problem, these studies evinced the necessity of further detailed studies on the Asian summer monsoon. At this juncture, a need arises to comprehend the multifarious complex mechanisms associated with the monsoon circulation in order to improve prediction in various spatio and temporal scales. Despite a few studies which focused on the Asian summer monsoon, they are limited by the scope of investigation as these are based on a few years of data sets. In recent years the National Centre for Environmental Prediction/National Centre for Atmospheric Research (NCEP/NCAR, hereafter NCEP) has made an excellent effort to generate a reanalysis of global data sets (KALNAY et al., 1996). The 52-year NCEP reanalysis provides a consistent and reliable data set for investigating dynamics and energetics of short-term climate phenomena. It provided for the first time a unique database to examine climate variability of circulation features and energetics of the monsoon. Recent studies (ANNAMALAI et al., 1999; SPERBER et al., 2000; RAJU et al., 2002) conducted making use of the NCEP reanalysis, confirm the usefulness of these data sets. As a result, understanding the dynamical mechanisms of the monsoon variability in time-scales ranging from weeks to years is an issue of considerable importance.

In the present study, the mean circulation features and energetics of the Asian summer monsoon are studied with 52-years (1948–1999) NCEP reanalysis data sets. Further, the mean circulation features and energetics associated with the composite of surplus and deficient monsoon seasons over India during the 1948–1999 period are examined.

2. Data and Analysis System

The NCEP/NCAR have cooperated in a project to produce a retrospective 52-year (1948–1999) record of global analyses of atmospheric fields supporting the needs of the research and climate monitoring communities. This effort involved the recovery of land surface, ship, rawinsonde, pibal, aircraft, and satellite and other observational data, quality control and immersion of these data in a data assimilation system that is kept unchanged over the reanalysis period. The reanalysis system is continuing with current data on a real time basis (Climate Data Assimilation System or CDAS), so that its products are available from 1948 to the present.

The NCEP/NCAR reanalysis system used a state-of-the-art data assimilation (3-D variational) with the horizontal resolution of T62 (about 210 km) and 28 sigma vertical levels. The model has five levels in the boundary layer and about seven levels above 100 hPa. The model is identical to the global system implemented operationally at NCEP except for the horizontal resolution T126 (105 km) (KANAMITSU, 1989, 1991). The analysis scheme is a three-dimensional variational scheme cast in spectral statistical interpolation (PARRISH and DERBER, 1992). The module contains complex quality control of rawinsonde data including time interpolation checks with confidence corrections of height and temperature (COLLINS and GANDIN, 1990, 1992). Optimal interpolation-based complex quality control is applied for all other data (WOOLEN, 1991; WOOLEN *et al.*, 1994). The model includes parameterization schemes of all major physical processes such as convection, large-scale precipitation, radiation, boundary layer physics, an interactive surface hydrology, and vertical and horizontal diffusion processes. The moist convection is represented by a simplified form of the Arakawa-Schubert parameterization scheme (PAN and WU, 1994) and clouds are diagnosed from the model which generated outgoing long-wave radiation (CAMPANA *et al.*, 1994). The NCEP model uses a three-layer soil scheme based on PAN and MAHRT (1987), in which the bottom layer is set to the annual mean climatological value. A detailed description of the NCEP/NCAR reanalysis project is described by KALNAY *et al.* (1996).

In this study, the daily averaged (00 and 12 UTC) reanalysis data set produced at NCEP with a horizontal resolution of 2.5° on a regular latitude/longitude grid are extracted for the monsoon domain. The basic meteorological fields considered for the study include geopotential height (z), wind (u and v), temperature (T) and specific humidity (q) at twelve pressure levels (1000, 925, 850, 700, 600, 500, 400, 300, 250, 200, 150 and 100 hPa). In order to avoid the problems with the divergent wind, the vertical velocity fields in this study have been computed from horizontal wind components (u and v) by using the kinematic method as suggested by O'BRIEN (1970). In this technique the divergence is adjusted to its vertically integrated value zero in the entire column of the atmosphere. The vertical velocity distribution obtained from the kinematic method delineates realistic Hadley circulation over the monsoon domain, compared to the archived field.

3. Methodology

The comprehensive analysis of dynamical features of the Asian summer monsoon is accomplished through the study of large-scale budgets of kinetic energy, vorticity, heat and moisture. The budget equations are obtained from the prognostic and diagnostic equations of the atmospheric model on simple mathematical transformations and represented below in the flux form with pressure as the vertical coordinate. The overbar in the budget equations denotes the composite seasonal mean value of a quantity, and prime quantities denote their corresponding deviations from the composite seasonal mean. In general, tropical circulations are driven by the mean flow, while extra tropical circulations are driven by both mean and eddy components. The eddy part in extra tropics is usually more significant than the mean flow. In this study the time mean large-scale balance equations are bifurcated into stable mean and transient eddy parts. The primary focus is on the mean component of the budgets as the tropical circulations are dominated by the mean component of flow.

Following HOLOPAINEN (1978), the kinetic energy budget equation is expressed as

$$\frac{\overline{\partial K_M}}{\partial t} + \nabla.(H_0 + H_1) + \frac{\partial}{\partial P} \overline{(K_M + \overline{V}V')\omega} = -\overline{V}.\nabla\overline{\phi} - C(K_M, K_T) + \overline{V}.F \qquad (1)$$

where

$$K_M = \frac{1}{2}\overline{V}^2 \text{ Kinetic energy of the mean flow,}$$

$$K_T = \frac{1}{2}\overline{V'^2} \text{ Kinetic energy of the eddy flow.}$$

Various notations used in the equation are given below:

$$H_0 = K_M\overline{V} \qquad ; \qquad H_1 = \overline{(\overline{V}.V')V'}.$$

H_0 and H_1 are kinetic energy fluxes due to mean and eddy component of flow, respectively.

$$C(K_M, K_T) = C_H(K_M, K_T) + C_v(K_M, K_T)$$

$$C_H(K_M, K_T) = -\frac{\overline{u'u'}}{a \cos\varphi}\frac{\partial\overline{u}}{\partial\lambda} - \frac{\overline{u'v'}\cos\varphi}{a}\frac{\partial}{\partial\varphi}\left(\frac{\overline{u}}{a \cos\varphi}\right) - \frac{\overline{u'v'}}{a \cos\varphi}\frac{\partial\overline{v}}{\partial\lambda} - \frac{\overline{v'v'}}{a}\frac{\partial\overline{v}}{\partial\varphi} + \overline{u'u'}\frac{\overline{v} \tan\varphi}{a}$$

$$C_V(K_M, K_T) = -\overline{u'\omega'}\frac{\partial\overline{u}}{\partial P} - \overline{v'\omega'}\frac{\partial\overline{v}}{\partial P}.$$

The first term on the left side of equation (1) designates the local tendency of kinetic energy. The second and third terms describe the horizontal and vertical flux divergences of kinetic energy, respectively. Similarly, the first term on the right side of the equation denotes the conversion of available potential energy to kinetic energy through the action of pressure forces (adiabatic generation of kinetic energy). The second term describes the exchange of energy between mean and transient flows that arises from the horizontal and vertical Reynolds stresses. The last term signifies the dissipation of kinetic energy by the turbulent frictional processes.
The vorticity budget equation is designated as

$$\frac{\overline{\partial\zeta}}{\partial t} + \nabla.(\overline{\zeta V}) + \beta\overline{v} + \frac{\partial(\overline{\zeta\omega})}{\partial P} = -\overline{(\zeta D)} - k.(\overline{\nabla\omega x}\frac{\partial v}{\partial P}) + \overline{Z}, \qquad (2)$$

where
$\zeta = \left(\frac{\partial v}{\partial x} - \frac{\partial u}{\partial y}\right)$ is the relative vorticity ; $D = \left(\frac{\partial u}{\partial x} + \frac{\partial v}{\partial y}\right)$ is the divergence.
The first term on the left-hand side of equation (2) denotes the local rate of change of vorticity. The second and third terms indicate the horizontal flux divergences of relative and planetary vorticity, respectively. The fourth term describes the vertical divergence flux of relative vorticity. Similarly, the first and second terms on the right-hand side evince the vorticity generation due to stretching and tilting, respectively. The final term designates the residue of vorticity (i.e., generation/dissipation of vorticity from subgrid scale processes).

The heat budget equation in the flux form can be written as

$$\frac{\overline{\partial(CpT)}}{\partial t} + \nabla.\overline{V}Cp\overline{T} + \frac{\partial}{\partial p}(\overline{\omega}\,Cp\,\overline{T}) - \overline{\omega\alpha} = \overline{Q}_H. \tag{3}$$

In the equation (3), the first term on the left denotes the local variation of enthalpy. The second and third terms designate the horizontal and vertical flux divergences of heat. The fourth term indicates the adiabatic conversion of available potential energy to kinetic energy. Similarly, the term on the right of equation (3) describes the diabatic heating, which is due to radiation, condensation, turbulent transfer, evaporation of falling raindrops and turbulent transfer of sensible heat.

The moisture budget equation in the flux form can be expressed as

$$\frac{\overline{\partial(Lq)}}{\partial t} + \nabla.\overline{V}L\overline{q} + \frac{\partial}{\partial p}(\overline{\omega}L\overline{q}) = \overline{Q}_M. \tag{4}$$

The first term on the left of equation (4) indicates the local tendency of moisture. The second and third terms designate the horizontal and vertical divergence fluxes of moisture, respectively. The right term evinces the diabatic contributions to latent heat energy or the moisture source/sink that arises from diabatic heating due to latent heat release and condensation as well as turbulent transfer of latent heat.

The vertical integration of all the budget equations with the boundary condition that vertical motion ($\omega = 0$) vanishes at the bottom and the top of the atmosphere leads to the elimination of all the terms representing the vertical flux divergences of various quantities. All last terms on the right of the budget equations (1 to 3) represent the contribution from the subgrid scale physical processes. These terms are evaluated implicitly in this study as residues of all the other terms in the respective budget equations. Although we evaluated all budget equations, the discussion in this paper is restricted to the terms which contribute significantly to the budgets that are largely responsible for the maintenance of the summer monsoon circulation. Further, Student's *t*-test has been applied to the basic parameters and energetics to identify the most significant zones (95% confidence level). The critical value of the Student's *t*-test with 95% confidence level is 1.75. We computed the Student's *t*-test by using the formula

$$t = \frac{\overline{X}_e - \overline{X}_d}{\sigma\left(\frac{1}{N_e} + \frac{1}{N_d}\right)},$$

where σ is the standard deviation, $\overline{X}_e$ and $\overline{x}$ are the mean of excess and deficient monsoons. N_e and N_d are the total number of excess and deficient monsoon seasons.

4. Results and Discussion

The Indian summer monsoon undergoes substantial variability in the amount of rainfall from one season to another. The average seasonal rainfall in summer monsoon (June-September) over India is 852 mm and standard deviation is 82 mm. Table 1 illustrates the excess (surplus) and deficient (drought) monsoon years during the period 1948–1999. The values of the Indian summer monsoon rainfall are taken from a data source of India Meteorological Department (IMD) and PARTHASAR-ATHY *et al.* (1994) to categorize the surplus and deficient monsoon years. The departure of the rainfall more (less) than 10% from the long-term mean is considered as surplus (deficient) monsoon. Based on the above criteria, 23 (18%) deficient monsoon years and 17 (13%) surplus monsoon years and the remaining 89 (69%) normal monsoon years were recorded during a 129-year (1871–1999) period. In this regard during the recent 52 years (1948–1999) under present study, 8 (15%) surplus and 11 (21%) deficient and 33 (64%) normal monsoon years are identified. Thus, no appreciable changes are found in the mode of occurrence of extreme monsoon events from 1948 to 1999. In this study the large-scale features of kinetic energy, vorticity, heat and moisture and its spatial variability in terms of mean and standard deviation are examined over the Asian summer monsoon using 52-year (1948–1999) NCEP reanalysis data sets. Further, the comprehensive analysis of contrasting dynamical features between eight surplus and eleven deficient monsoon year is also investigated with Student *t*-test at 95% confidence level to identify the statistically significant regions.

4.1 Precipitation, OLR and Net Tropospheric Moisture

The seasonal mean precipitation of NCEP reanalysis is illustrated in Figure 1. The climatology of rainfall (Fig. 1a) indicates maximum rainfall over north Bay of Bengal and the adjoining Indian region, the west Indian coast and the adjoining Arabian Sea. Maximum rainfall is noted over the equatorial Indian Ocean. This pattern is consistent with the rainfall climatology of RAO (1976) and XIE and ARKIN

Table 1

Surplus and deficient monsoon years during 1948–1999

Surplus Year	Deficient Year
1956 (15.12)	1951 (−13.54)
1959 (10.76)	1965 (−16.77)
1961 (19.70)	1966 (−13.18)
1970 (10.26)	1968 (−11.47)
1975 (12.95)	1972 (−23.39)
1983 (12.12)	1974 (−12.23)
1988 (12.80)	1979 (−16.96)
1994 (10.07)	1982 (−13.72)
	1985 (−10.86)
	1986 (−12.83)
	1987 (−18.20)

(The values in the bracket depict the percentage departure from the mean)

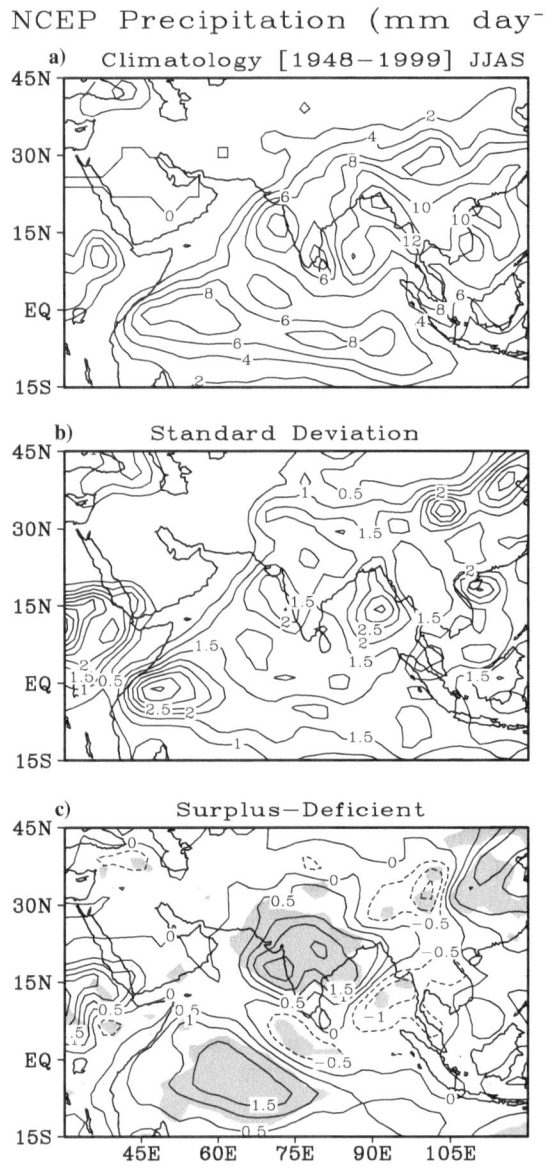

Figure 1

Geographical distribution of seasonal mean (JJAS) NCEP precipitation (mm day^{-1}) a) climatology (1948–1999), b) standard deviation, c) difference (surplus-deficient) [shaded region is 95% significant level].

(1997). Further, RAJU *et al.* (2002) stated that on a seasonal scale the NCEP precipitation appears to be reliable as regards the distribution of rainfall over India. The variability of rainfall measured in terms of its standard deviation from the climatological average (1948–1999) is shown in Figure 1b. It shows the maximum variability over the Bay of Bengal, the west coast and the western equatorial Indian Ocean. The difference of NCEP precipitation between surplus and deficient monsoon years is shown in Figure 1c. The statistically significant regions at 95% confidence level (based on Student's t-test) are shaded. The difference indicates more precipitation over Indian landmass, western equatorial Indian Ocean, small pockets over western Bay of Bengal and east Arabian Sea during surplus monsoon years. These regions are statistically significant. It may be noted that the negative values (less rainfall) during surplus years as compared to deficient years are observed over equatorial east Bay of Bengal off Myanmar and central China. The geographical distribution of outgoing long wave radiation (OLR) is presented in Figure 2. The low value of OLR corresponds to emission from higher levels and hence higher cloud tops (more rain when the clouds are deep). The climatological pattern (Fig. 2a) denotes maximum OLR over the Arabian Peninsula, Iraq and Iran where rainfall amounts are insignificant. The Bay of Bengal, the west Indian coast and the western equatorial Indian Ocean indicate lower OLR values. This is due to the fact that during the summer monsoon season, the lower OLR is associated with higher convective cloudiness and hence the precipitation. The maximum variability of OLR is observed over the equatorial Indian Ocean, north Africa and the south Indian peninsula (Fig. 2b). The difference of OLR between the composite of surplus and deficient monsoon years (Fig. 2c) reveals negative OLR over the equatorial Indian Ocean with its maximum in the western sector. These regions are statistically significant at 95% confidence level. It indicates that these regions are associated with low OLR values during surplus monsoon years. Further, the zones of low OLR values favourably agree with regions of excess precipitation (Fig. 1c). Thus, it is suggested that the equatorial convection, particularly in the western sector, plays an important role in the summer monsoon activity and hence precipitation over India. The geographical distribution of net tropospheric moisture is presented in Figure 3. The climatology of net tropospheric moisture (Fig. 3a) depicts the maximum moisture over the Indian subcontinent extending eastwards to southeast China through Myanmar, with minimum moisture over the Tibetan plateau. The maximum moisture zone is due to the low-pressure system that moves from the Pacific Ocean across Myanmar into the head of the Bay of Bengal, intensifies into depression and moves along the monsoon trough zone. The net tropospheric moisture shows the maximum variability in the monsoon trough zone and the south Arabian Sea (Fig. 3b). In the monsoon trough zone the variability could be due to the movement of monsoon lows/depression, while in the south Arabian Sea, it could possibly be due to pulses in the monsoon current associated with the strengthening and weakening of the monsoon. The difference of surplus and deficient monsoon years (Fig. 3c) indicates that most of the monsoon domain except a few small pockets is characterized by excess net tropospheric moisture during the surplus monsoon season, with a maximum over northwest India and the adjoining Arabian Sea. The

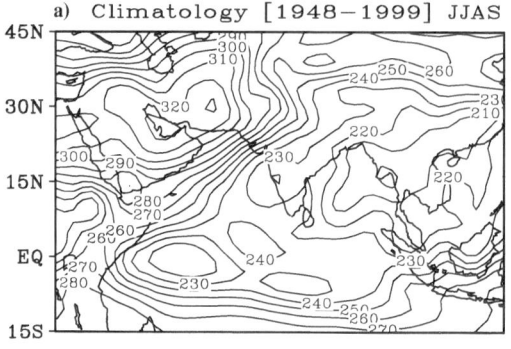

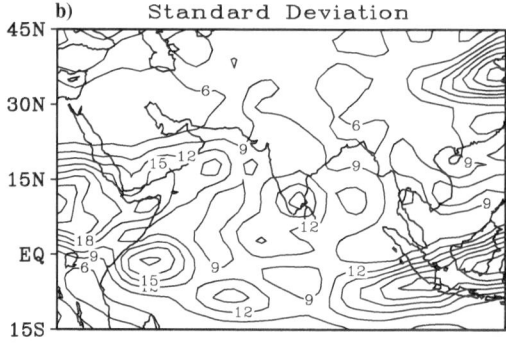

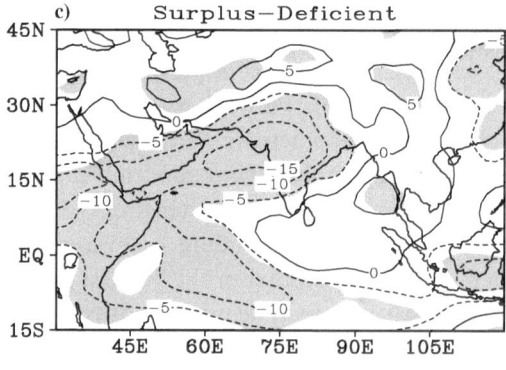

Figure 2

Geographical distribution of outgoing longwave radiation (W m^{-2}) for a) climatology (1948–1999), b) standard deviation, c) difference (surplus-deficient) [shaded region is 95% significant level].

statistically significant regions are observed over the Arabian Sea extending from the western equatorial Indian Ocean and Indian land mass except northeast India. These regions of statistical significance are in good agreement with regions of negative OLR (Fig. 2c) and excess rainfall (Fig. 1c).

Net Tropospheric Moisture (mm)

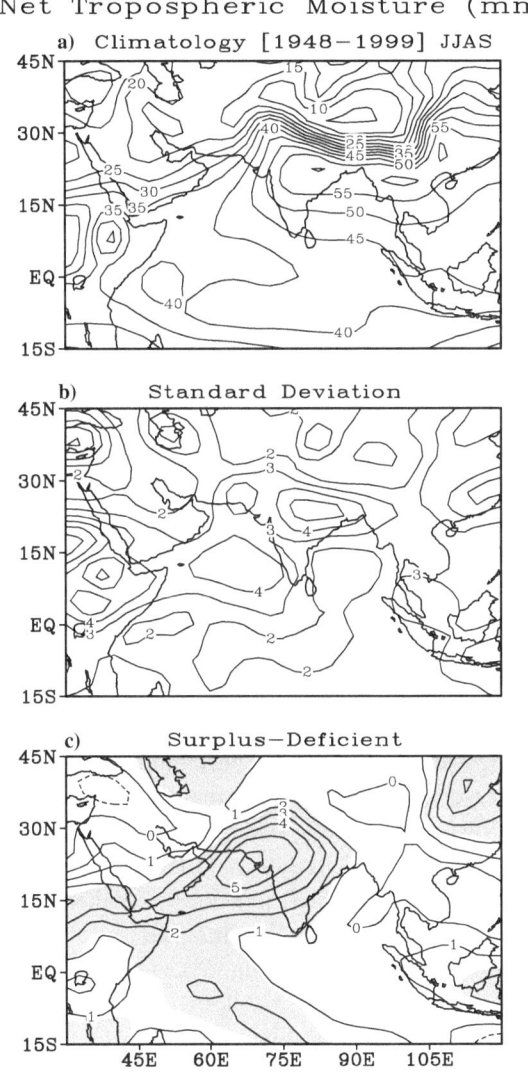

Figure 3
Geographical distribution of net tropospheric moisture (mm) for a) climatology (1948–1999), b) standard deviation, c) difference (surplus-deficient) [shaded region is 95% significant level].

4.2 Circulation Features

In order to delineate the predominant characteristic circulation features of Asian summer monsoon in the vertical plane, the sectorial mean cross sections of zonal wind, meridional wind over two longitudinal sectors; the Arabian Sea (45°E–75°E) and the Bay of Bengal (85°E–105°E) are considered. Figure 4 depicts the sectorial mean cross sections of zonal wind over Arabian Sea (left panel) and Bay of Bengal (right panels) sectors for JJAS climatology (Figs. 4a,d), standard deviation (Figs. 4b,e) and the difference between surplus and deficient years at 95% significant level (Figs. 4c,f). The significant features of zonal wind over the Arabian Sea (Fig. 4a) include low-level westerly jet (9 ms^{-1}) located around 10°N. The westerlies prevail up to 400 hPa level and above those strong easterlies with a core speed of 27 ms^{-1} around 150 hPa level in the tropics. The characteristic strong lower level easterlies due to southeast trades are monitored in the Southern Hemispheric tropics. In the Bay of Bengal sector (Fig. 4b), the low-level westerly jet is weaker as compared to the Arabian Sea sector that prevails up to 600 hPa, and above strong easterlies are observed. In extra tropics both sectors indicate strong westerlies with a maxima of 30 ms^{-1}. The maximum variability of zonal wind is noted at the upper levels both in the Arabian Sea and the Bay of Bengal sectors between the equator to 10°N. In addition, the Arabian Sea sector also shows substantial variability at 400 hPa over the equator. This variation is possibly due to a change of the wind from westerlies in the lower and middle troposphere to easterlies in the upper troposphere. The difference (surplus and deficient) indicates a stronger low-level westerly jet over the Arabian Sea sectors and weaker westerlies in the Bay of Bengal during surplus monsoon years (Figs. 4c,f). The Arabian Sea sector manifests significant negative differences between 20°N–30°N throughout the troposphere, with a maximum at 250 hPa. In the low level, the westerly jet was noticed between 10°N–20°N, and prevails up to 400 hPa. These regions are statistically significant at 95% confidence level. The mean meridional wind for the Arabian Sea and the Bay of Bengal sectors is illustrated in Figure 5. The salient features pertaining to the meridional component are a low-level convergent flow from the equatorial Indian Ocean to the Indian monsoon domain followed by a strong divergent flow from the monsoon region to the Southern Hemisphere confined to 300–100 hPa. The difference between surplus and deficient monsoon years (Figs. 5c,f) indicates a weakening of divergent flow in the upper level between 15°N–30°N in surplus years over the Arabian Sea sector (Fig. 5c). Over the Bay of Bengal sector (Fig. 5e) shows that a decrease in low-level convergence and upper level divergence between 10°S–10°N results in a decrease in convective activity which agrees favorably with OLR.

4.3 Kinetic Energy

The maintenance and intensity of the general circulation of the atmosphere depend on the balance between the generation and dissipation of the kinetic energy. The kinetic energy of the atmosphere is created through the conversion of available potential energy and eventually is dissipated through irreversible frictional processes. The monsoon circulation is maintained through the release of available potential energy for conversion to kinetic energy. The potential energy in the monsoon zone is provided by direct solar insolation and heating of the atmosphere over the land areas, by evaporation from oceans, developing cumulus cloud and the release of

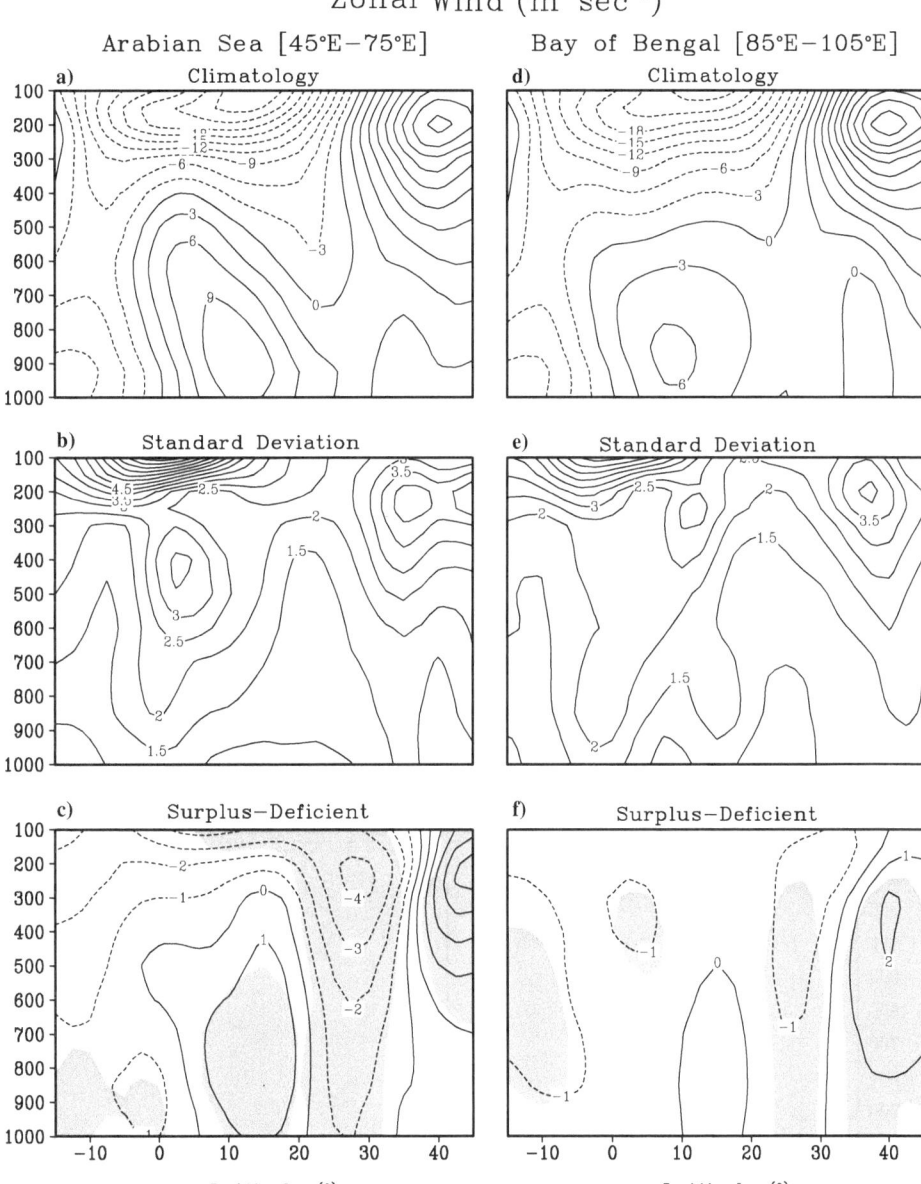

Figure 4
Height-latitude cross sections of zonal wind (m s^{-1}) for the Arabian Sea sector (left panels), a) climatology, b) standard deviation, c) difference (surplus-deficient) and Bay of Bengal sector (right panels), d) climatology, e) standard deviation, f) difference (surplus-deficient) [shaded region is 95% significant level].

Meridional Wind (m sec⁻¹)

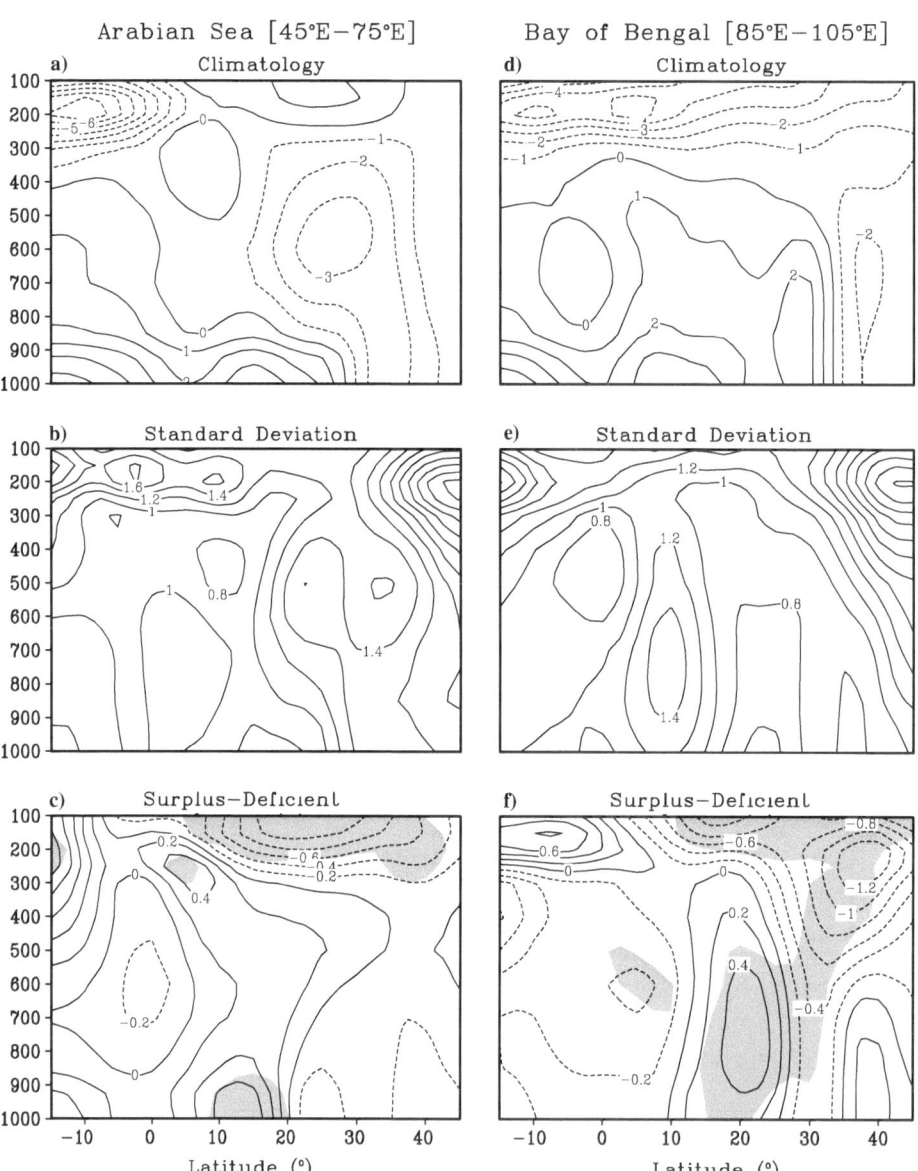

Figure 5
Height-latitude cross sections of meridional wind (m s⁻¹) for the Arabian Sea sector (left panels), a) climatology, b) standard deviation, c) difference (surplus-deficient) and the Bay of Bengal sector (right panels), d) climatology, e) standard deviation, f) difference (surplus-deficient) shaded region is 95% significant level].

latent heat. Though the potential energy is generated over the entire monsoon domain, the maximum production is expected to be around the monsoon trough zone (ANJANEYULU, 1971) where the maximum precipitation also takes place and is supported by the omega field. The local balance of kinetic energy is governed by three significant terms: namely horizontal flux, generation and dissipation of kinetic energy. In general, the dissipation of kinetic energy takes place through the surface friction and viscous stress within the atmosphere. In this section we examine horizontal flux and generation terms which play a major role in maintaining the monsoon circulation.

The geographical distribution of horizontal flux divergence of kinetic energy at 850 hPa and 150 hPa is shown in Figure 6. The top, middle and bottom panels depict the climatology (1948–1999), standard deviation and the difference between composite surplus and deficient monsoon years, respectively. At the lower level of 850 hPa (Fig. 6a), it is observed that the flux divergence of kinetic energy is over the Somali Coast and the flux convergence of kinetic energy is in the east Arabian Sea. It is associated with the low-level Somali jet entrance and exit regions, respectively. It is ascertained that the maximum variability lies in the same region of the Somali Coast and Arabian Sea (Fig. 6b). Over these regions the strong westerly (low-level jet) prevails during the summer monsoon. The difference between the surplus and deficient monsoon season with 95% significant level is depicted in Figure 6c. The flux divergence is seen over the Somali Coast and western Arabian Sea and convergence flux over the east Arabian Sea and Indian peninsula enhances in surplus years. This low-level jet is responsible for transporting the moisture along with the air mass from the Arabian Sea to the Indian land mass. The distribution of mean kinetic energy flux divergence at 150 hPa (Fig. 6d) over the summer monsoon region indicates that a zone of kinetic energy flux divergence extends over the south Asian region over spreading from the West Pacific to the eastern Arabian Sea with flux divergence maxima situated over the Bay of Bengal and flux convergence over the western Arabian Sea, adjoining the Arabian and eastern African regions. These zones of kinetic energy flux which transport maxima (minima) are situated at the respective entrance (exit) regions of TEJ. It is discovered that the entrance (exit) regions of TEJ are characterized by adiabatic production (destruction) of kinetic energy. Such a nature of kinetic energy production is necessary to maintain the strong outflow (inflow) of energy at the respective locations of TEJ (MOHANTY and RAMESH, 1994). Hence, in the maintenance of the summer monsoon circulation the adiabatic production of kinetic energy through the action of pressure force plays a very important role. Figure 6e shows the standard deviation of kinetic energy budget at 150 hPa. It can be seen that the horizontal flux of kinetic energy is observed as a large part of the variability over the Bay of Bengal, the southern Arabian Sea and the east African region. The difference between surplus and deficient monsoon years shows that the horizontal flux of kinetic energy at 150 hPa during the surplus monsoon season denotes the strong flux divergence over the Indian peninsula, Arabian Sea and

Horizontal Flux Divergence of KE (Wkg^{-1})

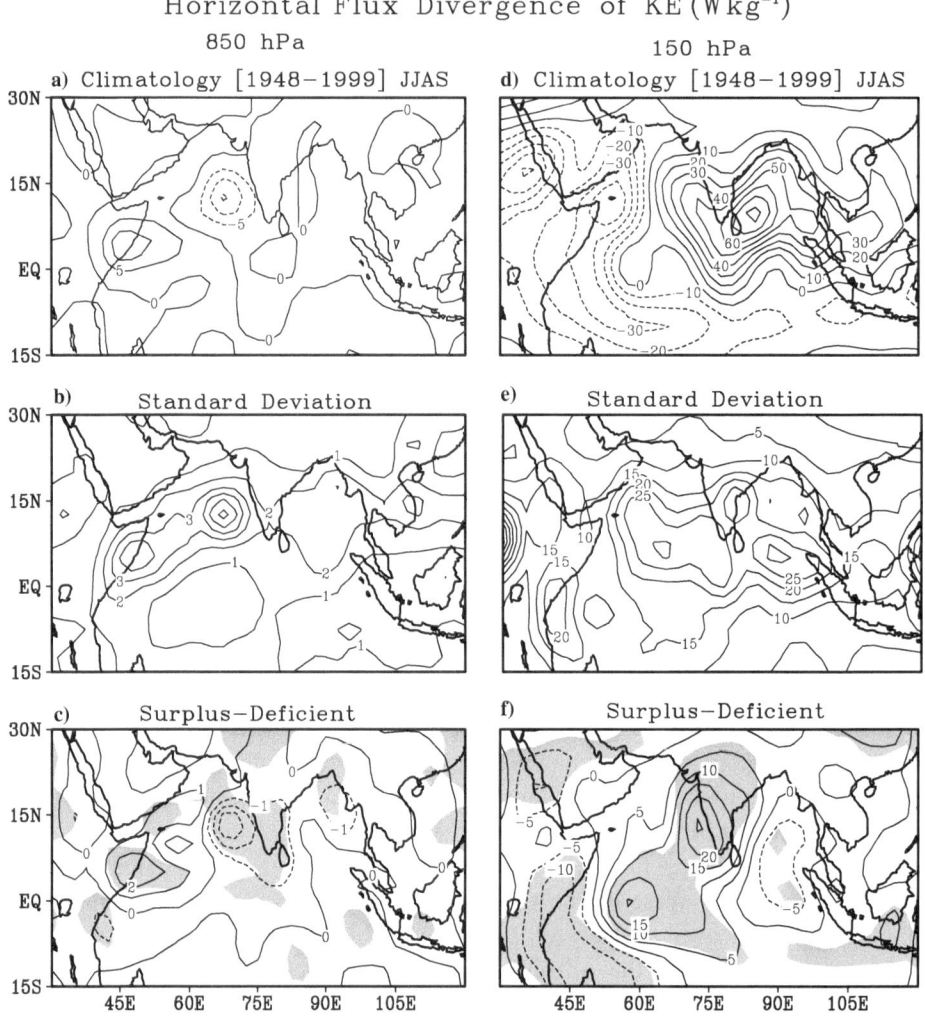

850 hPa   150 hPa

a) Climatology [1948–1999] JJAS   d) Climatology [1948–1999] JJAS

b) Standard Deviation   e) Standard Deviation

c) Surplus–Deficient   f) Surplus–Deficient

Figure 6

Geographical Distribution of Horizontal Flux Divergence of Kinetic Energy $(10^{-4}$ W kg$^{-1})$ for 850 hPa (left panels), a) climatology, b) standard deviation, c) difference (surplus-deficient) and 150 hPa (right panels), d) climatology, e) standard deviation, f) difference (surplus-deficient) [shaded region is 95% significant level].

western Indian Ocean, and the flux convergence over the east African and adjoining Indian Ocean enhances, signifying a large divergence of air mass from the Indian peninsula to the Southern Hemisphere. These regions indicate statistical significance at 95% confidence level (shaded region).

The kinetic energy is basically produced by the ageostropic component of the flow i.e., cross-isobaric flow. Positive magnitudes signify the generation of kinetic

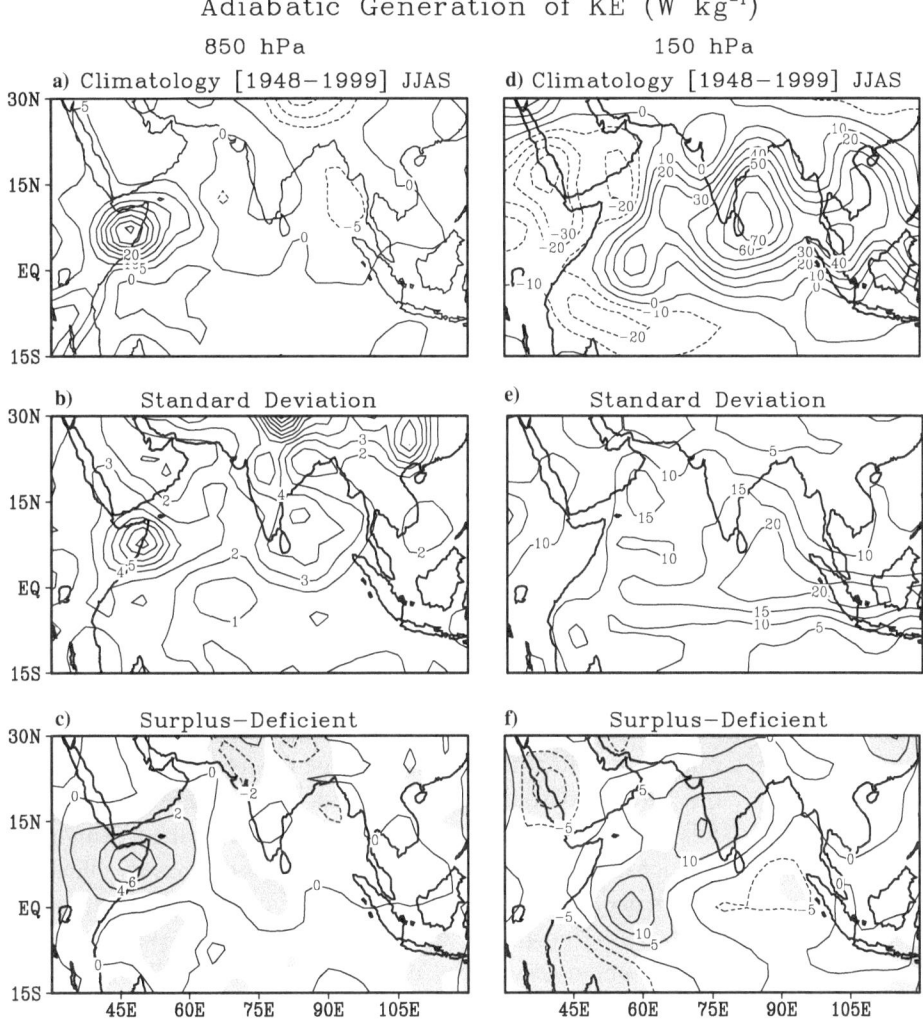

Figure 7

Geographical distribution of adiabatic generation of kinetic energy (10^{-4} W kg^{-1}) for 850 hPa (left panels), a) climatology, b) standard deviation, c) difference (surplus-deficient) and 150 hPa (right panels), d) climatology, e) standard deviation, f) difference (surplus-deficient) [shaded region is 95% significant level].

energy from the available potential energy, and negative magnitudes indicate the conversion of kinetic energy back to available potential energy. It is determined that the areas characterized by the flux divergence (convergence) of kinetic energy are the regions of strong generation (less/weak generation) of kinetic energy. The kinetic energy generation at 850 hPa (Fig. 7a) depicts maxima over the Somali Coast and is due to the strong ageostropic flow in that region. These production maxima are in agreement with the kinetic energy (horizontal) flux maxima. However, the maximum variability is also observed over the Somali Coast (Fig. 7b). The difference between surplus and deficient monsoon years (Fig. 7c) shows the generation of KE enhanced during surplus years as compared to deficient years over the Somali Coast. The generation of kinetic energy indicates maximum variability over the east equatorial Indian Ocean (Fig. 7e). The difference between surplus and deficient monsoon years (Fig. 7f) demonstrates that the generation of KE enhanced significantly during surplus years, extending from equatorial Indian Ocean to the Indian land mass.

Normally the zonal component contributes to generation in the extra tropics and dissipation in the tropics, and the meridional component *vice versa* (KUNG, 1971). The adiabatic generation of zonal kinetic energy is presented in Figure 8. In the lower tropospheric features (850 hPa) include the generation over the Indian peninsula and the adjacent Arabian Sea and the destruction of kinetic energy over the Bay of Bengal, western Arabian Sea and western Indian Ocean. The strong variability in zonal generation of kinetic energy can be seen over central India, east Bay of Bengal and Somali Coast (Fig. 8b). The difference between surplus and deficient monsoon seasons indicates destruction/negative generation over the Bay of Bengal, northwest India and western Indian Ocean, and the production over northeast Africa. These regions are significant at 95% confidence level. The zonal generation of kinetic energy at 150 hPa is represented in Figures 8d,e,f. The climatological features show that the generation over the Indian peninsula and adjoining Indian Seas of the Arabian Sea, Bay of Bengal and Indian Ocean. The destruction/weak generation is found over the northeast Bay of Bengal and northeast Africa. Further, the entire monsoon region exhibits strong variability, which is represented by standard deviation with the maximum over the Arabian Sea and Bay of Bengal. The difference between surplus and deficient monsoon years is illustrated in Figure 8f. The difference between surplus and deficient years demonstrates the generation over the western Indian Ocean and the destruction over the Bay of Bengal. The zones of maximum generation/destruction are statistically significant at 95% confidence level (shaded region). The meridional generation of kinetic energy at 850 hPa and 150 hPa is represented in Figure 9. In the lower levels at 850 hPa (Fig. 9, left panels), the climatological pattern depicts generation over the Somali Coast and the adjoining Arabian Sea. In addition, the positive generation is observed over the western Arabian Sea and Bay of Bengal. Further, the standard deviation of the meridional component (Fig. 9b) shows maximum variability over the Somali Coast and Bay of Bengal. These regions manifest higher magnitude in the surplus monsoon season. In the upper level at 150 hPa (Fig. 9, right panels), meridional generation of kinetic energy contributes to the production of kinetic energy over the Bay of Bengal and destruction over the western Indian Ocean and Arabian peninsula (Fig. 9d). The difference between surplus and deficient monsoon points out strong generation over the Indian peninsula and the adjoining Arabian Sea and Bay of Bengal. Destruction over the western Indian Ocean and north Arabian peninsula. The interesting feature

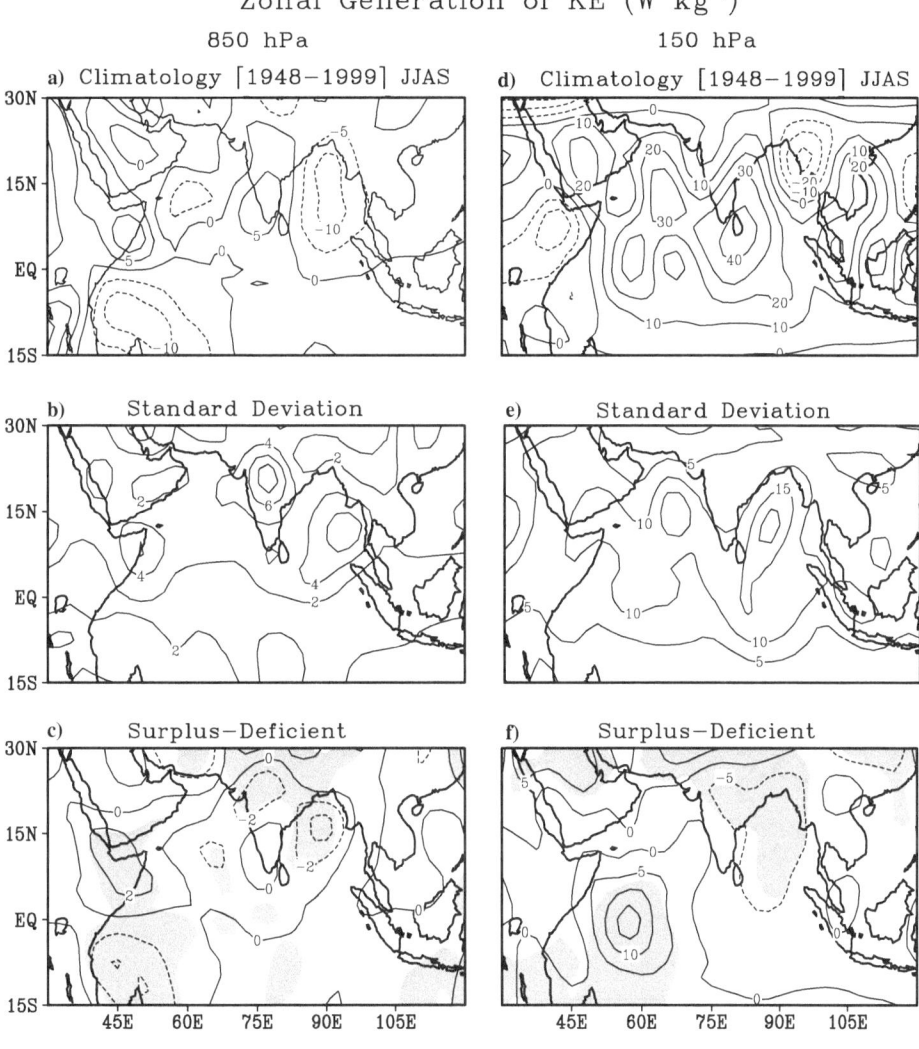

Zonal Generation of KE (W kg⁻¹)

Figure 8
Geographical distribution of zonal generation of kinetic energy (10^{-4} W kg⁻¹) for 850 hPa (left panels),
a) climatology, b) standard deviation, c) difference (surplus-deficient) and 150 hPa (right panels),
d) climatology, e) standard deviation, f) difference (surplus-deficient) [shaded region is 95% significant
level].

delineated by these two components over the monsoon domain is that both
contribute to generation, though the contribution by the meridional component is
significantly higher over the Somali Coast and Bay of Bengal.

    In this study it is found that the eastern Arabian Sea and southwest Bay of Bengal
maxima of kinetic energy production is maintained by the zonal component of the
ageostropic flow while that of the Bay of Bengal is maintained by the meridional

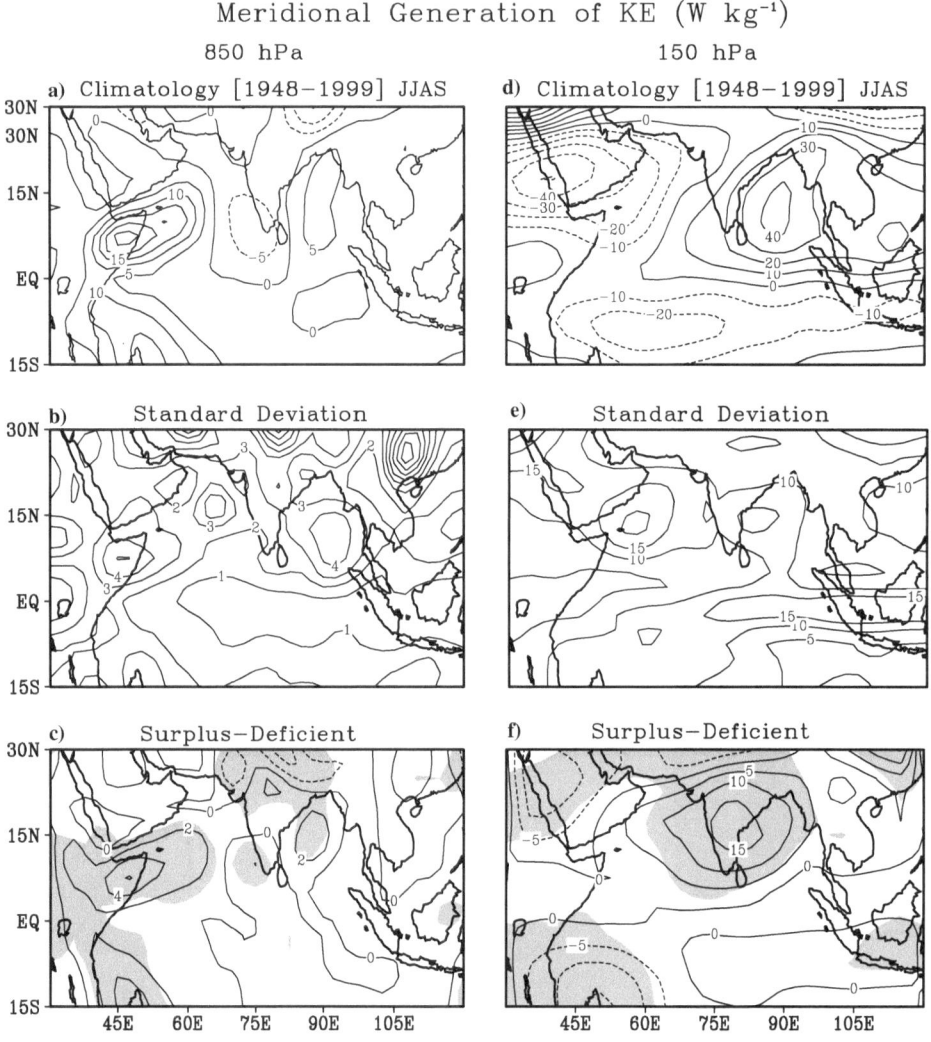

Figure 9
Geographical distribution of meridional generation of kinetic energy (10^{-4} W kg^{-1}) for 850 hPa (left panels), a) climatology, b) standard deviation, c) difference (surplus-deficient) and 150 hPa (right panels), d) climatology, e) standard deviation, f) difference (surplus-deficient) [shaded region is 95% significant level].

component of the ageostropic flow. The major part of the variability over the eastern equatorial Indian Ocean arises from the meridional generation of kinetic energy. It is seen that the zonal and meridional generations of kinetic energy show maximum variability over the Arabian Sea and Bay of Bengal, mainly due to fluctuation in the intensity and location of jets (entrance and exit regions).

4.4 Vorticity

The characteristic features of the monsoon are described with the vorticity budget. In general, the vorticity transport and generation terms are the most significant in the vorticity budget (HOLOPAINEN and OORT, 1981; CHU et al., 1981). DAGGUPATY and SIKKA (1977) stated that the lower tropospheric vorticity budget is balanced by advection of vorticity and frictional effects. In the lower troposphere, the divergence term contributes to the generation of cyclonic vorticity while in the upper troposphere it contributes to the generation of anti-cyclonic vorticity. The horizontal advection of planetary vorticity (Fig. 10) delineates positive advection over the Bay of Bengal and east Africa and adjoining west Indian Ocean. It manifests negative advection over the eastern equatorial Indian Ocean, south peninsular India and Arabian peninsular region. It is apparent that both these advections oppose each other over the summer monsoon. Also, over the south Indian peninsula and extending up to east Bay of Bengal considerable variability of horizontal advection of relative vorticity is exhibited. The difference between surplus and deficient years in horizontal advection of planetary vorticity (Fig. 10c) depicts flux divergence over the Indian peninsular region, western equatorial Indian Ocean, and east Asian region during the surplus monsoon season. The flux convergence is observed over east Arabia and the adjoining Arabian Sea in the surplus monsoon season. Further, these zones are identified as statistically significant.

The vorticity generation due to stretching is illustrated in Figure 11. It indicates that the generation of cyclonic vorticity dominates the Indian peninsula, Bay of Bengal and Arabian peninsula. However, the anti-cyclonic vorticity generation is observed over the equatorial Indian Ocean and the northwest sector of the Arabian Sea. The persistent convergence over the subcontinent sustains the circulation produced by interaction of the large-scale flow with the orographic barriers. This generation is crucial in order to sustain the cyclonic circulation in the lower troposphere. Further, the strong divergent circulation in the upper troposphere over the summer monsoon region is responsible for the generation of anti-cyclonic vorticity. The maximum variability of vorticity generation is detected over the north Indian region and southern Arabian Peninsula. The generation of vorticity (Fig. 11c) for the contrasting monsoon season shows that during the surplus monsoon season, there is a strong generation of vorticity over the entire Indian region. However, over the equatorial Indian Ocean and central Bay of Bengal weak generation/destruction of vorticity was observed. These regions are perceived at a 95% significant level.

4.5 Heat and Moisture

The mean climatological features of the summer monsoon are further analyzed through heat and moisture budgets. Also, the variability of the Asian summer monsoon is examined in terms of the standard deviation from the climatological average (1948–1999). The horizontal flux divergence of heat is depicted in Figure 12.

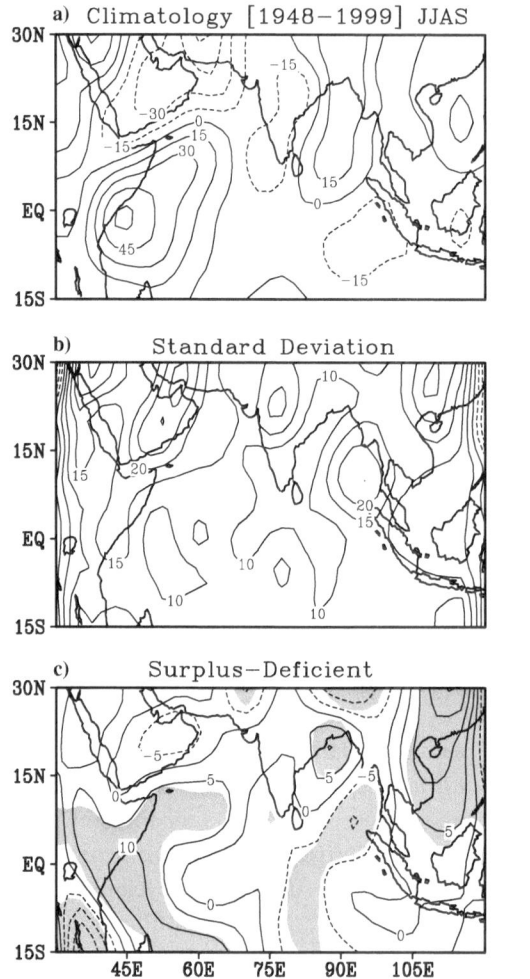

Figure 10
Geographical distribution of vertical integrated horizontal advection of planetary vorticity (10^{-8} Nm^{-3}), a) climatology, b) standard deviation, c) difference (surplus-deficient) [shaded region is 95% significant level].

The geographical distribution of horizontal flux divergence of heat (Fig. 12a) indicates that the summer monsoon domain is characterized by flux convergence with maxima over the Bay of Bengal and Arabian Sea. A zone of flux convergence of heat extends from the western Pacific to the Arabian peninsula across the Bay of Bengal and Arabian Sea. This is the region of monsoon trough/ITCZ. A zone of strong flux divergence is recognized off east Africa. The heat flux convergence in the monsoon trough/ITCZ region is essential to increase potential energy, which is available for conversion into kinetic energy and hence maintenance of the monsoon circulation.

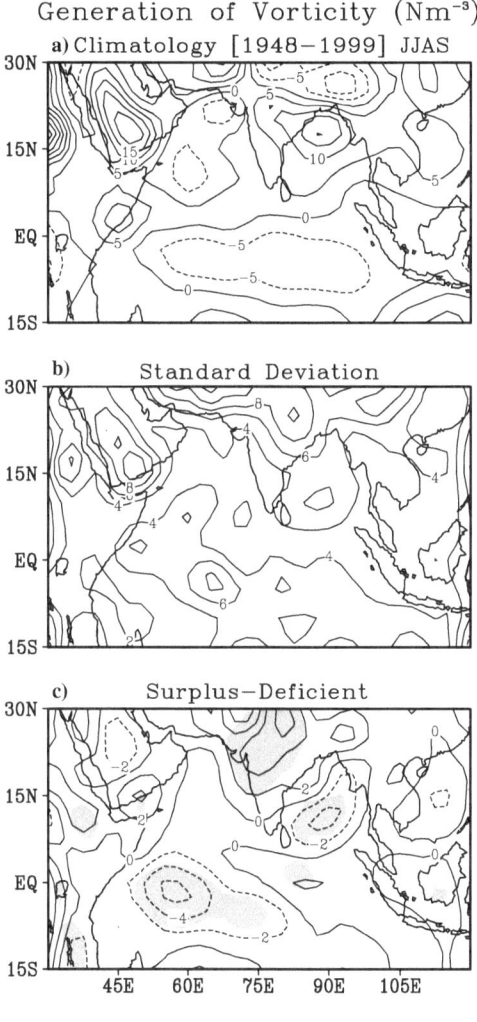

Figure 11
Geographical distribution of vertical integrated generation of vorticity (10^{-8} Nm^{-3}), a) climatology,
b) standard deviation, c) difference (surplus-deficient) [shaded region is 95% significant level].

These are consistent with earlier studies carried out using different analyses
(MOHANTY and RAMESH, 1994). The horizontal flux divergence of heat shows the
zone of maximum variability over the central Arabian Sea, Bay of Bengal, central
India and the western equatorial Indian Ocean. These zones observe maximum
variability in conversion of available potential energy to kinetic energy. The
difference of horizontal flux divergence of heat (Fig. 12c) is reflected of strong flux
convergence over western India and the adjoining Arabian Sea. This is a
characteristic feature of surplus monsoon. Further, strong convergence over east

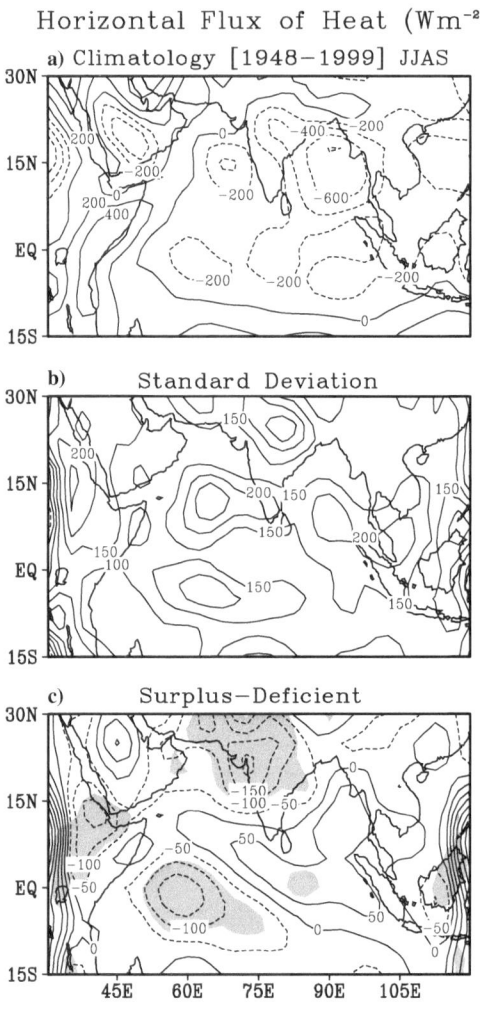

Figure 12
Geographical distribution of vertical integrated horizontal flux divergence of heat (Wm^{-2}), a) climatology,
b) standard deviation, c) difference (surplus-deficient) [shaded region is 95% significant level].

Africa is also a conducive feature for excess monsoon rainfall over India. The
interesting feature noted in this study is the strong convergence of heat flux over the
western equatorial Indian Ocean during surplus monsoon season. On the other hand,
convergence decreases over the eastern equatorial Indian Ocean and adjoining Bay of
Bengal.

The diabatic heating pattern (Fig. 13a) connotes an excess of convective activity
over the Bay of Bengal including the Indian peninsula and part of the eastern
Arabian Sea and south Indian Ocean, which indicates the predominant rising motion
and convective cloud formation over the summer monsoon region. During the

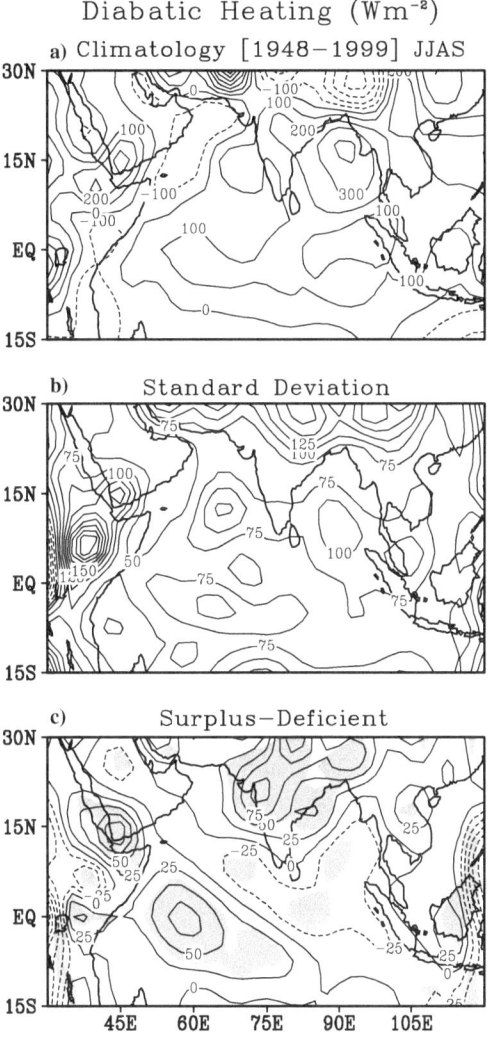

Figure 13
Geographical distribution of vertical integrated diabatic heating (Wm^{-2}), a) climatology, b) standard deviation, c) difference (surplus-deficient) [shaded region is 95% significant level].

summer monsoon, strong heating enhances the southwesterlies, and this may be augmented further by precipitation and latent heat release over the land area. These characteristics indicate the complexity of the diabatic forces and the additional role of dynamic factors that influence the summer monsoon. In addition, considerable variability of diabatic heating is observed over the Arabian Sea, the Bay of Bengal and the western equatorial Indian Ocean. The difference of diabatic heating pattern (Fig. 13c) connotes the decrease of convective activity in the deficient monsoon season and the increase of convective activity over the Indian region and western equatorial Indian Ocean during the surplus monsoon season. The maximum convective zones are correlated well with the precipitation centers. Further, these regions are statistically significant with 95% confidence level. The decreased diabatic heating over eastern equatorial Indian Ocean and adjoining east Bay of Bengal during surplus years is also consistent with precipitation and convective activity.

The geographical distribution of vertically integrated horizontal flux divergence of moisture (Fig. 14a) delineates that the entire monsoon region is characterized by strong flux convergence with maxima over the Bay of Bengal. The convergence of moisture is due to the monsoon trough and rapid cyclonic turning of low-level wind, which acts as a primary source for developing cumulus convection and ultimately sustains the monsoon circulation. PEARCE and MOHANTY (1984) studied the mean tropospheric moisture flux during May and June 1979 and showed that the buildup of the moisture flux over the summer monsoon domain could be attributed to the transportation from the south Indian Ocean. Further, the strong flux convergence zones are identified with excess diabatic heating. Corresponding to this heating, intense convective activity and rainfall are also observed over these zones. Thus, the moisture flux convergence contributes importantly in determining the diabatic heat patterns, which maintains the summer monsoon circulation. In addition, substantial variability of diabatic heating and horizontal flux divergence of moisture is noticed over the Arabian Sea, Bay of Bengal and western Indian Ocean. The difference (surplus-deficient) of vertically integrated horizontal flux divergence of moisture (Fig. 14c) indicates that during the surplus monsoon, excess convergence of moisture occurs over the east Arabian Sea, Indian landmass and the western equatorial Indian Ocean. These regions are statistically significant at 95% confidence level. A substantial decrease in convergence over the eastern equatorial Indian Ocean and adjoining Bay of Bengal is consistent with diabatic heating patterns and rainfall.

The seesaw pattern associated with most of the terms in the energetics over the equatorial western Indian Ocean and equatorial eastern Indian Ocean and adjoining Bay of Bengal may be due to the existence of the dipole phenomenon related to SST over the Indian Ocean (WEBSTER *et al.*, 1999 and SAJI *et al.*, 1999). The Indian Ocean climatology denotes maximum rainfall concentration over the tropical convergence zone of the Indonesian region. However, during the dipole mode event (positive SST anomaly over the western Indian Ocean and negative SST anomaly over the eastern Indian Ocean), rainfall decreases over the eastern Indian Ocean and increases over the western Indian Ocean. This pattern is dynamically consistent with divergence/convergence of patterns of wind shifts and outgoing long-wave radiation. In association with these aspects, during normal and deficient rainfall over India, the energetics regime over the Bay of Bengal seems to be intense. On the other hand, during the surplus monsoon season, the Arabian Sea delineates strong regime. During the dipole event the western Indian

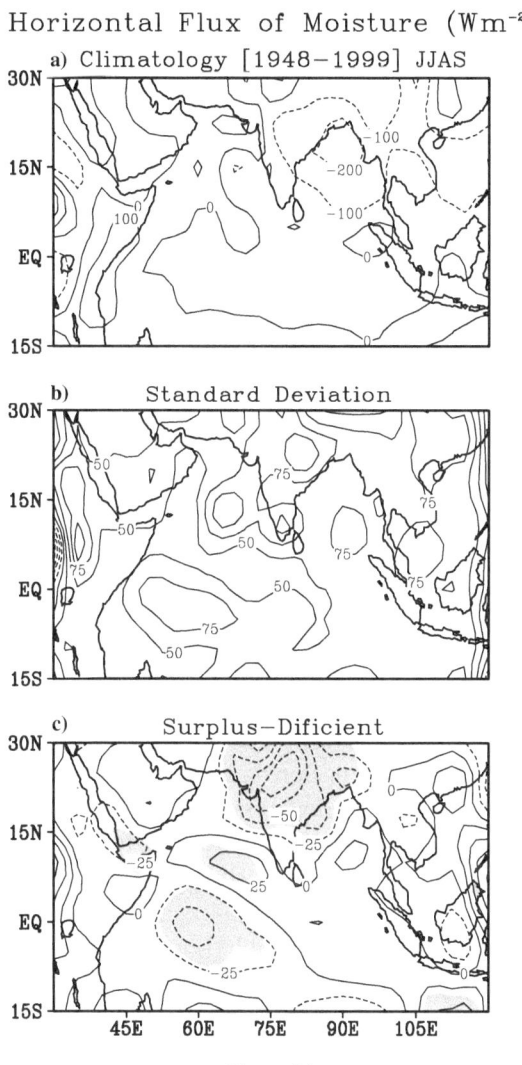

Figure 14
Geographical distribution of vertical integrated horizontal flux divergence of moisture (Wm^{-2}), a) climatology, b) standard deviation, c) difference (surplus-deficient) [shaded region is 95% significant level].

Ocean manifests anomalous warming. This warming is responsible for ample mass and moisture convergence. This anomalous warming produces an intense influx of heat and the formation of diabatic heat sources. In turn, the Indian subcontinent experiences maximum rainfall. However, during normal and drought monsoon the western Indian Ocean is relatively cooler. This inhibits the convergence of heat and moisture and the formation of diabatic heat sources. Contrary to that, the eastern Indian Ocean is warmer during drought and normal monsoon conditions over India. This warming is responsible for strong convective regime over the Bay of Bengal and off Indonesia. The outgoing long-wave radiation, SST and zonal wind patterns (WEBSTER *et al.*, 1999 and SAJI *et al.*, 1999) adduce this aspect.

5. Conclusions

The analysis of mean circulation features and energetics of the Asian monsoons renders the following broad conclusions.

The difference of OLR between the composite of surplus and deficient monsoon years shows negative OLR over the equatorial Indian Ocean with a maximum in the western sector. These zones of low OLR values are in good agreement with regions of excess precipitation. This suggests the possible relation between the equatorial Indian Ocean, particularly the western sector and the summer monsoon activity over India.

The entrance/exit regions of the TEJ are characterized by the production/ destruction of the kinetic energy, which is essential to maintain outflow/inflow prevailing at the respective location of the TEJ. The significant contribution originates from the meridional component over the Bay of Bengal to adiabatic generation of kinetic energy during the summer monsoon season.

The results indicate that the whole Indian Peninsula including the Bay of Bengal is unstable during the summer monsoon with the production of vorticity within the domain itself for maintaining the circulation. This production is manifested through subgrid-scale processes such as cumulus convection, unlike other regions where the balance is between the transportation and stretching term.

The summer monsoon evinces strong convergence of heat and moisture over the monsoon domain. In addition, considerable heat energy is generated through the action of adiabatic processes. The combined effect of these processes leads to the formation of strong diabatic heat sources in the region to maintain the monsoon circulation.

The interesting aspect presented in this study is that the large-scale budgets of kinetic energy, heat and moisture indicate excess magnitudes over the equatorial western Indian Ocean and Arabian Sea during the surplus monsoon. Conversely, the equatorial eastern Indian Ocean and adjoining east Bay of Bengal indicate excess magnitude during the deficient monsoon. During the normal and deficient monsoon the eastern Indian Ocean is relatively warmer compared to its western counterpart. The anomalous warming over the western Indian Ocean during the dipole formation is in principle responsible for intense energy flux transport over the Arabian Sea regime, leading to, surplus monsoon over India. In tandem with the warmer eastern Indian Ocean, during drought and normal monsoon season, the Bay of Bengal branch is intense. This is reflected in various budget terms considered in this study.

Acknowledgements

The authors sincerely acknowledge the NCAR for providing the NCEP/NCAR reanalysis data set. The authors express their thanks to two anonymous reviewers for their valuable comments enhancing the manuscript. Financial support from the Department of Science and Technology, Govt. of India is acknowledged.

REFERENCES

ANJANEYULU, T.S.S. (1971), *Estimates of Kinetic Energy over the Indian Monsoon Trough Zone*, Quart. J. Roy. Met. Soc. *97*, 103–109.

ANNAMALAI, H., SLINGO, J.M., SPERBER, K.R., and HODGES, K. (1999), *The Mean Evolution and Variability of the Asian Summer Monsoon: Comparison of ECMWF and NCEP-NCAR Reanalysis*, Mon. Wea. Rev. *127*, 1157–1186.

CAMPANA, K.A., HOU, Y.T., MITCHELL, K.E., YANG, S.K., and CULLATHER, R. (1994), *Improved Diagnostic Cloud Parameterization in NMC's Global Model*, Preprints, Amer. Meteor. Soc. 324–325.

CHU, J.H., YANAI, M., and SUI, C.H. (1981), *Effects of Cumulus Convection on the Vorticity Field in the Tropics. Part-I. The Large-Scale Budget*, J. Met. Soc. Japan, *59*(4), 535–546.

COLLINS, W.G., and GANDIN, L.S. (1990), *Comprehensive Hydrostatic Quality Control at the National Meteorological Centre*, Mon. Wea. Rev. *118*, 2754–2767.

COLLINS, W.G., and GANDIN, L.S. (1992), *Complex Quality Control of Rawinsonde Heights and Temperatures at the National Meteorological Center*, NMC office Note 390, 30 pp.

DAGGUPATY, S.M., and SIKKA, D.R. (1977), *On the Vorticity Budget and Vertical Velocity Distribution Associated with a Life Cycle of Monsoon Depression*, J. Atmos. Sci. *33*, 773–792.

HOLOPAINEN, E.O. (1978), *A Diagnostic Study of Kinetic Energy Balance of the Long-Term Mean Flow and the Associated Transient Fluctuations in the Atmosphere*, Geophysica *15*, 125–145.

HOLOPAINEN, E.O., and OORT, A.H. (1981), *Mean Surface Stress Curl over the Oceans as Determined from the Vorticity Budget of the Atmosphere*, J. Atmos. Sci. *33*, 773–792.

KALNAY, E., KANAMITSU, M., KISTLER, R., COLLINS, W., DEAVAN, D., GANDIN, L., IREDELL, M., SAHA, S., WHITE, G., WOOLLEN, J., ZHU, Y., CHELLIAH, M., EBISUZAKI, W., HIGGINS, W., JANOWIAK, J., MO, K.C., ROPELEWSKI, C., WANG, J., LEETMAA, A., REYNOLDS, R., JENNE, R., and Joseph, D. (1996), *The NCEP/NCAR 40-year Reanalysis Project*, Bull. Amer. Meteor. Soc. *77*, 437–471.

KANAMITSU, M. (1989), *Description of the NMC Global Data Assimilation and Forecast System*, Wea. Forecasting, *4*, 335–342.

KANAMITSU, M. (1991), *Description of the Global Data Assimilation and Forecast System*, Wea. Forecasting, *4*, 334–342.

KANAMITSU, M., and KRISHNAMURTI, T.N. (1978), *Northern Summer Tropical Circulation during Drought and Normal Rainfall Months*, Mon, Wea. Rev. *10*, 331–347.

KRISHNAMURTI, T.N., BEDI, H.S., and SUBRAMANIAM, M. (1989), *The Summer Monsoon of 1987*, J. Climate *2* (4), 321–340.

KRISHNAMURTI, T.N., BEDI, H.S., and SUBRAMANIAM, M. (1990), *The Summer Monsoon of 1988*, Meteorol. Atmos. Phys., *42*, 19–37.

KRUEGER, A.F., and WINSTON, J.S. (1975), *Large-Scale Circulation Anomalies over the Tropics during 1971—72*, Mon. Wea. Rev. *103*, 465–473.

KUNG, E.C. (1971), *A Diagnosis of Adiabatic Production and Destruction of the Kinetic Energy by the Meridional and Zonal Motion*, Quart. J. Roy. Meteor. Soc. *97*, 61–74.

KUNG, E.C., and TANAKA, H. (1983), *Energetics Analysis of the Global Circulation during the Special Observation Periods of FGGE*, J. Atmos. Sci. *40*, 2575–2592.

KUNG, E.C., and SMITH, P.J. (1974), *Problems of Large-scale Kinetic Energy Balance, a Diagnostic Analysis in GARP*, Bull. Amer. Meteor. Soc. *55*, 768–777.

Li, C., and Yanai, M. (1996), *The Onset and Interannual Variability of the Asian Summer Monsoon in Relation to Land-sea Thermal Contrast,* J. Climate *9*, 358–375.

Lorentz, E. N. (1955), *Available Potential Energy and Maintenance of the General Circulation,* Tellus *7*, 157–167.

Mohanty, U.C., and Ramesh, K.J. (1994), *A Study on the Dynamics and Energetics of the Indian Summer Monsoon,* Proc. Indian National Science Academy *60, A, 1,* 23–55.

Mohanty, U.C., Dube, S.K., and Sinha, P.C. (1982a), *On the Large-scale Energetics in the Onset and Maintenance of Summer Monsoon – I: Heat Budget,* Mausam *33,* 139–152.

Mohanty, U.C., Dube, S.K., and Sinha, P.C. (1982b), *On the Large-scale Energetics in the Onset and Maintenance of Summer Monsoon – II: Moisture Budget,* Mausam *33,* 285–294.

Mohanty, U.C., Dube, S.K., and Singh, M. P. (1983), *A Study of Heat and Moisture Budget over the Arabian Sea and their Role in the Onset and Maintenance of Summer Monsoon,* J. Meteorol. Soc. Japan *61,* 208–221.

Newell, R.E. (1970), *The Energy Balance of the Global Atmosphere, Global Circulation of the Atmosphere,* G.A. Corby Ed., Roy. Met. Soc., 42–90.

O'Brien, J.J. (1970), *Alternative Solutions to Classical Vertical Velocity Problem,* J. Appl. M, *9,* 197–203.

Oort, H. (1964), *On Estimation of the Atmospheric Energy Cycle,* Mon, Wea. Rev. *92,* 483–493.

Pan, H.L., and Mahrt (1987), *Interaction between Soil Hydrology and Boundary Layer Development,* Bound. Layer Meteor. *38,* 185–220.

Pan, H.L., and Wan-shu Wu (1994), *Implementing a Mass Flux Convective Parameterization Package for the NMC Medium-range Forecast Model,* Preprints, Amer. Meteor. Soc. 96–98.

Parthasarathy, B. Munot A.A., and Kothawale, D.R. (1994), *All India Monthly and Seasonal Rainfall Series 1871—1993,* Theor. Appl. Climatol. *45,* 217–224.

Parrish, D.F., and Derber, J.C. (1992), *The National Meteorological Center's Spectral Statistical Interpolation Analysis System,* Mon. Weather Rev. *120,* 1747–1763.

Pearce, R.P. (1979), *On the Concept of Available Potential Energy,* Quart. J. Roy. Meteor. Soc. *104,* 737–755.

Pearce, R.P., and Mohanty, U.C. (1984), *Onsets of the Asian Summer Monsoon 1979—1982,* J. Atmos. Sci. *41*(9), 1622–1639.

Ramesh, K.J., Mohanty, U.C., and Rao, P.L.S. (1996), *A Study on the Distinct Features of the Asian Summer Monsoon during the Years of Extreme Monsoon Activity over India,* Meteo. Atmos. Phys. *59* (3-4), 173–183.

Ramesh, K.J., Rao, P.L.S., and Mohanty, U.C. (1999), *A Study on the Performance of the NCMRWF Analysis and Forecasting System during of the Asian Summer Monsoon: Thermodynamic Aspects,* Pure Appl. Geophys. *154,* 141–162.

Raju, P.V.S., Mohanty, U.C., Rao, P.L.S., and Bhatla, R. (2002), *The Contrasting Features of the Asian Summer Monsoon during Surplus and Deficit Rainfall Years over India,* Int. J. Climatology *22*(15), 1897–1914

Rao, Y.P. (1976), *Southwest Monsoons,* Meteor. Monogr. No.1, Indian Meteorological Department, 1–367.

Rao, P.L.S., Mohanty, U.C., Raju, P.V.S., and Iyengar, G. (2003): *The Indian Summer Monsoon as Revealed by the NCMRWF System,* Proc. Indian Acad. Sci. (Earth and Planet. Sci.), *112*(1), 95–111.

Saji, N.H., Goswami, B.N., Vinayachandran, P.N., and Yamagata, T. (1999), A *Dipole Mode in the Tropical Indian Ocean,* Nature *401,* 360–363.

Sperber, K.R., Slingo, J.M., and Annamalai, H. (2000), *Predictability and the Relationship between Subseasonal and Interannual Variability during the Asian Summer Monsoon,* Quart. J. Roy. Meteor. Soc. *126,* 2545–2574.

Webster, P.J., Loschinigg, J.P., Moore, A.M., and Leben, R.R. (1999), *The Great Indian Ocean Warming of 1997–1998: Evidence of Coupled Oceanic-atmospheric Instabilities.* Nature. *401,* 356–360.

Woolen, J.S. (1991), *New NMC Operational OI Quality Control,* Preprints, Ninth Conf. on Numerical Weather Prediction, Denver, CO, Amer. Meteor. Soc, 24–17.

Woolen, J.S., Kalnay, E., Gandin, L., Collins, W., Saha, S., Kistler, R., Kanamitsu M., and Chelliah (1994), *Quality Control in the Reanalysis System.* Preprints, Amer. Meteor. Soc. 2417–2419.

XIE, P., and ARKIN, P. (1997), *Global Precipitation: A 17-year Monthly Analysis Based on Gauge Observations, Satellite Estimates and Numerical Model Outputs,* Amer. Meteor. Soc. *78*, 2539–2558.

(Received January 20, 2004, accepted April 30, 2004)

 To access this journal online:
http://www.birkhauser.ch

Pure appl. geophys. 162 (2005) 1543–1555
0033–4553/05/091543–13
DOI 10.1007/s00024-005-2682-y

© Birkhäuser Verlag, Basel, 2005

❙ Pure and Applied Geophysics

Radiative Feedbacks and Monsoon Circulations: A View from Simplified Models

Hervé Le Treut,[1] and Gilles Bellon[1]

Abstract—The modulation of radiative processes by changes in water vapor and cloudiness is at the origin of important feedbacks which control climate variability as well as climate changes. These feedbacks are especially active in the intertropical area, where it is possible to diagnose a combination of partially compensating positive and negative feedbacks. The characteristics and the strength of those feedbacks is closely associated with the dynamical regimes in which they develop. Reverse changes in dynamical patterns may cause a modulation of the radiative processes. A first approach to these problems is to distinguish between two ascending and subsiding circulation patterns. This bimodality of the circulation is well established in the tropical area, and favors the use of simplified models as an appropriate tool to carry out a first-order quantification of these processes. In particular, this combination of radiative and dynamical feedbacks characterizes the development of the monsoons and their variability. Simple conceptual models can thus serve to characterize some of the factors which will affect the intraseasonal variations of the monsoon.

Key words: Monsoon, radiative-convective, idealized model.

1. Introduction

Radiative feedbacks are among the main driving factors which affect climate change and climate variability at all time scales, and their role may be especially important in the tropics. The importance of radiative feedbacks has been emphasized by the study of future climate changes. The difficulty diagnosing and modeling these radiative processes (IPCC, 2001; Bony *et al.*, 2004) is linked with the large number of parameters which are involved: temperature lapse rate, water vapor content, cloud type, cloud height, cloud cover, cloud water phase, cloud water content, and it explains the substantial uncertainty which affects the current scenarios of future climate change. Radiative feedbacks are characterized by intricate compensation effects between both positive and negative feedbacks, and the resulting uncertainty also affects the study of climate fluctuations at the intraseasonal time scale.

[1] CNRS, Laboratoire de Météorologie Dynamique /IPSL, 24 rue Lhomond, Ecole Normale Supérieure, F-75231 Paris Cedex 05, France. E-mail: letreut@lmd.ens.fr; bellon@lmd.ens.fr

For many years the diagnostics of radiative feedbacks in climate models has been carried out without any explicit linkage to the role played by atmospheric dynamics. The approach originally introduced by HANSEN et al. (1984) or WETHERALD and MANABE (1988) to quantify those feedbacks has used off-line radiative diagnostics in which each perturbed parameter — cloud cover, water vapor, temperature, — was substituted from a perturbed to an unperturbed simulation, in order to determine its relative importance. These results have emphasized the important role of water vapor (LI and LE TREUT, 1992) and the role of cloud modeling assumptions in the simulated climate sensitivity. However they have provided a limited insight into why models are different from each other. A more mechanistic approach of the impact of those physical processes requires a better understanding of how they interact with changes in the atmospheric dynamics.

One of the first studies to explicitly describe the possible linkage between radiative processes and dynamics has been that of PIERREHUMBERT (1995). His work opposed the moist ascending regions of the Hadley cells, which are very often unstable (due to a possible run-away greenhouse effect associated with water vapor), and the dry subsidence areas, which are stabilized by strong radiation to space. In this simple approach, the link between radiative feedbacks and dynamical aspects is two-fold: on the one hand, the definition of dry or moist areas is of course closely associated with the location of the Hadley-Walker cells; on the other hand, reversely, these circulations must satisfy constraints linked with the conservation of radiative energy, in particular because the Hadley-Walker circulations are slow, largely stationary at the time-scale of a few weeks and have therefore the time to achieve at least an approximate equilibrium with the radiative sources or sinks of energy.

The simple conceptual approach proposed by PIERREHUMBERT (1995) may therefore serve as a guide to investigate a number of meteorological systems. The studies by BONY et al. (1997), in which the properties of clouds are studied according to mean dynamical regimes in the tropical area may also be viewed as an extension of the same idea to cloud feedbacks, with a more precise definition of the dynamical regimes than simply distinguishing between ascending and subsiding areas.

The role of cloud and water vapor feedbacks in the development of monsoon may also be approached with similar ideas, which of course cannot provide an adequate description of the full complexity of the real situation, but may provide partial explanations. During monsoon episodes, the circulation is organized following tropical cells, characterized by well-defined ascending areas, and more diffuse subsiding areas. KRISHNARMURTI (1971) has shown that when monsoon is active, warmer air ascends over the land, while relatively colder air descends over oceanic areas. SIKKA and GADGIL (1980) have shown that two zones of maximum cloudiness can be observed; one over the Indian continent and the other over the equatorial area. Part of the monsoon intraseasonal variability may then be interpreted as the competition between two tropical cells, characterized by distinct locations of their respective ascending branch. Active phases of the monsoon are associated with the development of a convergence zone in North India and a weakening of the equatorial convection, whereas breaks are characterized by its suppression by large-scale subsidence, while the equatorial convergence zone is enhanced. As noted by WEBSTER et al. (1998), these two extremes can be considered as quasi-equilibria and the intraseasonal variability of the Asian monsoon can be seen as a "dynamic tension between two quasi-equilibrium states."

Furthermore, breaks and active phases of the monsoon are associated with modifications of the circulation and convection over the West Pacific: while the active composites show intensification the convection around 10°N east of Vietnam, break composites tend to suggest a weakening of the Hadley cell in the West Pacific (GADGIL, 2003).

Some GCM studies have confirmed this role of radiative processes: SHARMA et al. (1998) have carried out sensitivity experiments using the LMD GCM, for the years 1987–1988. By modifying systematically the prescription of clouds used for radiation computations, they were able to show a significant impact of radiation on the development of monsoon, as well as emphasize the necessary consistence of cloud prescription with the simulated dynamics. The same competition was noted between two cells whose active ascending zones were located near the equator or near 10°N. Cloud radiative effects were able to affect this competition.

The objective of this paper is to describe certain radiative processes which may be involved in both the near-equilibrium maintenance of those cells, and their competition. In Section 2, we show why the bimodality of the tropical circulation, with strongly different ascending and subsiding areas, may be possible in near equilibrium radiative conditions — a key justification for using simple radiative (or radiative-convective) models of the tropical cells. A two-column model first described by BELLON et al. (2003) is introduced. In Section 3 we use the model to describe some of the key mechanisms which may regulate the competition between two cells. The last section discusses implications to the monsoon studies followed by concluding remarks.

2. Bimodality of the Radiative Equilibrium in the Tropics: A Simple Model Approach

2a. The View from Simplified 1-D Radiative Models

Satellite images of the water vapor field in the upper troposphere have shown the strong contrast that may exist in the intertropical area, between very moist areas (more than 80% humidity), which may be immediately adjacent to very dry areas (less than 20% humidity). The role of the atmospheric dynamics in determining those contrasts shows most clearly on instantaneous images, however the strong gradient which separates moist and dry areas can also be found in monthly distributions of the water vapor for planetary-scale systems (ZHANG et al., 2003).

A first intriguing feature characterizing this bimodal nature of the intertropical climate is that it is not reflected in the temperature field. Atmospheric temperatures are on the contrary very uniform throughout the entire intertropical area, as is required to maintain quasi-geostrophy, in the presence of the Hadley-Walker cells.

This issue has been investigated using a very simplified, almost analytic model of the radiative equilibrium (LI *et al.* 1997). The model is a 1-D vertical model, describing the temperature profile which is established following radiative equilibrium only, in a column characterized by a set of given absorption and reflection coefficients for the solar and terrestrial components. Using this model, IDE *et al.* (2001) have shown that the radiative-equilibrium equations could explain how similar near-equilibrium energy balance conditions could be maintained in two areas sharing the same temperature profile, but very different water vapor profiles. The search of equilibria is carried out by computing the infrared transmissivities which we know reflect primarily the presence of water vapor. It is computed as the quantity which permits the attainment of a radiative equilibrium, in conditions where the vertical profiles of temperature and solar absorption are specified within the atmosphere. The solar absorption has a slight dependence on the humidity, and neglecting temporarily the role of clouds, it is not a factor which opposes dry and moist areas. It is found that, for temperatures higher than a certain threshold, there are two possible radiative equilibria of the column: one equilibrium corresponds to a vertical profile of very low infrared transmissivities and therefore to very moist conditions, while the other is on the contrary characterized by a vertical profile of very high infrared transmissivities and therefore very dry conditions (Fig. 1). It should be noted that, although temperature within the atmosphere is prescribed to be the same for both solutions, the surface temperature is different, a feature which also connects well with the real world, where the observed uniformity of the atmospheric temperature throughout the tropics breaks down near the surface.

These results are obtained in the context of a highly simplified approach, in which the role of clouds is ignored and the computed equilibria are radiative rather than radiative-convective, as would be more realistic. Nonetheless the simplicity of the approach also induces a pronounced robustness of the results: the simple and generic equations that have been used can describe at least conceptually many aspects of a more complex reality. To better demonstrate the significance of these results, LI *et al.* (2001) have used results from the LMD GCM and/or observed reanalysis, as atmospheric temperature input to their simplified 1-D-model. They find that, very consistently, the two infra-red transmissivities which ensure radiative equilibrium for these observed conditions closely relate to those simulated by the GCM in moist and dry conditions, respectively.

This illustrates how the Hadley-Walker circulation can exist in near-equilibrium radiative conditions, adapting itself to both the rather uniform temperature field that exists throughout the intertropical belt, the very contrasted water vapor distribution, with very moist areas adjacent to very dry areas.

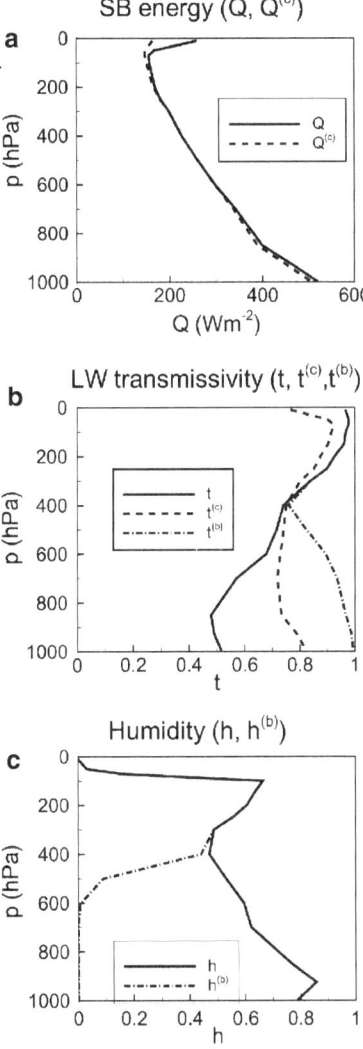

Figure 1

The temperature profile (featured as $Q = \sigma T^4$, with Q^c the threshold which allows for two transmissivities to co-exist at radiative equilibrium), the two associated transmissivities (with the one profile which would be obtained at $Q = Q^c$), and the corresponding atmospheric relative humidities, as diagnosed in the model of IDE *et al.* (2001).

2b. *A two-column Model of the Tropical Atmosphere*

Such results justify to study of the quasi-equilibrium of the tropical circulation with respect to radiative forcing. In what follows we will use a model developed by BELLON *et al.* (2003). It is a two-column model of the intertropical climate which, although conceptually simple, and inspired by PIERREHUMBERT (1995)'s work, introduces a number of additional feedbacks. In particular the model explicitly introduces a coupling between the two columns, in the form of water and energy exchanges. The ascending column occupies a prescribed fraction of the tropical area. In the moist ascending part of the model, the vertical profile of temperature and humidity corresponds to the moist adiabat, and a simple representation of convective clouds is taken into account. The representation of the dry subsiding area includes a formulation for a cloud-topped atmospheric boundary-layer. A schematic of the model is provided in Fig. 2.

The model is also coupled to a simple ocean mixed layer, and includes a comprehensive representation of atmospheric processes, and in particular a representation of clouds and convection.

For a more detailed description of the model the reader is referred to the paper of BELLON et al. (2003). In the following paragraphs however, we develop two aspects of its formulation which have critical consequences for the results of Section 3: The treatment of cloud-radiation, and the prescription of simplified atmospheric dynamics.

2c. Cloud Feedbacks

How cloud radiative forcing may affect the simple view of the radiative equilibrium in the Tropics which we have described above, is still a debated topic. The longwave and shortwave components of the convective-cloud forcing seem to cancel each other over the warm pool (KIEHL and RAMANATHAN, 1990), a feature which is at the basis of the simplified approach of PIERREHUMBERT (1995) in which cloud effects are ignored. HARTMANN et al. (2001) explain the near cancellation of the two components forcing by dynamical feedbacks.

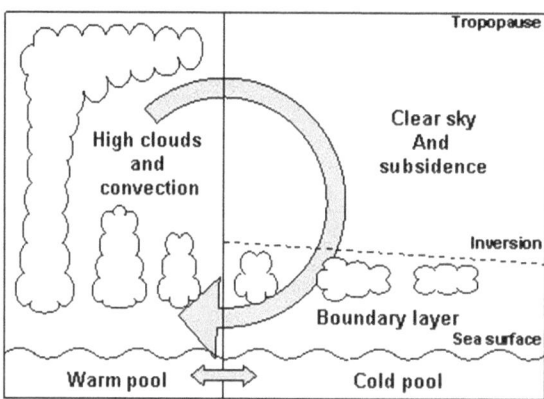

Figure 2

Schematics of the BELLON et al. (2003) model. The two columns are linked by energy and water transport, based on a diagnostic of the radiation balance above the boundary layer in the subsiding branch, and on the gradient of SST between the two boxes.

This cancellation is however not a universal and constant feature of the tropical climate. It is for example not the case over the Asian monsoon region (RAJEEVAN and SRINIVASAN, 2000), where the convective anvils appear to cool the atmosphere. Also, different theories about the sign and amplitude of the net forcing, and its possible changes, have been developed, such as the still controversial "Iris" effect (LINDZEN *et al.*, 2001; LIN *et al.*, 2002), which claims that the extension of cirrus clouds would decrease with an increase of SST due to the increase of precipitation efficiency with temperature. In the present study we have chosen to ignore those effects, and the model radiative code is build to ensure a zero-cloud radiative forcing at the top of the atmosphere, for those convective clouds which develop in the ascending branch. LARSON *et al.* (1999) showed that, in this type of model, the convective cloud radiative forcing is a second-order regulator of the climate. The cloud feedback can certainly modulate the response of the system, however the core of this response is expected to be independent of the convective cloud radiative forcing.

In fact, the climate sensitivity is likely to depend more on the radiative forcing of the subsiding area, and thus on the boundary-layer cloudiness. Simple models have proven the importance of stratus clouds radiative feedback (MILLER, 1997), while more elaborate models tend to show that the shallow-cumulus trade-wind regions are crucial to the response of the climate to a perturbation (BONY *et al.*, 2004). Subsiding areas play a key role in the study described in Section 3 and we have ensured that our model is able to generate stratus clouds in the subsiding area, with a prescribed albedo.

2d. Treatment of Dynamical Feedbacks

The bimodality of the tropical atmosphere, with moist conditions associated with ascending motions, also opens the way to another set of possible feedbacks which correspond to changes in the atmospheric circulation. Indeed these changes in circulation are likely if the response of the climate to some external perturbation (such as an increase of the incoming solar radiation, or a change in greenhouse gases) is different for each dynamical regime.

The most superb example of the dynamical feedback which affects the energy cycle of a convective system in quasi-equilibrium is perhaps the so-called WISHE (Wind Induced Surface Heating Enhancement) mechanism proposed by YANO and EMANUEL (1991), also known as evaporation-wind mechanism: variations of large-scale circulation can enhance the convection through a wind-driven increase of surface fluxes. The change in the respective areas corresponding to ascending and descending atmospheric motions in the tropics is another potentially important feedback.

There are two ways in which dynamical feedbacks may be investigated with our simple model. One approach is to try to diagnose certain components of the atmospheric circulation, e.g., those which are closely associated with energy conservation. For example a measure of the atmospheric overturning associated with a tropical cell can be derived from the radiative heat loss in subsiding areas above the boundary layer. The gradient of surface temperature between the ascending and subsiding areas also act as a factor that accelerates the Hadley cell. A representation of both effects is included within the model.

Another approach is to test its sensitivity to prescribed parameters. For instance, by increasing the surface of the ascending motions by about 10%, thus favoring a larger extent of the areas where the climate is self-destabilizing, as opposed to the subsidence areas where it is stabilized by a strong infrared emission toward space, BELLON et al. (2003) create a warming of 1.6 °C, almost equivalent to the effect of a CO_2 doubling (2.2 °C using the same model).

BELLON et al. (2003) have shown that a strong interaction exists between the areal feedback and other dynamical processes, such as the changes in the surface evaporation, (a climatic WISHE mechanism) which may result from changes in the surface winds.

3. Application to Monsoon

As explained in the introduction, monsoons dynamics is at least partly character-ized as a situation where two adjacent tropical cells compete. There is therefore a need to investigate which are the main processes that may regulate this competition. A full investigation would require a more complex model with at least three columns being explicitly described. However the two-column model of BELLON et al. (2003), because it includes those dynamical features which are most directly related to mean energy conservation, may be used for a preliminary investigation of the involved mechanisms.

In the broad and simplified picture of the Indian monsoon which we have depicted above, two ascending branches are located in distant areas, and share very largely a common subsiding branch. Most naturally the energy exchanges which occur within this partially common subsiding area will constitute a decisive link between the two cells: the convective branch of an active cell imports water vapor from the subsiding branch in the low troposphere and exports dry energy in the subsiding free-troposphere; the intensity of the circulation is controlled by the radiative cooling of the subsiding branch, and the change in the energy budget of the subsiding region will therefore have a crucial impact on the intensity of the circulation. We therefore use the model of BELLON et al. (2003) to describe the sensitivity of a tropical cell to perturbations in the energy balance of its subsiding branch.

Two kinds of perturbations are considered. In a first experiment a prescribed transport of sensible heat is added to the free-tropospheric component of the subsiding area (e.g., is added above the boundary layer). This additional flux may be positive (energy gain by the subsiding area) or negative (energy loss). The SST, precipitation and subsidence rate in the clear-sky region increase with a gain of energy (Fig. 3). The simulated impact on the SST is largely linear (Fig. 3a), but the complexity of the model response appears clearly in the strong nonlinearity of the precipitation response (Fig. 3b), or in the diagnostics of the cell overturning, as measured by the subsidence rate associated with radiative cooling in its descending branch (Fig. 3c). This latter parameter would constitute the main link with another competing cell: the nonlinearity of the response shows the potential complexity of this competition.

A second set of results corresponds to the prescription of a latent heat flux in the boundary layer of the subsiding area. This is a crude representation of the perturbation which one cell may cause to another one by modifying the low-level flow or perturbing the evaporation over the subsiding area, also changing the subsequent water vapor transport toward the ascending branch. The horizontal transport associated with SST gradients then plays an important role which further amplifies the nonlinearity of the model response especially in terms of SST. The results of this sensitivity study approximate those obtained by varying the sensible heat horizontal transport (Fig. 4): whereas the response of the temperature to an input of latent energy is close to linear, the response in the precipitation and intensity of the cell exhibits a strong nonlinearity.

The nonlinearity is linked to the response of the boundary layer in the subsiding column of the model. For a large input of energy (more than 20 Wm^{-2}), the shallow convection intensifies and extracts more energy from the ocean via the evaporation. The thermodynamically-controlled circulation is thus strongly enhanced. The threshold of 20 Wm^{-2} and slope of the curves above this threshold are certainly dependent on the parameterization of the cloud–topped boundary layer. Although the deepening of the boundary layer is expected to be a strong feedback (BONY et al., 2004).

If a cell is extremely active, its subsiding branch is not sufficient to provide all the humidity needed for deep convection and to radiatively compensate for the runaway greenhouse effect in its ascending branch. The energetic behavior of a very active cell is thus likely to be dominated by the behavior of its convective part. It would therefore tend to export heat in the free troposphere and import water vapor in the lower levels. Therefore, the competition between two cells could be roughly represented by the net effect of the loss of latent energy in the boundary layer and the gain of sensible heat in the free troposphere. The resulting effect will at least modulate the intensity the circulation and precipitation, nonetheless it might also determine its regime (normal or heavily precipitating), because of the non-linearity of the response.

4. Conclusions and Perspectives

We have shown that the study of the energetic equilibrium of tropical cells is relevant to explain certain features of monsoon circulations. This direct impact of

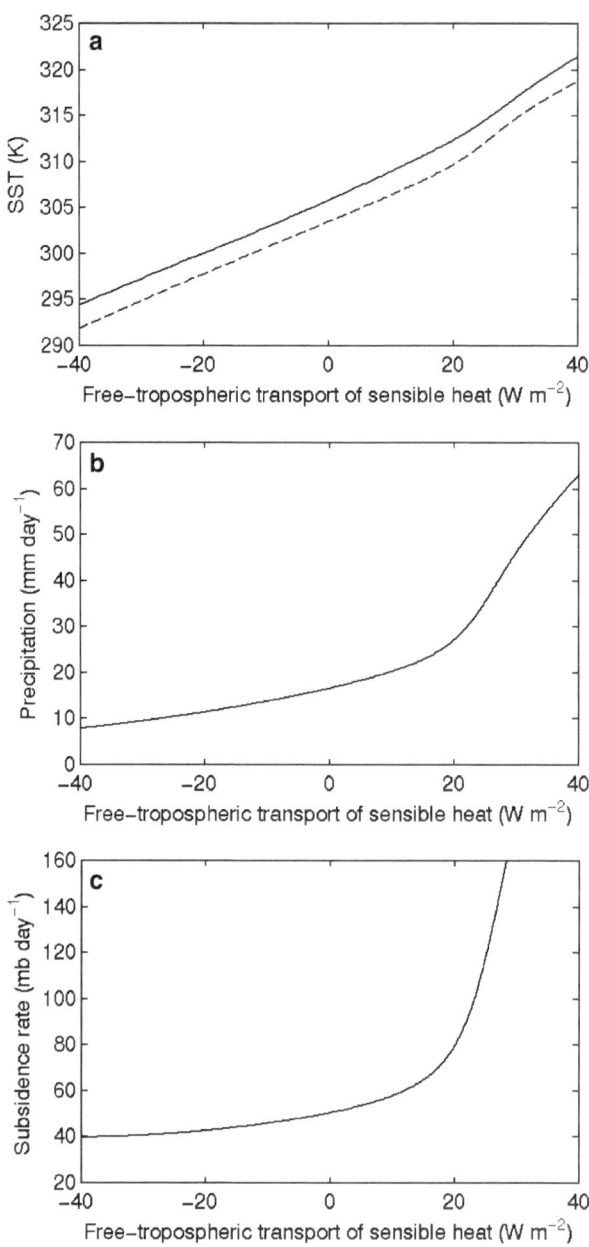

Figure 3

Response of the model to a prescribed sensible heat flux added to the subsiding branch above the boundary layer (e.g., within the troposphere): a) Sea-surface temperature in the convective area (full line), or in average (dashed line); b) precipitation; c) subsidence rate diagnosed in the subsiding branch (also see text).

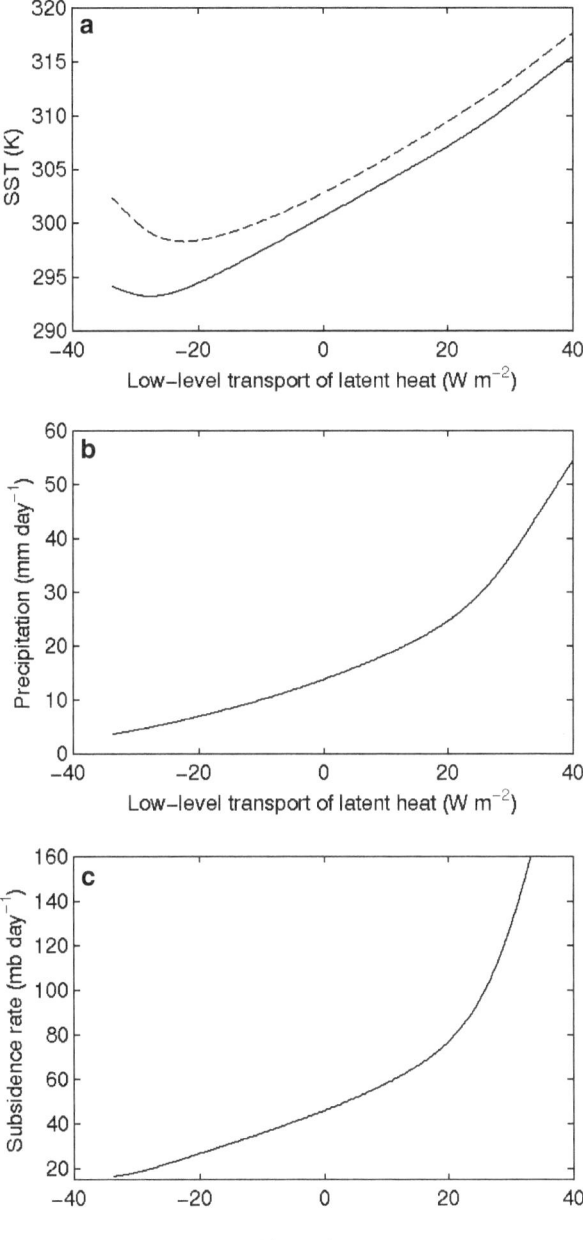

Figure 4
Same as Figure 3, but for a latent heat flux added to the boundary layer of the subsiding branch.

radiative heating on the development of monsoon may depend in a very complex manner on a combination of strongly interacting radiative and dynamical feedbacks.

Our model results suggest that these processes will give rise to a strongly nonlinear competition between the tropical cells participating in monsoon dynamic. The boundary layer processes are the primary cause of these nonlinearities. Although the model is limited to the description of stationary patterns, we may expect such nonlinearities in the competition between tropical cells to reinforce the complexity of the associated monsoon fluctuations, producing for example abrupt transitions between different atmospheric regimes, and therefore providing intrinsic limitations to monsoon prediction.

This preliminary study suggests new areas for further investigations of the monsoon and its variability using simplified models. These models can provide insight into the dynamical and radiative mechanisms maintaining or destabilizing these quasi-equilibria, including cloud effects. In addition, simplified models could provide a useful tool to study the interplay between the large-scale circulation over the Indian Ocean/Subcontinent and that over the West Pacific. A strong limitation of these approaches, however, is that they are restricted to equilibrium conditions and cannot consider the transition between different equilibria: A larger hierarchy of models, with transient models of a higher level of complexity, is also necessary.

Finally the approach described in this paper may serve as a conceptual guide to analyze the substantial amount of new satellite data that will soon become available. In addition to data which are already available such as those from METEOSAT and the Tropical Rainfall Measuring Mission (TRMM), the large variety of instruments that have been or will be launched as part of the AQUA-train (radar, lidar, high-resolution spectrometer,...) will dramatically enhance our capacity to describe the Earth radiative budget and the hydrological cycle. Studying the energetics of the tropical atmosphere for specified dynamical regimes on one side, studying the response of those dynamical regimes to specified thermodynamical forcing on the other side, should help distinguish the specific role of each of those processes.

Acknowledgements

This paper is based on an invited lecture given at the occasion of the SIVOM meeting held in Munnar, Kerala, in October 2003. It summarizes certain aspects of a work which has been developed over several years and published in some of the cited references. The many discussions and contributions from L. Li, M. Ghil, K. Ide and O.P. Sharma are gratefully acknowledged.

REFERENCES

BELLON, G., LE TREUT, H., and GHIL, M. (2003), *Large-Scale and Evaporation-Wind Feedbacks in a Box Model of the Tropical Climate*, Geophys. Res. Lett. *30*, 10.1029/2003GL017895.

BONY, S., LAU, K.M., and SUD, Y.C. (1997), *Sea Surface Temperature and Large-Scale Circulation Influences on Tropical Greenhouse Effect and Cloud Radiative Forcing*, J. Climate *10*, 2055–2077.

BONY, S., DUFRESNE, J.-L., LE TREUT, H., MORCRETTE, J.-J., and SENIOR, C. (2004), *On Dynamic and Thermodynamic Components of Cloud Change*, Clim. Dyn. *22(2/3)*, 71–76.

GADGIL, S. (2003), *The Indian Monsoon and its Variability*, Ann. Rev. Earth Planet. Sci. *31*, 429–467.

HANSEN, J.E., LACIS, A., RIND, D., RUSSELL, G., STONE, P., FUNG, I., RUEDY, R., and LERNER, J., *Climate sensitivity: Analysis of feedback mechanisms*. In *Climate Processes and Climate Sensitivity* (eds. Hansen, J.E. and Takahashi, T.) (American Geophysical Union, Washington D.C. 1984) pp. 130–163.

HARTMANN, D. L., Moy, L. A., and Fu, Q. (2001) *Tropical convection and the energy balance at the top of the atmosphere*, J. Climate *14*, 4495–4511.

IDE, K., LE TREUT, H., LI, Z.-X., and GHIL, M. (2001), *Atmospheric Radiative Equilibria, Part II: Bimodal Solutions for Atmospheric Optical Properties*, Climate Dynamics *18*, 29–49.

INTERGOVERMENTAL PANEL ON CLIMATE CHANGE, *Climate Change 2001: The Scientific Basis* (Cambridge University Press, Cambridge, 2001).

KIEHL, J.T. and RAMANATHAN, V. (1990), *Comparison of Cloud Forcing Derived from the Earth Radiation Budget Experiment with that Simulated by the NCAR Community Climate Model*, J. Geophys. Res. *95,(D8)*, 11, 679–11, 698.

KRISHNAMURTI, T.N. (1971), *Observational Study of the Tropical Upper Tropospheric Motion Field During the Northern Hemisphere Summer*, J. Appl. Meteor. *10*, 1066–1096.

LARSON, K., HARTMANN, D.L., and KLEIN, S.A. (1999), *The Role of Clouds, Water Vapor, Circulation, and Boundary Layer Structure in the Sensitivity of the Tropical Climate*, J. Climate *12*, 2359–2374.

LI, Z.-X., IDE, K., LE TREUT, H., and GHIL, M. (1997), *Radiative Equilibrium: Investigations Using a One-dimensional Analytical Model*, Clim. Dyn. *13*, 429–440.

LI, Z.-X. and LE TREUT, H. (1992), *Cloud-Radiation Feedbacks in a General Circulation Model and their Dependence on Cloud Modelling Assumptions*, Clim. Dyn. *7*, 133–139.

LIN, B., WIELICKI, B., CHAMBERS, L., HU, Y., and XU, K.-M. (2002), *The Iris Hypothesis: A Negative or Positive Cloud Feedback?*, J. Clim. *15*, 3–7.

LINDZEN, R.S., CHOU, M.D., and HOU, A. (2001), *Does the Earth Have an Adaptive Iris?*, Bull. Am. Met. Soc. *82*, 417–432.

MILLER, R. L. (1997), *Tropical thermostats and low cloud cover*, J. Climate 10, 409–440.

PIERREHUMBERT, R.T. (1995), *Thermostat, Radiator Fins and the Local Runaway Greenhouse*, J. Atmos. Sci. *52*, 1784–1806.

RAJEEVAN, M. and SRINIVASAN, J. (2000), *Net Cloud Radiative Forcing at the Top of the Atmosphere in the Asian Monsoon Region*, J. Climate *13*, 3650–3657.

SHARMA, O.P., LE TREUT, H., SEZE, G., FAIRHEAD, L., and SADOURNY, R. (1998), *Interannual Variations of Summer Monsoons: Sensitivity to the Cloud Radiative Forcing*, J. Climate *11*, 1883–1905.

SIKKA, D.R. and GADGIL, S. (1980), *On the Maximum Cloud Zone and the ITCZ over Indian Longitudes During the Southwest Monsoon*, Mon. Wea.. Rev. *108*, 1840–1853.

WEBSTER, P.J., PALMER, T., YANAI, M., TOMAS, R., MAGANA, V., SHUKLA, J., and YASUNARI, A. (1998), *Monsoons: Processes, Predictability and the Prospects for Prediction*, J. Geophys. Res. *103*, 14,451–14,510.

WETHERALD, R.T. and MANABE, S. (1988), *Cloud Feedback Processes in a General Circulation Model*, J. Atmos. Sci. *45*, 1397–1415.

YANO, J.-I. and EMANUEL, K.A. (1991), *An Improved WISHE Model of the Equatorial Atmosphere and its Coupling with the Stratosphere*, J. Atmos. Sci. *48*, 377–389.

ZHANG, C., MAPES, B.E., and SODEN, B.J. (2003), *Bimodality in Tropical Water Vapour*, Quart. J. Roy. Meteorol. Soc. *129*, 2849–2866.

(Received March 29, 2004, accepted July 9, 2004)

 To access this journal online:
http://www.birkhauser.ch

Pure appl. geophys. 162 (2005) 1557–1586
0033–4553/05/091557–30
DOI 10.1007/s00024-005-2683-x

❚ Pure and Applied Geophysics

The Global Warming Debate: A Review of the State of Science

M.L. Khandekar,[1] T.S. Murty,[2] and P. Chittibabu[3]

Abstract—A review of the present status of the global warming science is presented in this paper. The term global warming is now popularly used to refer to the recent reported increase in the mean surface temperature of the earth; this increase being attributed to increasing human activity and in particular to the increased concentration of greenhouse gases (carbon dioxide, methane and nitrous oxide) in the atmosphere. Since the mid to late 1980s there has been an intense and often emotional debate on this topic. The various climate change reports (1996, 2001) prepared by the IPCC (Intergovernmental Panel on Climate Change), have provided the scientific framework that ultimately led to the Kyoto protocol on the reduction of greenhouse gas emissions (particularly carbon dioxide) due to the burning of fossil fuels. Numerous peer-reviewed studies reported in recent literature have attempted to verify several of the projections on climate change that have been detailed by the IPCC reports.

The global warming debate as presented by the media usually focuses on the increasing mean temperature of the earth, associated extreme weather events and future climate projections *of increasing frequency of extreme weather events worldwide*. In reality, the climate change issue is considerably more complex than an increase in the earth's mean temperature and in extreme weather events. Several recent studies have questioned many of the projections of climate change made by the IPCC reports and at present there is an emerging dissenting view of the global warming science which is at odds with the IPCC view of the cause and consequence of global warming. Our review suggests that the dissenting view offered by the skeptics or opponents of global warming appears substantially more credible than the supporting view put forth by the proponents of global warming. Further, the projections of future climate change over the next fifty to one hundred years is based on insufficiently verified climate models and are therefore not considered reliable at this point in time.

Key words: Carbon dioxide, global warming, land use effects, sea level, extreme weather events, solar influence.

1. Introduction

Studies and discussions of global warming as well as initiation of ice ages remained mainly as a scientific problem in the 19th century and most of the 20th century. Starting in the mid to late 1980s, this debate has spilled over into the media, the public and in the political arena as well. The debate has become

[1] Consulting Meteorologist, Unionville, Ontario, Canada. e-mail: mkhandekar@rogers.com
[2] Department of civil engineering, University of Ottawa, Ottawa, Canada
[3] W.F. Baird Associates Coastal Engineers Ltd., Ottawa, Ontario, Canada

emotionally charged with the proponents and opponents of global warming dug in their rigid stance.

The present global warming debate appears to have been accelerated following the publication of the IPCC (1996) report on the science of climate change which included a phrase that *the balance of evidence suggests a discernible human influence on climate.* This sacramental phrase has caused considerable controversy among atmospheric scientists, environmentalists and policymakers, as evidenced in a number of scientific commentaries and articles (AVERY *et al.*, 1996; MASOOD, 1996; SEITZ, 1996; SINGER *et al.*, 1997; KONDRATYEV, 1997) that appeared soon after publication of the IPCC 1996 report. These commentaries and articles questioned the link between the observed warming of the earth's surface and the increasing concentration of greenhouse gases in the atmosphere. The global warming debate has also sparked publication of a number of books and monographs in the last five years; noteworthy among the recent books are *The Satanic Gases* by MICHAELS and BALLING (2000), *Global Warming: The Hard Science* by HARVEY (2000), *Taken by Storm* by ESSEX and McKITRICK (2002) and *The Greenhouse Delusion* by GRAY (2002). The book by Harvey presents a supporting view of the global warming science while the other three present a dissenting view of the science.

The present manuscript is arranged as follows: Evolution of the earth's atmosphere, the natural greenhouse effect, the case for global warming, the case against global warming, earth's temperature variation in geological and historical times, urbanization and land-use change, impact of solar variability and sun's brightness, sea-level variations, extreme weather events and finally summary and conclusions.

2. Evolution of the Earth's Atmosphere

It is generally believed that when the earth evolved about 5 billion years ago (BY), all materials that made up the atmosphere and the oceans were contained inside the earth. The atmosphere began about 4.5 BY ago, as a mixture of water vapor, hydrogen, hydrogen chloride, carbon monoxide, carbon dioxide and nitrogen (GRAY, 2002). Through interaction with surface rocks and living organisms, it gradually reached its present composition, some 280 million years (MY) ago and has remained, more or less, unchanged. During the past 4.5 BY to 280 MY, the most important transformation was the conversion of much of the carbon dioxide (CO_2) into oxygen by abundant plant life, particularly during the carboniferous period, when most of our coal and oil deposits were formed (GRAY, 2002).

The following Table 1 lists four principal gases of the dry atmosphere and their proportional amounts in a well-mixed atmosphere. Besides these four gases, there are several other trace gases like Helium, Methane (CH_4), Nitrous Oxide (N_2O), Krypton Hydrogen, etc., whose proportion by volume is too small to be of any significance in the present discussion. The gases methane and nitrous oxide along with carbon dioxide are referred to as the greenhouse gases (GHG) whose radiative properties are a subject of intense study at present.

In Table 1, we have not considered the highly variable component of the atmosphere, namely the water vapor which is the most important greenhouse gas and can influence the earth's mean temperature structure significantly. There is an upper limit to the quantity of water the atmosphere can hold in gaseous (vapor) form at any temperature. Theoretically, water vapor could increase to a maximum of 5% of the total atmosphere (by volume) at the highest temperature measured near the earth's surface; however, in reality a value of 2% of water vapor in the atmosphere is considered a high value representative of a very humid atmosphere. As we discuss below, the presence of water vapor in the atmosphere produces a greenhouse effect, making the earth-atmosphere system warm and comfortable enough for animal and plant life.

3. The Natural Greenhouse Effect

Sun's radiation is mostly in short wavelengths and passes through the atmosphere without much absorption, except the ultraviolet part of the solar radiation which is absorbed by ozone in the stratosphere. Solar radiation heats the earth and the oceans and they in turn emit radiation back to space in longer wavelengths, hence known as longwave radiation. Some of the gases in the atmosphere, including water vapor and CO_2 absorb this longwave radiation from the earth (and oceans) and in this process, maintain an annual global surface temperature of about 14 °C to 15 °C. Since this phenomenon is somewhat similar to keeping plants warm in a greenhouse, is generally referred to as the greenhouse effect. In the actual greenhouse, vertical mixing is limited by the glass panes, whereas the atmospheric greenhouse gas effect reduces radiative loss to space through the absorption and then reemission downward for longwave radiation by CO_2, O_3 and water vapor. Without this natural greenhouse warming, the earth's annual average surface temperature would

Table 1

The four principal gases of the dry atmosphere, below 25 km (PETTERSEN, 1958)

Gas	Symbol	Percent by Volume
Nitrogen	N_2	78.09
Oxygen	O_2	20.95
Argon	A	0.93
Carbon Dioxide	CO_2	0.03

be about $-18\ °C$ to $-19\ °C$. Thus the natural greenhouse effect contributes about 33 °C to the earth's annual average surface temperature. The greenhouse effect extends through the troposphere and stratosphere.

Figure 1a adapted from RAMANATHAN (1998) shows the global energy balance for annual mean conditions. Here, the earth receives 343 units $(W.m^{-2})$ at the top of the atmosphere, while globally averaged longwave emission by the earth's surface is 395 ± 5 units. At the top of the atmosphere, a total of 237 ± 3 units are lost to space (often identified as OLR—Outgoing Longwave Radiation). Thus the intervening atmosphere and clouds cause a reduction of $(395-237) = 158 \pm 7$ units of energy which is the *Greenhouse Effect*. It is this natural greenhouse effect that keeps the earth's mean temperature a comfortable 289 K (15 °C), or about 33 °C warmer than it would be without this greenhouse blanket. Figure 1b shows global average clear-sky vs. average cloudy-sky radiation budget based on ERBE (Earth Radiation Budget Experiment) data from earth-orbiting satellites. The ERBE data separated the clear-sky radiation from average cloudy-sky radiation based on a five-year period (1985–1989) and the difference, about $-18\ W/m^2$, is attributed to cloud radiative forcing by RAMANATHAN *et al.* (1989a). This average "cloudy-sky" forcing is considerably larger than the $2.45\ W/m^2$ forcing due to increased concentration of GHGs (IPCC, 2001) and, as we shall discuss later, this can be a significant source of uncertainty in the hypothesis on greenhouse gas-induced climate change.

4. The Case for Global Warming

The concept that the earth's atmosphere acts somewhat like the glass of a greenhouse, letting through the sunlight (shortwave light rays) while retaining a portion of the longwave radiation emanating from the earth's surface was first introduced by the French mathematician JOSEPH FOURIER (1827). This concept was further expounded by TYNDALL (1861) who carried out sophisticated experiments to study the infrared radiative properties of water vapor and carbon dioxide and demonstrated that water vapor is the most important greenhouse gas.

▶

Figure 1a

Global energy balance for annual mean conditions. The top-of-the-atmosphere estimates of solar insolation (343 units), reflected solar radiation (106 units) and outgoing longwave radiation (237 units) are obtained using satellite data. Refer to the text for more details (from RAMANATHAN, 1988).

Figure 1b

Global average clear-sky radiation budget (left panel) and average cloudy fluxes and cloud radiative forcing (right panel) from ERBE data. Outgoing arrows denote OLR (Outgoing Longwave Radiation) and incoming arrows denote incoming solar radiation. Values shown are for 5-year averages between 1985 and 1989 with uncertainties in the fluxes of about $\pm\ 5\ W/m^2$ (from RAMANTHAN *et al.*, 1989b).

(a)

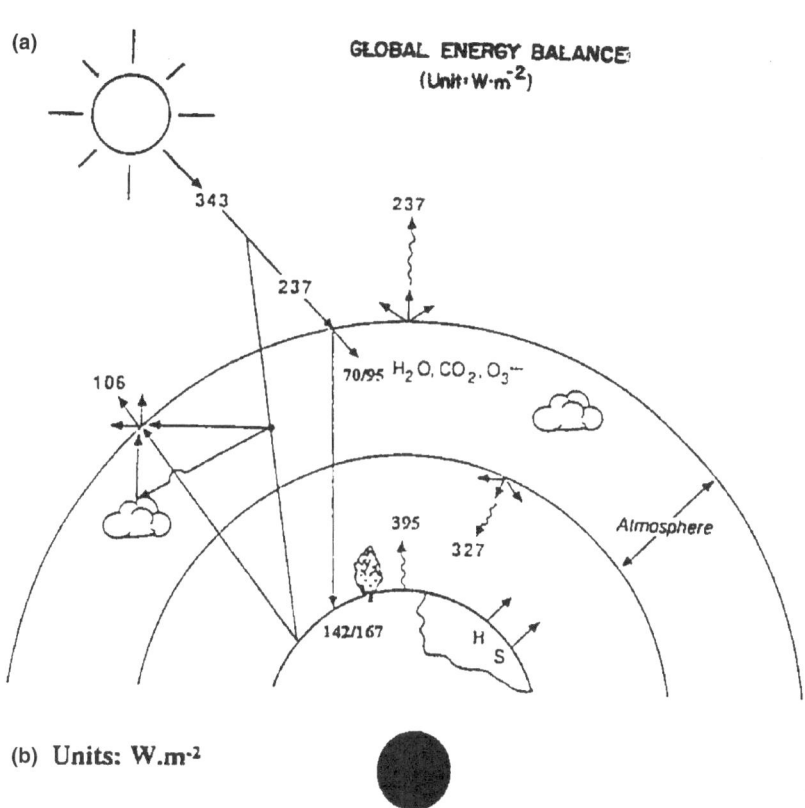

GLOBAL ENERGY BALANCE
(Unit: W·m^{-2})

(b) Units: W.m^{-2}

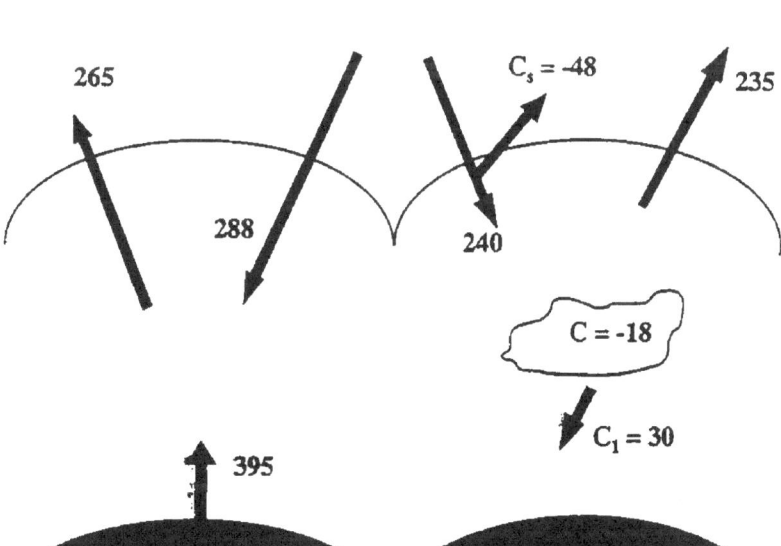

The first estimates of how changes in the global concentration of "carbonic acid" (a primary greenhouse gas, now more commonly referred to as carbon dioxide) might affect mean global surface temperature were made by a Swedish chemist Svante Arrhenius more than 100 years ago. ARRHENIUS (1896) demonstrated that an increase in the atmospheric concentration of CO_2 by a factor of two would lead to a heating of the earth's temperature by 5 to 6 °C. Arrhenius' work was followed by the studies of American geologist CHAMBERLIN (1899) whose work was focused on the role of CO_2 in the formation of glacial periods in geological times. The studies of Arrhenius and Chamberlin received scant support from the atmospheric science community of that time, since there was a general consensus that the absorption of longwave radiation (emanating from the earth) by water vapor was so strong that the absorption by carbon dioxide was considered negligible.

In 1938, British engineer CALLENDAR (1938, 1940) demonstrated through laboratory experiments that CO_2 does indeed have absorption bands outside of those dominated by water vapor and that increased CO_2 concentration could have significant global effects on the surface temperature of the earth. Callendar also speculated for the first time that humans could have a significant influence on the atmospheric CO_2 concentration, but estimated that it would take several centuries of continued industrial emission to achieve a doubling of concentration. In an important paper, the well-known American geophysicist Roger Revelle proposed that "humans are carrying out a large-scale geophysical experiment through worldwide industrial activity that could lead to a build-up of CO_2 larger than the rate of CO_2 production from volcanoes" (REVELLE and SUESS, 1957). Revelle was instrumental in establishing the first station for long-term monitoring of atmospheric CO_2 at Mauna Loa (Hawaii) and in launching an accelerated international research program on the potential human influence on the climate system.

The most direct and visible evidence of global warming is the change of ocean heating, and not the (earth's) surface air temperature record, even though it is the latter that has received the most attention from IPCC and the media. A more appropriate unit to measure warming (or cooling) is Joules, and not degrees Celsius. A recent paper by PIELKE (2003) analyzes the heat storage system in the earth system and points out how the utilization of surface temperature as a monitor of the earth system climate change is not particularly useful in evaluating the heat storage changes to the earth system. However, since surface air temperature is the most commonly used and most easily understood variable in the present global warming debate, we will refer to air temperature record as the basis for our discussion.

Through a careful analysis of a vast amount of land-ocean surface data, mean temperature variation of the earth's surface over a long period (1860–2000) has been prepared by JONES *et al.* (1999, 2001) and this temperature variation as shown in Fig. 2 has become the benchmark for the present global warming debate. For the present discussion here, we focus our attention on the 20th century temperature variation: During the last 100 years, the earth's temperature increased rather steeply (by ∼0.5 °C) from 1910 through 1945, then decreased (by ∼0.2 °C) from 1945 through 1975 and since about 1977 the mean temperature has increased by about 0.3 °C. It is this recent temperature increase of about 0.3 °C or more that has become the focus of the present global warming debate. The bottom part of Fig. 2 shows earth's temperature changes for the last millennium, based on MANN *et al.* (1998, 1999). This temperature variation is estimated using a number of proxy data like tree-ring widths, ice core data from Greenland, etc. Mann *et al.* conclude recent warming of the Northern Hemisphere as unprecedented and the 1990s are likely the warmest decades in 1000 years. As we shall discuss later, this temperature variation of the last millennium has become a subject of vigorous debate at present.

The most recent IPCC report on the climate change (IPCC, 2001) states that *the atmospheric concentration of carbon dioxide has increased from 280 ppm (parts per million) from 1750 to 367 ppm in 1999 (31% increase). Today's carbon dioxide concentration has not been exceeded during the past 420,000 years and likely not during the past 20 million years.* The IPCC also refers to a substantial recent increase in other greenhouse gases (GHG) namely, methane and nitrous oxide which have increased by 145% and 15%, respectively in the last 250 years. These greenhouse gases have added a total direct radiative forcing of about 2.45 $W.m^{-2}$ which, according to IPCC, has led to an increase in the mean surface temperature of the earth. Further, the recent steep increase (∼ 0.16 °C per decade) in the mean surface temperature is being *directly linked* to the increased greenhouse gas emission of the last 25/30 years.

A number of climate models developed at various national and international institutions in North America, Europe and elsewhere have simulated the mean surface temperature increase and its intimate link to the increasing concentrations of greenhouse gas emissions. Noteworthy among the climate modeling studies are those reported by BOER *et al.* (1992, 2000), HANSEN *et al.* (2001), MANABE *et al.* (1990), MANABE and STOUFFER, (1996), MEEHL *et al.* (2003) and MITCHELL *et al.* (1995). These modeling studies have simulated the earth's temperature change over the twentieth century and in particular the recent temperature increase, using gradually increasing levels of greenhouse gases.

The case for global warming and its link to GHG resides in the claim (by IPCC) that the increase in the mean surface temperature of the earth cannot be explained by the natural variability of the atmosphere-ocean system alone. Further, the recent increase in the mean temperature is unprecedented and can be explained (only) through climate model simulations which demonstrate the purported link between mean temperature and the increasing concentrations of GHG in the atmosphere.

Variations of the Earth's surface temperature for:

the past 140 years

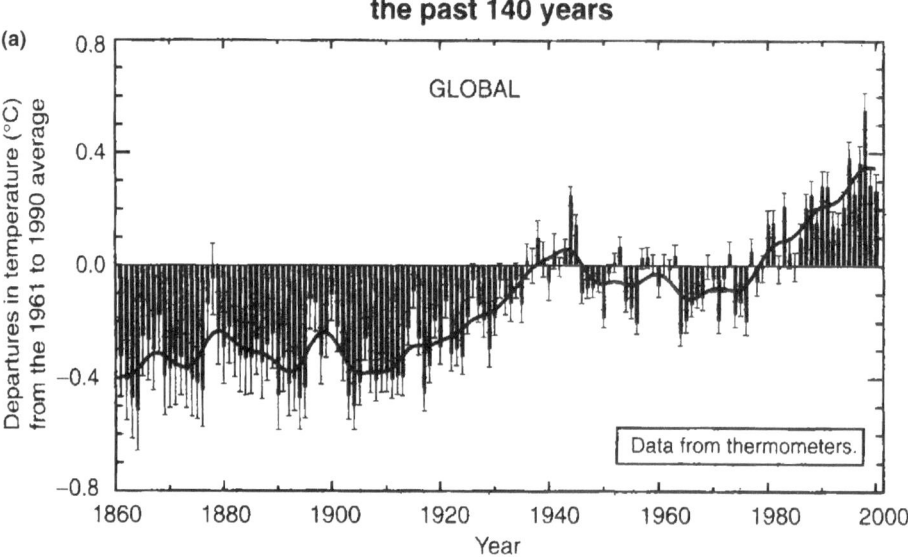

the past 1,000 years

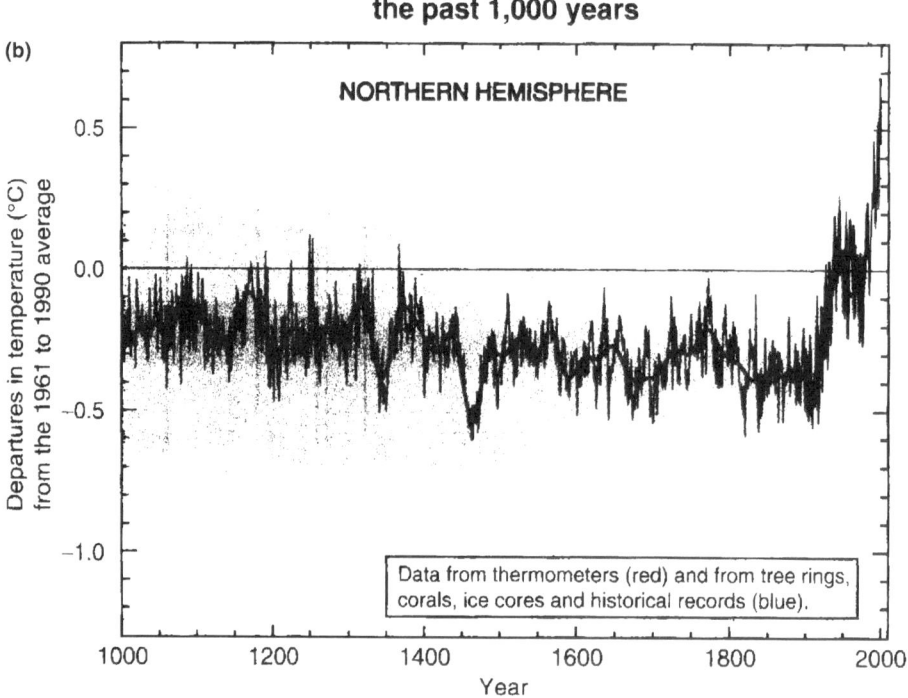

5. The Case against Global Warming

The case for global warming as presented above appears convincing and seemingly governed by a simple but attractive physical argument that more CO_2 in the atmosphere will trap more outgoing longwave radiation and thus the earth's surface will eventually become warm enough to make a case for "global warming." As mentioned before, the global warming and associated climate change issues are governed by many complex mechanisms and it is imperative to more closely examine these mechanisms before making definitive conclusions about cause and consequence of global warming.

Among the important issues that we discuss here are: **1.** Mean temperature calculation and the impact of urbanization and land-use change; mean temperature changes on regional and hemispheric scale and their variation in the context of large-scale atmospheric circulation patterns and other mechanisms. **2.** The impact of solar variability on climate change (and global warming) in geological times as well as on shorter time scales of a few hundred to a few thousand years and **3.** Consequences of global warming in terms of the "increase in mean sea-level and in extreme weather events world-wide." Within each of these three main issues there are numerous related other issues which make the simple warming/CO_2 link questionable, and consequently the warming/extreme weather link becomes tenuous at best.

In the last few years, several studies have brought into sharper focus some of the issues mentioned above: The urbanization and land-use change impact is now considered as providing a climate forcing which may be equal to or even stronger than the GHG forcing. The solar variability on geological as well as on shorter time scales appears to provide a significant influence on the mean temperature change. Further, the large-scale atmospheric circulation patterns and their decadal changes appear to provide significant impact on the mean temperature calculation. Finally the consequence of global warming in terms of increasing the frequency of extreme weather events and sea-level rise is fraught with considerable uncertainty due to the lack of good data on extreme weather events over a long period of time and due to other reasons of natural variability. When all these issues are taken together, a strongly dissenting view of the global warming science appears to emerge.

Figure 3 shows mean monthly temperature anomalies over global, Northern and Southern Hemisphere land-areas from January 1990 through January 2004. These temperature anomalies are obtained as departures from the 1971–2000 base period

◄

Figure 2
(a) Earth's mean temperature variation from 1860 to 2000, with respect to 1961–1990 average. (b) Mean temperature variation over the Northern Hemisphere for the past 1000 years with respect to 1961–1990 average (from IPCC, 2001).

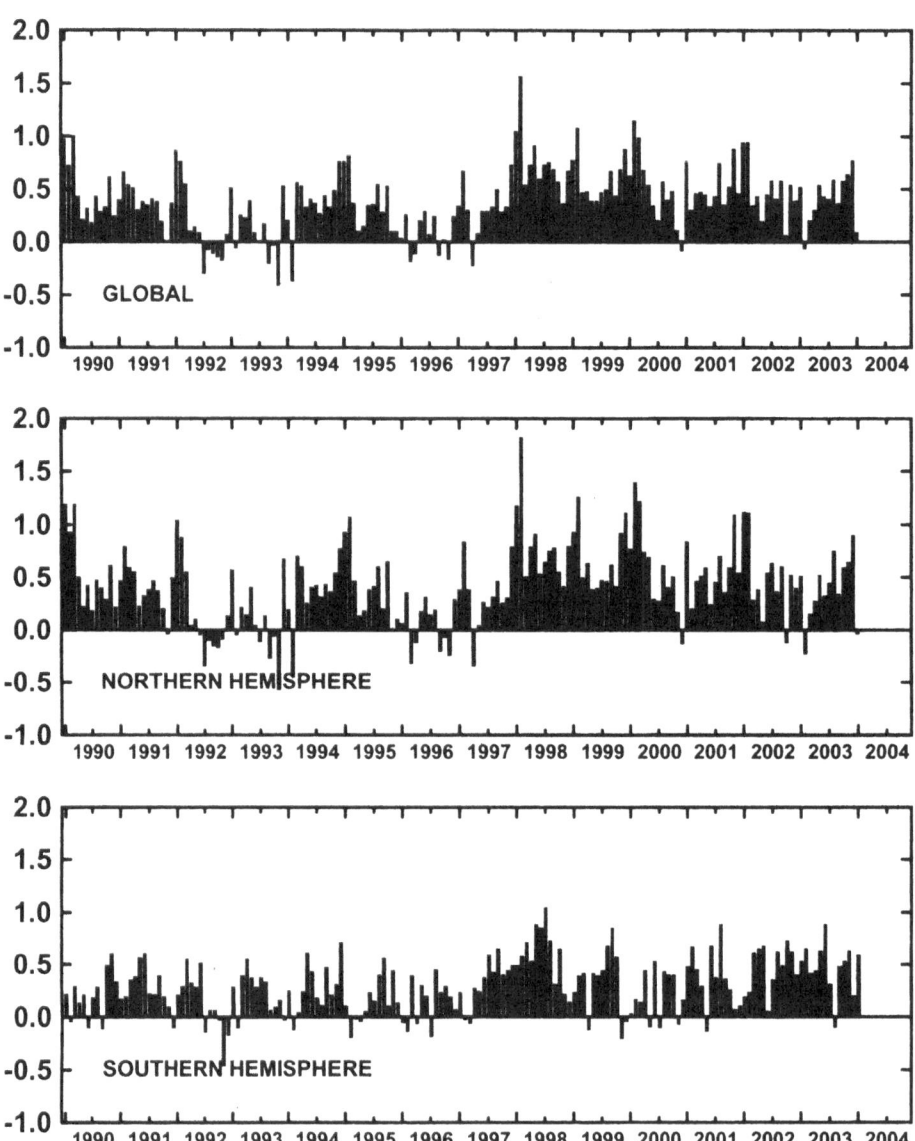

Figure 3

Monthly global (top), Northern Hemisphere (middle) and Southern Hemisphere (bottom) surface
temperature anomalies in °C (land areas only) from January 1990-present, computed as departures from
the 1971–2000 base period means. (from CLIMATE DIAGNOSTICS BULLETIN, January 2004).

means and show how the global temperature anomaly (land-areas) which peaked in
1998 after the 1997 El Niño, has steadily declined in recent years. Similar decrease in
mean temperature anomalies for northern and southern land-areas is also seen in the
same Figure. These temperature anomalies show how mean temperature calculation

can be significantly influenced by natural climatic events like El Niño (KUMAR et al., 2001).

6. Earth's Temperature Variation in Geological and Historical Times

Earth's temperature does not change monotonically. It rises and falls in highly irregular cycles and the amplitude of the change is highly variable. Earth's climate has been changing in geological as well as in historical times due to natural processes and it is instructive to take a closer look at the earth's temperature variation. While there is no universal agreement on the earth's climate history, there is some consensus on the following temperature history, which is based on a variety of data sources, mostly proxy records for the geologic past, with weather diaries for the past few centuries and instrumental records for the 20th century and part of the 19th century. The following is a partial list of proxy records: boreholes, glaciers, coral reefs, tree rings, sediments, pollen, insects, sea organisms, river flow, dune migration, stalactites, crop amounts, etc. The list provided here is in random order and does not reflect the order of importance of various proxy records.

The earth's temperature history reverting to 500 MY (Million Years) is given below:'

- Earth was warmer than now at −500 MY, −390 MY, −250 MY, and −100 MY and colder than now at −445 MY, −310 MY, −170 MY and −35 MY (VIEZER et al., 2000; KUMP, 2000).
- −490 to −443 MY (Ordovician glaciation). Colder than present (SHAVIV and VIEZER, 2003).
- −145 MY (Cretaceous). Very warm. Speculation that there was no ice on the planet, even at the poles (ENVIRONMENT CANADA, 2003).
- −43 MY (Eocene). Very warm. CO_2 levels then were less than during the glaciation at −114,000 years (ENVIROTRUTH, 2003).
- −17 MY (Miocene) very warm. CO_2 levels then were less than present levels (ENVIROTRUTH, 2003).
- −2 MY very warm. Forests almost extended towards the North Pole (ENVIRO-TRUTH, 2003).
- −1.6 MY to now. Thirty-three glacial advances (ice ages) and retreats; earth was much colder than at −2 MY (ENVIROTRUTH, 2003). Periodic and rapid fluctuations from cooler to warmer periods are referred to as interglaciations. Reasons cited are: continental drift, changes in ocean configurations, changes in atmospheric and ocean circulations, natural wobbles in earth's orbit (called Milankovich cycles) and variations in solar energy.
- −125,000 years. Very warm in Europe. Hippopotami and other animals, now confined mainly to Africa in natural habitat, existed in Northern Europe (ENVIROTRUTH, 2003).

- −114,000 years. Beginning of the most recent glacial period. Very cold. High CO_2 levels (ENVIROTRUTH, 2003).
- −50,000 years. Very cold. Most of North America was covered by ice, some places up to 1.5 km thick (ENVIRONMENT CANADA, 2003).
- −15,000 years. Earth was emerging from the last ice age. Temperatures in Greenland rose by 9 °C in 50 years (WEART, 2003).
- −12,000 years. In Europe, temperatures varied from warmer than present to the coldest during the ice age in a few decades and then bounced back. In Greenland, temperatures rose by 8 °C in a single decade (WEART, 2003).
- −11,000 years. Last ice age ended. Since then temperatures have been fluctuating (ENVIRONMENT CANADA, 2003).
- −7,000 to −4,000 years. 1 to 3°F warmer than now (ENVIRONMENT CANADA, 2003; BRIFFA, 2000).
- −5,000 years. Cooling of 2 °C globally. 6 °C cooling in the Arctic and only 0.5 °C in lower latitudes (ENVIROTRUTH, 2003).
- −2,000 years. Tree-ring records from Siberia suggest no temperature change except three episodes (i) Medieval Warm Period (MWP), (ii) Little Ice Age (LIA) and (iii) high temperatures of 20th Century with peak in 1940 (CENTER FOR THE STUDY OF CARBON DIOXIDE AND GLOBAL CHANGE, 2003).
- 800 to 1300 A.D. MWP (Medieval Warm Period): 1 to 2 °C warmer than present. Warmest period was during 900 to 1100 A.D (SOON and BALIUNAS, 2003; VILLALBA, 1990, 1994).
- 1000 A.D. Very warm in the Arctic. Sailing activity reported where there is a permanent ice pack now (THOMPSON *et al.*, 2000; BRIFFA, 2000 and also LAMB, 1972a, b; VILLALBA, 1990, 1994).
- 1350 – 1800 A.D. LIA (Little Ice Age): Average temperature dropped by 1.5 °C in 100 years. Coldest period of the LIA was during 1550 to 1700 (JONES *et al.*, 1998; VILLALBA, 1990, 1994).
- 1860 – to present time: This period of instrument data has provided the most detailed description of the earth's climate with a steep temperature increase between 1910–1940, a moderate decrease from 1945 to 1970 and the present warming from about 1975 (LAMB, 1972a, b; DEGAETANO and ALLEN, 2003; GRAY, 2002.

In the context of the earth's climate through the last 500 million years, the recent (1975–2000) increase in the earth's mean temperature does not appear to be unusual or unprecedented as claimed by IPCC and many supporters of the global warming hypothesis. According to MANN *et al.* (1998, 1999), *the 20th century is likely the warmest century in the Northern Hemisphere and the 1900s were the warmest decades with 1998 as the warmest year in the last 1000 years.*

Several recent studies have questioned the conclusion of Mann *et al.*, which has become the pivotal issue in the global warming debate at present. A paper by SOON and BALIUNAS (2003) examines a large number of proxy records and concludes that the 20th Century is probably not the warmest or a uniquely extreme climatic period of the last millennium. Another recent paper by MCINTYRE and MCKITRICK (2003) recalculates and reconstructs the Mann *et al.* temperature curve of the last 1000 years shown in Fig. 2 (this curve has been dubbed "the Hockey Stick Curve") using all available data and improved quality control. The recalculated curve by McIntyre and McKitrick is shown in Fig. 4. This Figure suggests that the 20th century is unexceptional when compared to the 15th century (1400–1500 A.D.) which according to McIntyre and McKitrick, could be warmer than the 20th century. Several other studies (e.g., LAMB, 1965; GROVE, 1996, 2001; OGILVIE and JONSSON, 2001) have suggested that the MWP (Medieval Warm Period) and the LIA (Little Ice Age) were global scale climatic anomalies and not just regional phenomena as concluded by IPCC (2001). The debate concerning the scale and extent of MWP and LIA continues at present.

The structure of the observed warming has become another point of debate in recent years. According to MICHAELS *et al.* (2000), the observed warming of the last fifty (1950–2000) years is mostly confined to the dry, cold anticyclones of Siberia and northwestern North America during the winter season. Further, on a seasonally weighted basis, a relatively small area (12.8%) contributed over half of the annual warming, while in the winter season of the Northern Hemisphere, 26% of the area

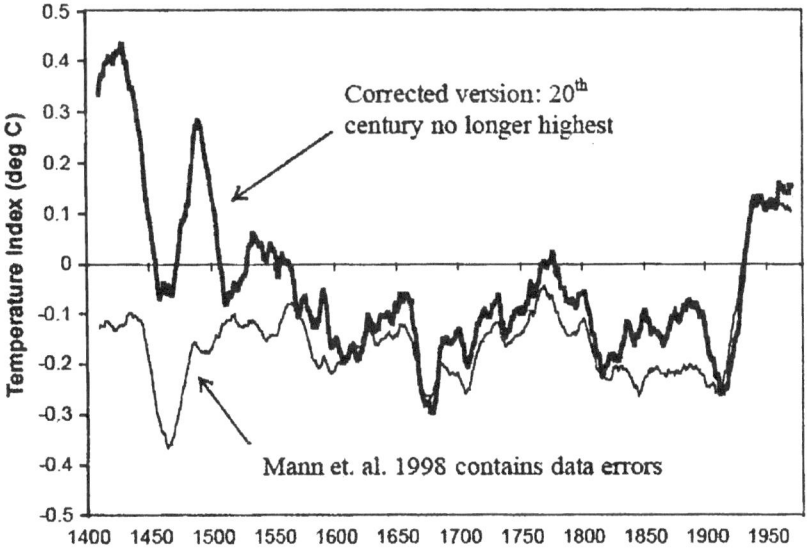

Figure 4
Temperature change (with respect to present mean temperature) over the last six centuries by Mann *et al.* (thin line) and as recalculated by McIntyre and McKitrick (thick line). See text for further discussion (from MCINTYRE and MCKITRICK, 2003).

accounts for 78% of the warming. The spatial patterns of observed warming, according to MICHAELS *et al.* (2000) are not consistent with that projected by many of the climate models as reported in the IPCC documents (e.g., IPCC, 1996). In another recent paper by JONES and MOBERG (2003), the observed warming of the 20th century is identified as having occurred during two distinct periods, 1920–1945 and 1975–2000. JONES and MOBERG further document that the recent (1975–2000) warming is statistically significant only at 19% of the grid locations; these grid locations being predominantly distributed over heavily populated and industrial areas of the earth. Thus it can be argued that the recent warming could be due to land use change and related economic activity and need not be linked to the atmospheric GHGs.

Finally, the warming that is being debated at present appears to be restricted to the lowest layer of the earth's atmosphere, approximately 1.5 km atmospheric layer above the earth's surface. Extensive analysis of satellite-derived surface-temperature data of the recent 22 years (1978–1999) by CHRISTY *et al.*, (1995, 1998, 2000) shows that in the troposphere (850–200 hPa layer), temperature trends range from −0.03 to +0.04 °C per decade (see Fig. 5), too small to be statistically significant from zero. Thus the troposphere has not warmed appreciably with respect to the earth's surface, as documented by Christy and coworkers. The tropospheric temperature changes have been recalculated in a recent paper by SANTER *et al.* (2003). A commentary by PIELKE SR. and CHASE (2004) discusses these calculations in additional details. These and other papers suggest that the

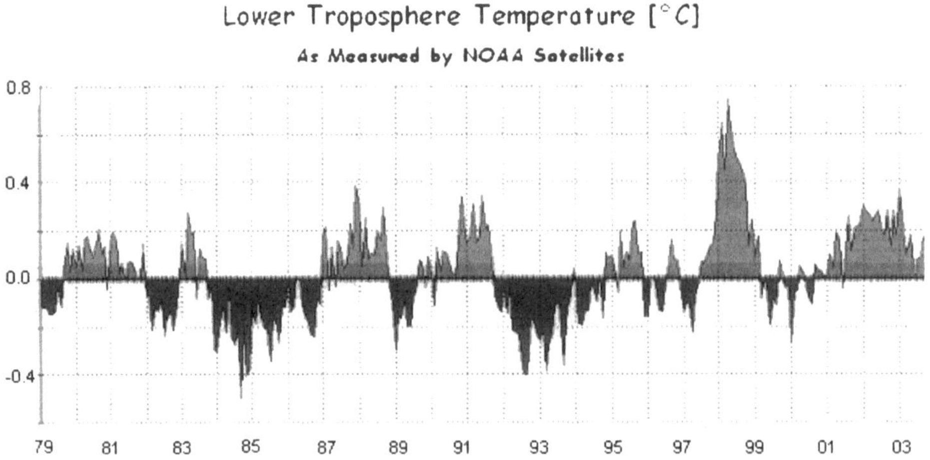

Figure 5

Lower tropospheric temperature as measured by NOAA satellite. The abscissa is years and the ordinate is temperature. The average trend in temperature change over 22 years (1979–2000) is estimated to be between −0.03°C to +0.04°C per decade. Note the high positive spike in temperature change during the strong El Niño years of 1997/98.

troposphere has warmed in the recent 25 years, however the warming remains significantly less than the modeled warming of the lower-to-mid troposphere. (CHASE *et al.*, 2004).

7. *Urbanization and Land-Use Change*

Assessing the impact of urbanization and land-use change on the mean temperature calculation is a challenging task. The classical studies of MITCHELL (1961) and OKE (1973) suggest that an urban heat island effect could be significant even for towns with a population of a few thousand people. For large cities, the urban heat island effect has been shown to be as high as 10 °C (temperature difference between the city center and a remote suburban location) in studies by OKE (1973) and others. The IPCC (1990) identified the urban heat island and its potential impact on surface air temperature as the most serious source of systematic error in the mean temperature calculation. In the context of the present global warming debate, the urbanization impact has been assessed as about 0.05 to 0.06 °C over one hundred years (JONES *et al.*, 1990; IPCC, 2001). This value appears too small in view of similar calculations for individual large cities or local regions.

Several recent studies (FUJIBE, 1995; GALLO *et al.*, 1996, 1999; HANSEN *et al.*, 2000, DEGAETANO and ALLEN, 2002; KALNEY and CAI, 2003) have taken a closer look at the land-use change impact on mean temperature calculations and these and other studies strongly suggest the impact to be significantly more than the value 0.06 °C used by IPCC (2001). The study by Fujibe shows trends in mean temperature at several Japanese stations (where long-term temperature data of 100 years or longer are available) of between 2 to 5 °C per hundred years at large cities and about 1 °C per hundred years at medium-sized cities. Another study by HINGANE (1996) estimates rising temperature trends of 0.84 °C and 1.39 °C per hundred years in the mean surface temperature calculated for Bombay (Mumbai) and Calcutta (Kolkata), two of the largest cities in India. Hingane used mean temperature data from Mumbai and Kolkata in conjunction with two neighboring stations to estimate the temperature trend due to urbanization as shown in Figure 6. Figure 6 shows how urbanization impact on temperature can gradually grow with time as two of the largest cities in India have grown in population and industrial development over 100 years.

Based on the study by KARL (1993) and others, it is now recognized that urbanization and land-use change influence minimum temperature, which in the last 100 years has risen faster than the maximum temperature at most locations, and this has led to a decrease in the DTR (Diurnal Temperature Trend). The recent study by GALLO *et al.* (1999) developed an innovative approach to identify rural vs. urban locations over the conterminous USA by using a night-light index from a satellite-based device. Their study (GALLO *et al.*, 1999) found that the

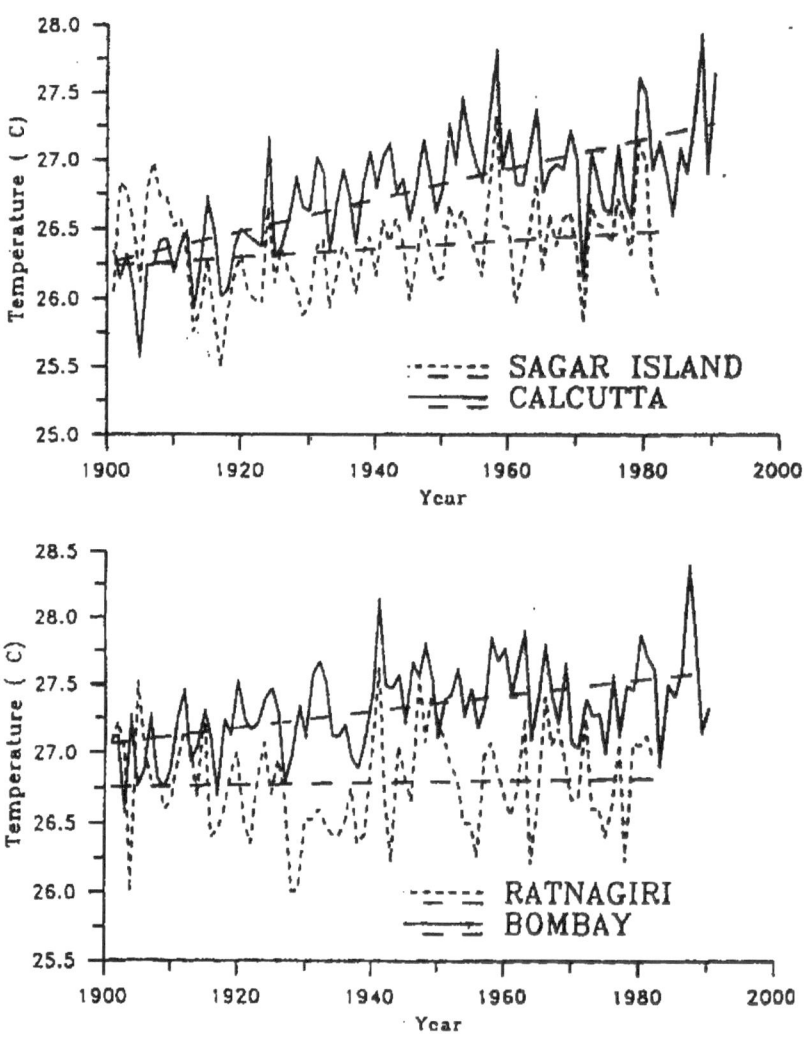

Figure 6

Long-term variation of surface-air temperature at two industrialized cities in India in relation to temperature variation at adjacent nonindustrial towns. **Top:** Calcutta vs. Sagar Island. **Bottom:** Bombay vs. Ratnagiri (from HINGANE, 1996).

decreasing trend in the DTR was smaller at rural stations than at urban stations by about 0.45 °C per 100 years. The study by HANSEN *et al.* (1997) suggests the urbanization impact of at least 0.1 °C on the earth's mean temperature calculations over a 100-year period. A more recent study by HANSEN *et al.* (2001) concludes that *local human effect (urban warming) can be identified even in suburban and small-town surface air temperature records.* Another recent study by KALNEY and CAI (2003) used the NCAR (National Center for Atmospheric

Research, Boulder, USA) reanalysis for the 1950–2000 period and reconstructed the surface temperature by extrapolating the various tropospheric level data to the surface level. Kalney and Cai obtained the urbanization and land-use change impact of 0.27 °C per century and about 0.18 °C for the recent 25-year period; this value (0.18 °C) being highly significant.

In another recent study by OOKA (2002), the urbanization impact on the temperature variation over Metropolitan Tokyo is documented using a network of stations in and around Tokyo for which data are available from 1870 through 2002. It is of interest to note that the temperature has risen by about 4 °C in the central part of Tokyo from its surroundings during the 150-year period. This increase of about 4 °C in the central part of Tokyo can most assuredly be attributed to urbanization and land-use change impact in and around Tokyo. Another recent study by DEGAETANO and ALLEN (2002) makes a detailed analysis of temperature structure at many locations in the USA and obtains trends in the occurrence of maximum and minimum temperatures which are significantly influenced by urbanization. In the most recent study, PETERSON (2003) carefully analyzes rural/ urban temperature differences for several clusters of stations in the USA and demonstrates that the differences between rural and urban sites are insignificant when homogeneity and other corrections are applied. Peterson's analysis is, however, based on only three years' data (1989–1991) and may not apply over longer durations as shown by Hingane, Fujibe and Ooka for locations in India and Japan, respectively.

Since urbanization is decidedly linked to economic activity, another recent study (MCKITRICK and MICHAELS, 2004) has investigated the global temperature histories at over 200 individual stations in 93 countries, to identify the impact of local economic activity like income, GDP (Gross Domestic Produce) growth rates, coal use, etc. McKitrick and Michaels find that the recent (1979–2000) temperature trends are influenced by economic activity and sociopolitical characteristics of the region surrounding individual stations.

Besides the urban/rural influence, the impact of land-use change and landscape dynamics on the climate system is being increasingly recognized and studied. In a landmark paper by PIELKE et al. (2002), the impact of land-use change and associated landscape dynamics on the climate system has been documented. It is further concluded that a more complete indication of human contribution to climate change will require the climatic influence of land-surface conditions and other processes to be included. Many of these processes will have strong regional effects that are not represented in a globally averaged metric. The study by PIELKE et al. brings out an important 'climate forcing' in the radiation budget of the earth-atmosphere climate system. Pielke has further suggested (PIELKE, 2002) that the climate forcing by land-use change and landscape dynamics can overwhelm the GHG forcing in the future. Other related studies reported in recent literature (CHASE et al., 1996; RADDATZ, 2003a,b) suggest how land-cover changes in terms of agricultural practices over various regions and subsequent leaf-area index can influence the global climate system. These and related papers have added a new dimension to the global warming debate.

In summary, the impact of urbanization and land-use change on the mean temperature calculation appears considerably more significant than that which has been assumed to date. There is a definite need to reanalyze the mean surface temperature calculation and to determine the "true warming" due to GHG (CO_2–induced) forcing only. The above discussion also points to a need for adequately incorporating the impact of land-use/land-cover change in present and future climate models, as pointed out by PIELKE *et al.* (2002).

8. *The Impact of Solar variability and Sun's Brightness*

The IPCC 1996 report did consider solar irradiance change over the last 100 years (and earlier) and its possible impact on global warming. However, it was concluded that the solar irradiance variation in the past century is likely to have been considerably smaller than the anthropogenic radiative forcing, and consequently its impact on global warming and climate change was considered to be insignificant. Several studies (e.g., HOYT and SCHATTEN, 1993; LEAN *et al.*, 1995; LEAN and RIND, 1998; SOLANKI and FLIGGE, 1998) published in the last ten years have attempted to reconstruct solar irradiance variations over the last 300 years or more and these reconstructed solar irradiance values have been examined in conjunction with earth's mean temperature changes. These studies now suggest that solar forcing can be significant and may have contributed significantly to the observed warming of the earth's surface. According to LEAN *et al.* (1995) and LEAN and RIND (1998), solar forcing may have contributed to about half of the observed warming of 0.55 °C since 1860 and about one third of the warming since 1970. Lean and Rind further state that since 1970, GHG forcing has been significantly larger than solar forcing and consequently the solar forcing signal in the global warming data cannot be easily isolated. Other studies, notably by RIND and OVERPECK (1993), SOON *et al.* (1996) and POSMENTIER *et al.* (1998) have suggested that up to 78% of earth's warming between 1885 and 1987 can be accounted for by an increase in the sun's irradiance.

A few other studies reported in recent literature have attempted to analyze the solar brightness through geological times and have strongly suggested the dominant role of the sun in driving the earth's climate over geological times. Notable among these studies are by VIEZER *et al.* (2000) and SHAVIV and VIEZER (2003). These studies now suggest that increased CO_2 levels in geological times were not linked with increased temperature. The earth's climate history listed earlier (section 6) suggests no correlation between warmer periods in earth's climate and CO_2 levels. For more than 90% of earth's history the mean temperature of the earth was warmer than present (ENVIROTRUTH, 2003). According to SHAVIV and VIEZER (2003) the earth's temperature fluctuated from slightly below −3 °C to slightly less than +3 °C from the present mean temperature over the last 500 million years (see Fig. 7). Their study which is based on the so-called "sea shell thermometer" also establishes that (Fig. 7) there is absolutely no correlation between atmospheric CO_2 levels and temperature. Shaviv and Viezer further show that the earth's temperature correlates well (Fig. 7) with changes in cosmic ray flux.

According to BENESTAD (2002) it is inappropriate to claim that response to solar activity can explain all the 20th century warming, as it is to dismiss it as making a negligible contribution. Any mechanism, linking solar activity and climate, must involve a forcing agent that penetrates at least into the stratosphere, if not into the

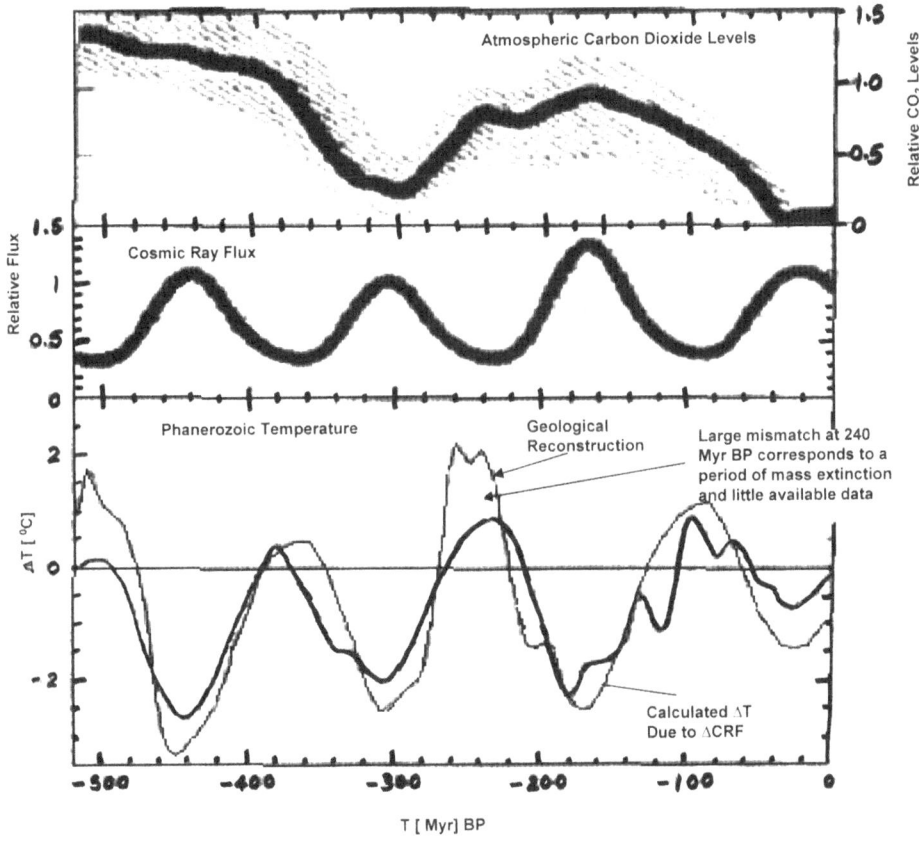

Figure 7
Top panel: CO_2 levels (determined from reconstructed partial pressure of atmospheric carbon dioxide). **Middle panel:** Cosmic ray flux (determined from a diffusion model that takes into account the geometry and dynamics of the spiral arms of the galaxy). **Bottom panel:** Temperature anomaly (determined from reconstructed sea-water temperatures based on various proxy records). The abscissa is years (in millions) BP (Before Present). (Simplified from SHAVIV and VEIZER, 2003 and PATTERSON, 2003).

troposphere. Benestad further proposes that solar ultraviolet irradiance can provide forcing which can influence the earth's climate. Observations show that this forcing does affect the climate at the 30 hPa (~25 km) level. The question then is: Can this forcing somehow affect the dynamics and climate at the 1000 hPa level? However, there are two other solar agents that appear more promising, in view of their ability to produce changes in atmospheric ionization. These two agents are solar wind and galactic cosmic ray flux affecting clouds. These two (solar) forcing mechanisms are present in the stratosphere as well as the troposphere with modulation amplitudes of about 10%. SVENSMARK and FRIIS-CHRISTENSEN (1997) demonstrate that there is a direct connection between earth's cloud cover and cosmic ray flux and this can influence earth's climate more effectively than increasing CO_2 levels.

It has been argued by LANDSBERG (1974) and RANDALL *et al.* (1984) that merely one to four percent increases in marine stratocumulus cloud cover can offset any warming due to a doubling of CO_2. Svensmark and Friis-Christensen further document an increase in solar irradiance of about 1.5 W/m^2 over a short period from 1986 to 1990. Further, studies by BARLOW and LATHAM (1983) and by DICKINSON (1975) show that secondary ions produced by cosmic rays can provide condensation nuclei which can enhance cloud cover. Thus solar irradiance change can influence earth's climate far more effectively over a short period of time than increased CO_2 forcing over a considerably longer period as claimed by IPCC.

Finally, two new studies deserve attention here: A statistical analysis of satellite-derived tropospheric (and stratospheric) temperature anomaly and solar irradiance has been carried out by KARNER (2002) in an attempt to identify solar influence on the tropospheric temperature trends. Karner's study demonstrates that global average tropospheric temperature anomaly and solar irradiance anomaly behave similarly and show antipersistency for scales longer than two months. A precise definition of "antipersistency" is provided by KARNER (2002). Karner's study suggests a cumulative negative feedback in the earth's climate system, contrary to the suggestion by MITCHELL (1989) of a positive feedback. The study further emphasizes the dominant role of solar irradiance variability and lends no support to the hypothesis of anthropogenic climate change. In a more recent investigation, FOUKAL (2003) suggests that slow variation in solar luminosity can provide a missing link between sun and climate. Foukal's study suggests that solar impact on the earth's climate may be driven by variable output of ultraviolet radiation or plasmas and fields via more complex mechanisms than direct forcing of tropospheric temperature.

In summary, the impact of solar variability and sun's brightness on the earth's climate has been brought into sharper focus in several recent studies. These studies further suggest a much stronger solar impact on earth's climate than previously believed. Some of the recent studies (e.g., SHAVIV and VIEZER, 2003; BENESTAD, 2002) suggest a definite link between solar variability and cloud cover which can significantly influence earth's mean temperature. The role of solar variability and the sun's brightness on earth's climate has not been fully incorporated in most climate modeling studies at present.

9. Sea-level Variations

This is not a review of sea-level variation, but rather an evaluation of the suggestion that global warming may be causing (accelerated) sea-level rise currently and in the future.

Sea-level measurements are even more biased than weather stations (GRAY, 2002). They are mainly near Northern Hemisphere ports, and are subject to local and short and long-term geological changes which are difficult to allow for. Sites in remote, low population places, such as the smaller Pacific Islands, show no evidence of recent sea-level changes.

Figure 8 shows that the sea level is more or less steady in Tuvalu, one of the small atolls in South Pacific. According to the National Tidal Facility (NTF) of Australia, the historical record shows no visual evidence of any acceleration in the sea level trends. They suggest that coastal degradation and sinking islets in Funa Futi were the result of environmental conditions, and not due to sea-level rise.

Other Pacific islands showing no detectable change in sea level are:

Tarawa, Kribati for 24 years
Kanton Island for 28 years

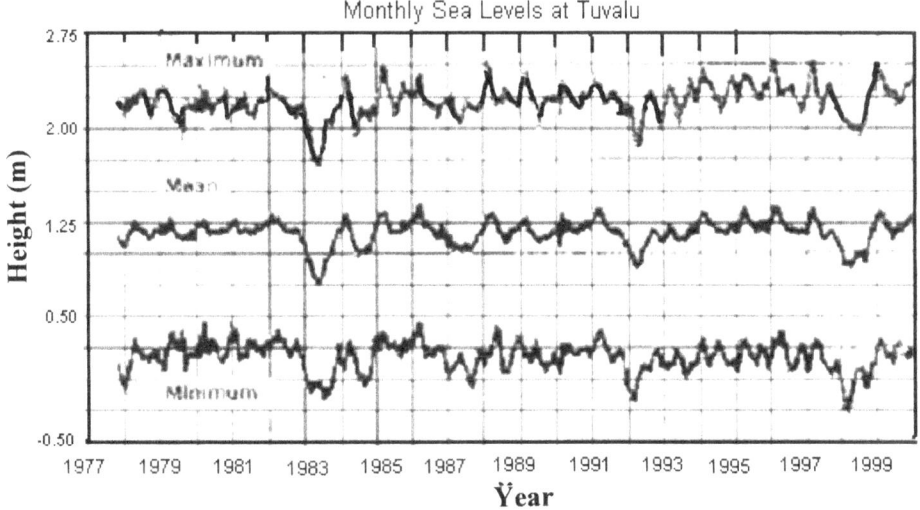

Figure 8
Sea levels at Tuvalu in the South Pacific as measured by tide gauges for the period 1978 to 2000 (from the National Tidal Facility of Australia)

Nauru for 26 years
Johnston Island for 50 years
Honiara, Solomons, for 26 years
Saipan for 22 years

Many others exhibit a stable period followed by a sudden jump, most likely due to hotel or airport construction, or a hurricane. Most of them also show no mean temperature increase over the period. The El Niño events of 1983 and 1998 show low readings.

According to IPCC (2001), *"the sea level rise in the 20^(th) century is in the range of 1.0 to 2.0 mm per year with a central value of 1.5 mm per year. No significant acceleration in the rate of sea-level rise during the 20^(th) century has been detected.* The study by BALTUCK *et al.* (1996) concludes that *it is very probable that the rising sea level is due to natural causes and not due to man's contribution to the greenhouse effect.*

Sea level has been rising naturally since the end of the last ice age and this has not accelerated recently. The total rise has been over 120 meters and is still proceeding at a rate of about 18 cm per century. An inspection of sea-level data does not show accelerated rise during the last fifty years, when the mean temperature of the earth increased by about 0.5 °C (ENVIROTRUTH, 2003).

Ongoing sea-level rise is due to the slow melting of Antarctic ice sheets that have been gradually disappearing for about 18,000 years; the date of the last glacial maximum. As far as we can tell from geological data, only temperature variations on a millennial time scale can affect this rate. Climate fluctuations lasting decades or even centuries are too short to affect appreciably this rate of melting. So unless another ice age commences in the meantime, sea level is bound to continue rising at about the same rate as present.

It is also important to understand that just as the melting of ice cubes in a glass of water does not cause the glass to overflow, the melting of polar sea ice will not result in ocean level changes. Only if massive quantities of inland Antarctic and Greenland glaciers melted, would sea levels raise enough to submerge coastal settlements. This did not happen 5,500 years ago, when the earth was three degrees warmer. Sea level was only two meters higher than now at 120,000 yeas ago, when temperatures were almost six degrees warmer than now.

Ordinarily, small island nations like the Maldives and Barbados are not threatened by such a rise. This is because these island countries are built entirely on coral and coral fragments. This coral is continually and quickly growing upwards and, unless something disastrous happens to the natural environment in a region, no sea-level rise is fast enough to get ahead of coral growth. The Maldivian reefs have been coping with increasing sea level for the past few thousand years and were able to keep up even when the ocean was rising ten times faster than it is now, about 10,000 years ago.

10. Extreme Weather Events

An important consequence of global warming is the possibility of increased incidences of extreme weather events worldwide. The most recent document on climate change (IPCC, 2001) has identified a number of extreme weather events which are expected to be observed with increased frequency during the 21st century. Increased incidences of some of the extreme weather events (e.g., frequent heavy precipitation, increased hot spells in summer, increased summer continental drying and associated risk of drought) are purported to have been observed and detected during the latter half of the 20th century. The earlier climate change document (IPCC, 1996) summarizes that *warmer temperature will lead to a more vigorous hydrologic cycle; this translates into prospects for more severe droughts and/or floods in some places and less severe droughts and/or floods in other places. Several models indicate an increase in precipitation intensity, suggesting a possibility for more extreme rainfall events.*

The global warming/extreme weather link has been investigated in numerous studies in recent years. Besides scientific studies, a number of informal articles and commentaries have appeared in news and print media (popular science magazines, etc.). Also, news treating extreme weather events worldwide (e.g., hurricane landfalling, outbreak of tornadoes, summer heat waves, large forest fires) is reported on television and in newspapers with suggestions of a possible link to global warming. According to UNGER (1999), American television viewers are three times more likely to see a story on severe weather today than they were only thirty years ago. The growing news coverage of extreme weather events and their socio-economic impact has created a perception that extreme weather events are on the rise at present and are due to the increasing mean temperature of earth.

When carefully analyzed, the link between global warming and extreme weather is more a perception than reality. A number of peer-reviewed studies cited in IPCC (2001) appear to suggest global warming/extreme weather link, however a close inspection of available data does not provide any evidence of such a link at this point in time. In a survey article, BALLING and ČERVÉNY (2003) analyze a number of severe weather events in the conterminous USA and find no upward trend in their frequency in recent years. Balling and Červény further analyze damage from severe weather which has increased in recent years in the USA, however, when the damage in terms of loss of human life and property is adjusted to inflation, population growth and wealth, this upward trend in damage disappears. A similar conclusion is arrived at in another comprehensive study by CHANGNON (2003) who has documented that *shifting economic impact from weather extremes in the United States is a result of societal change and not global warming.* Elsewhere in North America, KHANDEKAR (2000, 2002) has analyzed some of the synoptic scale events such as Atlantic and Pacific hurricanes and (USA/Canada) East Coast winter storms. Khandekar has also examined other extreme weather events like heat waves, intense thunderstorms/

tornadoes, ice storms in eastern Canada and winter blizzards on the Prairies. Based on a careful assessment of available data over Canada and the Canadian Prairies, Khandekar concludes that extreme weather events are not increasing in Canada or on the Prairies (south of 70°N) at this point in time. Some of the extreme weather events like Prairie winter blizzards are definitely on the decline (LAWSON, 2003) in frequency as well as in intensity. KHANDEKAR (2002) has further examined available data in the Arctic and sub-Arctic regions of Canada and found evidence of an increase in "heavy precipitation" events in the Canadian Arctic. However, the present database in the Canadian Arctic is limited in terms of length of time as well as in spatial coverage; consequently, a definitive conclusion regarding an increasing/ decreasing trend cannot be made currently. In another paper by KUNKEL (2003a), extreme precipitation trends are analyzed over North America and it is concluded that there is a definite increase in extreme precipitation trends in some regions of the USA (e.g., Great Lakes and Northeast) but not over Canada. In a subsequent paper KUNKEL (2003b) analyzes newly available data on rainstorms and concludes that the recent increase in extreme precipitation events in USA may be similar to what was observed more than 100 years ago and could thus be related to natural variability and not necessarily due to anthropogenic influence.

Elsewhere, extreme rainfall analysis has been made by ZHAI *et al.* (1999) for China and by RAKHECHA and SOMAN (1994) for summer Monsoon rains over India. Both these studies conclude no significant increasing trend in extreme rainfall events of 1-to-3-day duration. For southeast Asia, interannual rainfall variability is primarily governed by the ENSO (El Niño-Southern Oscillation) phase and does not reveal an increasing/decreasing tendency in recent years (KHANDEKAR *et al.*, 2000; KRIPLANI and KULKARNI, 1997). A comprehensive analysis by GROISMAN *et al.* (1999) finds a 20% increase in the probability of summer daily precipitation amount of over 25.4 mm (1 inch) in a few northern European countries like Norway and Poland, but such increasing trend has not been reported elsewhere in Europe. Over Australia, heavy rainfall events have increased in some areas, although this increase is not significant. For South Africa, a recent study by FAUCHEREAU *et al.* (2003) concludes that some regions have experienced a shift toward more extreme rainfall events in recent decades; however this increasing trend appears to be due to a closer link to ENSO phase in recent years and not due to anthropogenic influence.

In summary, the global warming/extreme weather link appears to be tenuous at best at this point in time. As pointed out by KHANDEKAR (2003), there is a definite need to closely examine this link using all available data in different parts of the world. A recent review paper by KARL and TRENBERTH (2003) states that *human influences are large enough to exceed the bounds of natural climate variability.* This statement and many other examples of "extreme weather" events given by Karl and Trenberth are debatable and require close examination.

11. Summary and Conclusions

During the long geological history of the earth, there was no correlation between global temperature and atmospheric CO_2 levels. Earth has been warming and cooling at highly irregular intervals and the amplitudes of temperature change were also irregular. The warming of about 0.3 °C in recent years has prompted suggestions about anthropogenic influence on the earth's climate due to increasing human activity worldwide. However, a close examination of the earth's temperature change suggests that the recent warming may be primarily due to urbanization and land-use change impact and not due to increased levels of CO_2 and other greenhouse gases.

Besides land-use change, solar variability and the sun's brightness appear to provide a more significant forcing on earth's climate than previously believed. Recent studies suggest solar influence as a primary driver of the earth's climate in geological times. Even on a shorter time scale, solar irradiance and its variability may have contributed to more than sixty percent of the total warming of the 20th century. The impact of solar activity like cosmic ray flux on the earth's cloud cover has not been fully explored and may provide an additional forcing to the earth's mean temperature change.

There appears to be no intimate link between global warming and worldwide extreme weather events to date. Increasing economic impact due to extreme weather events in the conterminous USA appears to be a result of societal change in wealth and population and not due to global warming. Outside of USA, very few studies have been reported thus far which make a meaningful analysis of economic impact of extreme weather events. There has been no accelerated sea-level rise anywhere during the 20th century.

Our review suggests that the present state of global warming science is at an important cross road. There is a definite need to reassess the science and examine various issues that have been discussed and analyzed here.

Acknowledgments

The authors would like to express their sincere appreciation to Ms. Maria Latyszewskyj, chief librarian at Environment Canada's library in Downsview (Ontario), for providing access to the library facilities. The authors also appreciate assistance from Ms. Brenda Bruce at Baird Associates in Ottawa. Professors Ross McKitrick (The University of Guelph, Ontario, Canada) and Pileke Sr. (Colorado State University, Fort Collins, USA) kindly provided links to some of their recent papers. One of the authors (MLK) would like to express his gratitude to his wife Shalan for her technical assistance on the home computer.

REFERENCES

ARRHENIUS, S. (1896), *On the Influence of Carbonic Acid in the Air upon the Temperature of the Ground*, Philosophical Mag. and J. Sc. *41*, 237–276.

AVERY, S. K., TRY, P. D., ANTHES, R. A., and HALLGRAN, R. E. (1996), *An Open Letter to Ben Santer*, Bull. Amer. Met. Soc. *77*, 1961–1966.

BALLING, Jr. R. C., and ČERVÉNY, R. (2003), *Compilation and Discussion of Trends in Severe Storms in the United States: Popular Perception vs. Climate Reality*, Natural Hazards *29*, 103–112.

BALTUCK, M., DICKEY, J., DIXON, T., and HARRISON C. G. A. (1996), *New Approaches Raise Questions About Future Sea Level Change*, EOS *1*, 385–388.

BARLOW, A. K., and LATHAM, J. (1983), *A Laboratory Study of the Scavenging of Sub-microscale Aerosol by Charged Raindrops*, Q. J. of Royal Met. Soc. *109*, 763.

BENESTAD, R., *Solar Activity and Earth's Climate* (Springer, New York 2002).

BOER, G. J., MCFARLANE, N. A., and LAZARE, M. (1992), *Greenhouse Gas-induced Climate Change Simulated with the CCC Second-generation General Circulation Model*, J. Climate *5*, 1045–1077.

BOER, G.J., Flato, G.J., Reader, M.C. and Ramsden, D. (2000), *A Transient Climate Change Simulation With Greenhouse Gas and Aerosol Forcing: Experimental Design And Comparison with the Instrumental Record for the 20ᵗʰ Century*, Climate Dyn. *16*, 405–426.

BRIFFA, K. R. (2000), *Annual Climate Variability in the Holocene: Interpreting the Message of Ancient Trees*, Quaternary Sci. rev. *19*, 87–105.

CALLENDAR, G. S. (1938), *The Artificial Production of Carbon Dioxide and its Influence on Temperature*. Q. J. of Royal Met. Soc. *64*, 223.

CALLENDAR, G. S. (1940), *Variation in the Amount of Carbon Dioxide in Different Air Currents*, Q.J. Royal Met. Soc. *66*, 395.

CENTER FOR THE STUDY OF CARBON DIOXIDE AND GLOBAL CHANGE (2003), *Was Late 20th Century Warming Unprecedented over the Past Two Millennia?* Internet. www.co2science.org/journal/2003/vol6n34c4.htm.

CHAMBERLIN, T. C. (1899), *An Attempt to Frame a Working Hypothesis to the Cause of Glacial Periods of an Atmospheric Basis*, J. Geology *7*, 751–787.

CHANGNON, S. A. (2003), *Shifting Economic Impacts from Weather Extremes in the United States: A Result of Societal Changes, not Global Warming*, Natural Hazards *29*, 273–290.

CHASE, T. N., PIELKE Sr., R. A., KITTEL, T. G. F., NEMANI, R. R., and RUNNING, S. W. (1996), *The Sensitivity of a General Circulation Model to Global Changes in Leaf Area Index*, J. Geophys. Res. *101*, 7393–7408.

CHASE, T. N., PIELKE Sr., R. A., HERMAN, B., and ZENG, X. (2004), *Likelihood of Rapidly Increasing Surface Temperatures Unaccompanied by Strong Warming in the Free Troposphere*, Climate Res. *25*, 185–190.

CHRISTY, J. R., SPENCER, R. W., and MCNIDER, R. T. (1995), *Reducing Noise in the MSU Daily Lower Tropospheric Global Temperature Data Sets*, J. Climate *8*, 888–896.

CHRISTY, J. R., SPENCER, R. W., and LOBL, E. S. (1998), *Analysis of the Merging Procedure for the MSU Daily Temperature Time Series*. J. Climate *11*, 2016–2041.

DEGAETANO, A. T., and ALLEN, R. J. (2002), *Trends in Twentieth-Century Temperature Extremes across the United States*, J. Climate, *15*, 3188–3205.

CHRISTY, J. R., SPENCER, R. W., and BRASWELL, W. D. (2000), *MSU Tropospheric Temperatures: Dataset Construction and Radiosonde Comparison*, J. Atmos. & Oceanic Technol. *17*, 1153–1170.

DICKINSON, R. E. (1975), *Solar Variability and the Lower Atmosphere*, Bull. Amer. Met. Soc. *58*, 1240.

ENVIRONMENT CANADA (2003), *CO₂/Climate Report* (summer issue, 32 pages, 2003), *Downsview, Ontario, Canada.*

ENVIROTRUTH (2003), Internet: *www.john-daly.com,index.htm*

ESSEX, C., and MCKITRICK, R., *Taken by Storm: The Troubled Science and Politics of Global Warming* (Porter Books, Toronto, Canada, 2002).

FAUCHEREAU, N., TRZASKA, S., ROUAULT, M., and RICHARD, Y. (2003), *Rainfall Variability and Changes in South Africa during the 20ᵗʰ Century in the Global Warming Context*, Natural Hazards *29*, 139–154.

FOUKAL, P. (2003), *Can Slow Variations in Solar Luminosity Provide Missing Link Between the Sun and Climate?* EOS *84*, 205–208.

FOURIER, J. B. J., *Théorie Analytique de Chaleur, Paris*, (translated in 1878 by Alexander Freeman), *The Analytical Theory of Heat* (Cambridge University Press, New York 1827).

FUJIBE, F. (1995), *Temperature Rising Trends at Japanese Cities during the Last Hundred Years and their Relationships with Population, Population Increasing Rates and Daily Temperature Changes*, Papers in Meteorology and Geophys. *46*, 35–55.

GRAY, V. (2002), *The Greenhouse Delusion: A Critique of "Climate Change 2001*, Multi-Science Pub. UK, 95 pp. www.multi-science.co.uk

GALLO, K. P., EASTERLING, D. R. and PETERSON, T.C., (1996) *The influence of Land Use/Land Cover on Climatological Values of the Diurnal Temperature Range*. J. Climate, *9*, 2941–2944.

GALLO, K. P., OWEN T.W., EASTERLING, D. R. and JAMASON, P. F. (1999) *Temperature Trends of the U.S. Historical Network Based on Satellite-Designated Land Use/Land Cover*. J. Climate *12*, 1344–1348.

GROISMAN, P.,YA, KARL, EASTERLING, KNIGHT, JAMASON, HENNESSY, SUPPIAH, PAGE, WIBIG, FORTU-NIAK, RAZUVAEV, DOUGLAS, FORLAND, and ZHAI (1999), *Changes in the Probability of Heavy Precipitation: Important Indicators of Climate Change*, Climatic Change 32, 243–283.

GROVE, J. M., *The century time-scale*. In *Time-scales and Environmental Change* (eds. Driver and Chapman), (Routledge, London 1996) pp. 39–87.

GROVE, J. M., *The onset of Little Ice Age*. In *History and Climate-memories of the Future?* (eds. Jones, Ogilivie, Davis, and Briffa) (Kluwer, New York 2001) pp. 153–185.

HANSEN, J., SATO, M., LACIS, A., and RUEDY, R. (1997), *The Missing Climate Forcing*. Phil Trans. R. Soc. London. *352*, 231–240.

HANSEN, J., RUEDY, R., LACIS, A., SATO M., NAZARENKO, N., TAUSNEV, N., TAGEN, I., and KOCH, D., *Climate modeling in the global warming debate*. In *Climate Modeling:Past, Present and Future* (ed. Randall. D.) (Academic Press, USA 2000).

HANSEN, J., RUEDY, R., SATO, M., IMHOFF, M., ESATERLING, D., PETERSON, T., and KARL, T. (2001), *A Closer Look at United States and Global Surface Temperature Change*, J. Geophy. Res. *106*, *D20*, 23,947–23,963.

HARVEY, L. D. D., *Global Warming: The Hard Science* (Prentice Hall, Canada 2000).

HINGANE, L. S. (1996), *Is a Signature of Socio-economic Impact written on the Climate?* Climatic Change *32*, 91–102.

HOYT, D. V., and SCHATTEN, K. H. (1993), *A Discussion on Plausible Solar Irradiance Variation*, J. Geophys. Res. *98*, 18895–18906.

IPCC, *The Science of Climate Change, Contribution of Working Group I to the Second Assessment Report of the IPCC* (eds. J. T. Houghton *et al.*) (Cambridge University Press, New York 1996).

IPCC, *Climate Change (2001), The Scientific Basis* (Cambridge University Press 2001).

JONES, P. D., GROISMAN, P. YA, COUGHLAN, M., PLUMMER, N., WANG, W.C. and KARL, T. R. (1990) *Assessment of Urbanization Effects in Time Series of Surface Air Temperature over Land*. Nature, *347*, 169–172.

JONES P. D., BRIFFA, K. R., BARNETT, T. P., and TETT, S. F. B. (1998), *High Resolution Paleoclimatic Records for the Last Millennium: Interpretation, Integration and Comparison with General Circulation Model Control-run Temperatures*, Holocene. *8*, 455–471.

JONES, G. S., SIMON, F. B., TETT, and STOTT, P. A. (2003), *Causes of Atmospheric Temperature Change 1960–2000: A Combined Attribution Analysis*, Geophys. Res. Lett. *30*, 1228, 32–1 to 32–4.

JONES, P. D., NEW, M., PARKER, D. E., MARTIN, S., and RIGOR, I. G. (1999) *Surface Air Temperature and its Changes over the Past 150 Years*, Rev. Geophys. *37*, (2), 173–199.

JONES, P. D., OSBORNE, T. J., BRIFFA, K. R., FOLLAND, C. K., HORTON, H., ALEXANDER, L. V., PARKER, D. E., and RAYNER, N. (2001), *Accounting for the Sampling Density in Grid-box Land and Ocean Surface Temperature Time Series*, J. Geophys. Res. *106*, 3371–3380.

JONES, P. D., and MOBERG, A. (2003), *Hemispheric and Large Scale Surface and Temperature Variations, An Extensive Revision and an Update to 2001*, J. Climate. *16*, 206–223.

KALNAY E., and CAI, M. (2003), *Impact of Urbanization and Land-use Change on Climate*, Nature. *423*, 528–531.

KARL, T., and TRENBERTH, K. E., (2003), *Modern Global Climate Change*, Science *302*, 1719–1723.

KARL, T. R. (1993) *A New Perspective on Global Warming.* EOS, 74, 25. p. 28

KARNER, O. (2002), *On Nonstationarity and Antipersistency in Global Temperature Series,* J. Geophys. Res. *107,* 1–11.

KHANDEKAR, M. L. (2000), *Uncertainties in Greenhouse Induced Climate Change,* Report prepared for Alberta Environment, March 2000, ISBN 0-7785-1051-4, 50 pages, Edmonton, Alberta.

KHANDEKAR, M. L. (2002), *Trends and Changes in Extreme Weather Events: An Assessment with Focus on Alberta and Canadian Prairies.* Report prepared for Alberta Environment, Oct. 2002, ISBN:0-7785-2428-0, 56 pages, Edmonton, Alberta.

KHANDEKAR, M. L. (2003), *Comment on WMO Statement on Extreme Weather Events,* EOS *84,* 428.

KHANDEKAR, M. L., MURTY, T. S., SCOTT, D., and BAIRD, W. (2000), *The 1997 El Niño, Indonesian Forest Fires and the Malaysian Smoke Problem: A Deadly Combination of Natural and Man-made Hazard,* Natural Hazards *21,* 131–144.

KONDRATYEV, K. (1997), *Comments on "Open Letter to Ben Santer",* Bull. Amer. Met. Soc. *78,* 689–691.

KRIPLANI, R. H., and KULKARNI, A. (1997), *Rainfall Variability over Southeast Asia: Connection with Indian Monsoon and ENSO Extremes—New Perspectives,* Int. J. Climatology *17,* 1155–1168.

KUMAR, A., WANG, W., HOERLING, M. P., LEETMAA, A., and JI, M. (2001), *The Sustained North American Warming of 1997 and 1998,* J. Climate *14,* 345–353.

KUMP, L. R. (2000). *What Drives Climate?* Nature, *408,* 651–652.

KUNKEL, E. K. (2003a), *North American Trends in Extreme Precipitation,* Natural Hazards *29,* 291–305.

KUNKEL, E. K. (2003b), *Temporal Variation of Extreme Precipitation of USA,* Geophys. Res. Lett. *30,* 1900–1903.

LAMB, H. (1965), *The Early Medieval Warm Period and its Sequel,* Paleogeogr. Paleoclim. Paleoecology. *1,* 13–37.

LAMB, H. H., *Climate: Present, Past and Future, 3 volumes.* (Methuen, London 1972a).

LAMB, H. H., *Weather, Climate and Human Affairs: A Book of Essays and other Papers* (Routledge, London 1972b).

LANDSBERG, H. E. (1974), *Man-made climate changes.* In *Proc. Symp. Phys. and Dyn. Climatology of the World Meteorol. Org.* 347, 262.

LAWSON, B. D. (2003), *Trends in Blizzards at Selected Locations on the Canadian Prairies,* Natural Hazards *29,* 121–138.

LEAN, J., BEER, J., and BRADLEY, R. (1995), *Reconstruction of Solar Irradiance since 1610: Implications for Climate Change,* Geophys. Res. Lett. *22,* 3195–3198.

LEAN, J., and RIND, D. (1998), *Climate Forcing by Changing Solar Radiation,* J. Climate *11,* 3069–3094.

LOMBORG, B., *The Skeptical Environmentalist* (Cambridge University Press 2001) 515 pp.

MANABE, S., BRYAN, K., and SPELLMAN, M. J. (1990), *Transient Response of a Global Ocean-atmosphere Model to a Doubling of Atmospheric Carbon Dioxide,* J. Phys. Oceanography *20,* 722–749.

MANABE, S., and STOUFFER, R. J. (1996) *Low-frequency Variability of Surface Air Temperature in a 1000-year Integration of a Coupled Atmosphere-ocean-land Surface Model,* J. Climate *9,* 376–393.

MANN, M., BRADLEY, R., and HUGHES, M. (1998), *Global-scale Temperature Patterns and Climate Forcing over the Past Six Centuries,* Nature *392,* 779–787.

MANN, M., BRADLEY, R., and HUGHES, M. (1999), *Northern Hemisphere Temperature during the Past Millennium: Inferences, Uncertainties and Limitations,* Geophys. Res. Lett. *26,* 759–762.

MANN, M., ANNAN, C., BRADLEY, R., BRIFFA, K., JONES, P., OSBORN, T., CROWLEY, T. M., HUGHES, M., OPPENHEIMER, M., OVERPECK, J., RUTHERFORD, S., TRENBERTH, K., and Wigley, T. (2003), *On Past Temperatures and Anomalous Late – 20th Century Warmth,* EOS *84,* 256–257.

MCINTYRE, S., and MCKITRICK, R. (2003), *Corrections to Mann et al (1998) Proxy Database and Northern Hemispheric Average Temperature Series,* Energy and Environ. *14,* 751–771.

MCKITRICK, R., and MICHAELS, P. J. (2004), *A Test of Corrections for Extraneous Signals in Gridded Surface Temperature Data.* Climate Research, *26,* 159–173. p. 29

MASOOD, E., (1996), *Sparks Fly over Climate Report,* Nature *381,* 639.

MEEHL, G. A., WASHINGTON, W. M., WIGLY, T. M. L., ARBLASTER, J. M. and DAI, A. (2003), *Solar and Greenhouse Gas Forcing and Climate Response in the Twentieth Century,* J. Climate *16,* 426–444.

MICHAELS, P. J., and BALLING Jr., R. C., *The Satanic Gases: Clearing the Air about Global Warming* (Cato Institute, Washington, USA 2000) 234 pp.

MICHAELS, P. G., KNAPPENBERGER, P. C., BALLING Jr., C., and DAVIS, R. E. (2000), *Observed Warming in Cold Anticyclones*, Climate Research *14*, 1–6.

MICHAELS, P. J. (2003), *The Political Science of Climate*, The Washington Times, July 29.

MITCHELL, J. M. (1961), *The Temperature of Cities*, Weatherwise. *14*, 224–229.

MITCHELL, J. F. B. (1989), *The Greenhouse Effect and Climate Change*, Rev. Geophys. *27*, 115–139.

MITCHELL, J. F. B., JOHNS, T. J., GREGORY, J. M., and TETT, S. F. B. (1995), *Climate Response to Increasing Levels of Greenhouse Gases and Sulfate Aerosols*. Nature *376*, 501–504.

OGILVIE, A. E., and JONSSON, T. (2001), *"Little Ice Age" A perspective from Iceland*, Climatic Change *48*, 9–52.

OKE, T. R. (1973), *City Size and the Urban Heat Island*, Atmos. Environ. *7*, 769–779.

OOKA, R. (2002), *Urban Warming and its Control, Newsletter of ICUS (International Center for Urban Safety Engineering, Institute of Industrial Science)* The University of Tokyo *2*, 1–3.

PATTERSON, T. (2003), *Celestial-Climate Connection Revealed*, Envirotruth, 7 pages.

PETERSON, T. C. (2003), *Assessment of Urban versus Rural in situ Surface Temperatures in the Contiguous United States: No Difference Found*, J. Climate *16*, 2941–2959.

PETTERSEN, S., *Introduction to Meteorology, 2nd Edition* (McGraw-Hill, New York 1958).

PIELKE, Sr., R. A. (2002), *Overlooked Issues in the U.S. National Climate and IPCC Assessment: An Editorial Essay*, Climatic Change *52*, 1–11.

PIELKE, Sr. R. A. (2003), *Heat Storage within the Earth System*. Bull. Amer. Met. Soc. *84*(3), 331–335.

PIELKE Sr., R. A., MARLAND, G., BELTS, R., CHASE, R. A., EASTMAN, T. N., NIBE, J. L., NEYOG, J. O., and RUNNING, S. W. (2002), *The Influence of Land Use Change and Land-scale Dimensions on the Climate System. Relevance to Climate-Change Policy Beyond the Radiative Effect of Greenhouse Gases*, Phil. Trans Roy. *360*, 1705–1719.

PIELKE, R. A. Sr., and CHASE, T. N. (2004), *Technical Comments on 'Contribution of Anthropogenic and Natural Forcing to Recent Tropospheric Height Changes'* Science, *303*, 1771b.

POSMENTIER, E. S., SOON, W. H., and BALIUMAS, S. L., *Relative impacts of solar irradiance variations and greenhouse changes on climate, 1880–1993. In (Bate, R. HS: Global Warming, the Continuing Debate, the European Science and Environment Forum (ESEF)* (Cambridge, 1998).

RADDATZ, R. L. (2003a), *Agriculture and Tornadoes on the Canadian Prairies: Potential Impact of Increasing Atmospheric CO_2 on Summer Severe Weather*, Natural Hazards *29*, 113–122.

RADDATZ, R. L. (2003b), *Aridity and the Potential Physiological Response of C_3 Crops to Doubled Atmospheric CO_2: A Simple Demonstration of the Sensitivity of the Canadian Prairies*, Boundary-Layer Meteorology *107*, 483–496.

RAKHECHA, P. R., and SOMAN, M. K. (1994), *Trends in the Annual Extreme Rainfall Events of 1 to 3-days Duration over India*, Theor. Appl. Climatology *48*, 227–237.

RAMANATHAN, V. (1988), *Trace-gas Greenhouse Effect and Global Warming*, AMBIO. *27*,187–197.

RAMANATHAN, V., BARKSTROM, B. R., and HARRISON, E. F. (1989d), *Climate and the Earth's Radiation Budget*, Physics Today, 22.

RAMANATHAN, V., CESS, R. D., HARRISON, E. F., MINNIS, P., BARKSTROM, B. R., AHMED, E., and HARTMAN, D. (1989b), *Cloud-radiative Forcing and Climate: Results from the Earth Radiation Budget Experiment*, Science *243*, 57–63.

RANDALL, D. A., COAKLY, J. A., FAIRALL, C. W., KROPFLI, R. A., and LENSCHOW, D. H. (1984), *Outlook for Research on Subtropical Marine Stratiform Clouds*, Bull. Amer. Met. Soc. *65*, 1290–1301.

REVELLE, R., and SUESS, H. E. (1957), *Carbon Dioxide Exchange between Atmospheric and Ocean and the Question of an Increase of Atmospheric CO_2 during the Past Decades*, Tellus *9*, 18–27.

RIND, D., and OVERPECK, J. T. (1993), *Hypothesized Causes of Decadal to Century Scale Climate Variability – Climate Model Results*, Quat. Sci. Rev. *12*, 357.

SANTER, B. D., WEHNER, M. E., WIGLY, T. M. L., SAUSEN, R., MEEHL, G. A., TAYLOR, K. E., AMMANN, C., ARBLASTER, J., WASHINGTON, W. M., BOYLE J. S., and BRUGGEMANN W. (2003), *Contributions of Anthropogenic and Natural Forcing to Recent Tropospheric Height Changes*, Science *301*, 25 July 2003, 479–483.

SEITZ, F. (1996), *A Major Deception on "Global Warming"*, Letter to Wall Street Journal, June 12, 1996, p. A16.

SHAVIV, N. J., and VIEZER, J. (2003), *Celestial Driver of Phanerozoic Climate,* GSA (Geol.Soc. of America) Today *13,* 4–10.

SINGER, F., BOE, B. A., DECKER, F. W., FRANK, N., GOLD, T., GRAY, W., LINDEN, H. R., LINDZEN, R., MICHAELS, P. J., NIRENBERG, W. A., PORCH, W., and STEVENSON, R. (1997), *Comments on "Open Letter to Ben Santer",* Bull. Amer. Met. Soc. *71,* 81–82.

SMAGORINSKY, J., ARMI, L., BRETHERTON, F. P., BYAN, K., CESS, R. D., GATES, W. L., HANSEN, J., KUTZBACH, J. E., and MANABE, S., *Carbon Dioxide and Climate: A Second Assessment* (National Academy Press, Washington, DC, 1982).

SOLANKI, S. K., and FLIGGE, M. (1998) *Solar Irradiance Flux since 1874 Revisited,* Geophys. Res. Lett. *25,* 341–344.

SOON, W., and BALIUNAS, SALLIE, (2003), *Proxy Climate and Environmental Changes of the Past 1000 Years,* Climate Res. *23,* 89–110.

SOON, W. H., POSMENTIER, E. S., and BALIUNAS, S. L. (1996), *Inference of Solar Irradiance Variability from Terrestrial Temperature Changes, 1880–1993, An Astrophysical Application of the Sun-Climate Connection,* The Astrophysical J. *472,* 891.

SVENSMARK, H., and FRIIS-CHRISTENSEN, E. (1997), *Variation of Cosmic Ray Flux and Global Cloud Coverage—A Missing Link in Solar Climate Relationships,* J. Atmos. Sol. Terr. Phys. *59,* 1225.

THOMPSON, L. G., YAO, T. E., MOSLEY-THOMPSON, E., DAVIS, M. E., HENDERSON, K. A., LIN, P. N. (2000), *A high-resolution Millennial Record of the South Asian Monsoon from Himalayan Ice Cores,* Science *289,* 1916–1919.

TYNDALL, J. (1861), *On the Absorption and Radiation of Heat by Gases and Vapours and the Physical Connection of Radiation, Absorption and Conduction,* Phil. Magazine *22,* 169–194.

UNGER, S. (1999) *Is Strange Weather in the Air? A Study of U. S. National Network News Coverage of Extreme Weather Events.* Climatic Change, *41,* 133–150.

VIEZER, GODDERIS and FRANCOIS (2000), *Evidence of Decoupling of CO2 and Global Climate During the Phanerozoic Eon,* Nature *408,* 698–701.

VILLALBA, R. (1990), *Climatic Fluctuations in Northern Patagonia during the last 1000 Years as Inferred from Tree-ring Records,* Quat. Res. *34,* 346–360.

VILLALBA, R. (1994), *Tree-ring and Glacial Evidence for the Medieval Warm Epoch and the Little Ice Age in Southern South America,* Climate Change *26,* 183–197.

WEART, S.: *The Discovery of Global Warming* (Harvard University Press, USA, 2003)

ZHAI, P., SUN, REN, LIU, GAO, and ZHANG (1999), *Changes of Climate Extremes in China,* Climatic Change *42,* 203–218.

(Received November 19, 2003; accepted June 21, 2004)

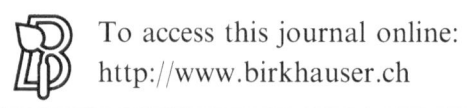

To access this journal online:
http://www.birkhauser.ch

Pure appl. geophys. 162 (2005) 1587–1606
0033–4553/05/091587–20
DOI 10.1007/s00024-005-2684-9

Pure and Applied Geophysics

Mountain Climates and Climatic Change: An Overview of Processes Focusing on the European Alps

MARTIN BENISTON

Abstract—This contribution provides an overview of the intricacies of mountain climates, particularly as they pertain to the European Alps. Examples will be given of issues that are related to climatic change as observed in the Alps during the course of the 20th century, and some of the physical mechanisms that may be responsible for those changes. The discussion will then focus on the problems related to assessing climatic change in regions of complex topography, the potential shifts in climate during the 21st century that the alpine region may be subjected to, and the associated climate-generated impacts on mountain environments.

Key words: Climate, climatic change, modeling, North Atlantic Oscillation, snow, mountain regions.

1. Introduction

Significant orographic features occupy close to 25% of continental surfaces (KAPOS *et al.*, 2000) and, although only about 26% of the world's population resides within mountains or in the foothills of the mountains (MEYBECK *et al.*, 2001), mountain-based resources indirectly provide sustenance for over half. This is principally because over 40% of the global population lives in the watersheds of rivers originating in the planet's different mountain ranges (BENISTON, 2000).

Although mountains differ considerably from one region to another, one common feature is the complexity of their topography. Orographic features include some of the sharpest gradients found in continental areas. Related characteristics include rapid and systematic changes in climatic parameters, in particular temperature and precipitation, over very short distances (BECKER and BUGMANN, 1997); greatly enhanced direct runoff and erosion; systematic variation of other climatic (e.g., radiation) and environmental factors (e.g., differences in soil types). Mountains in many parts of the world are susceptible to the impacts of a rapidly changing climate, and provide interesting locations for the early detection and study of the

[1]Department of Geosciences, University of Fribourg, Chemin du Musé 4, CH-1700, Fribourg, Switzerland. E-mail: Martin.Beniston@unifr.ch

signals of climatic change and its impacts on hydrological, ecological, and societal systems.

This paper will present an overview of the intricacies of mountain climates, particularly as they pertain to the European Alps. The Alps are without question one of the best-endowed mountain ranges in terms of data availability for studies related to climate and the environment. Examples will be given of issues that are related to climatic change as has already occurred in the Alps during the course of the 20th century, and the physical mechanisms that are responsible for the observed changes. In a final part, the potential shifts in climate during the 21st century in the alpine region will be reviewed, along with the possible impacts on mountain environments that these changes could generate.

2. Specificities of Mountain Climates

Climate is the principal factor governing the natural environment of mountains on short time scales, and characterizes the location and intensity of biological, physical and chemical processes. Mountain climates are determined by four major factors, namely continentality, latitude, altitude, and features related to topography itself (BARRY, 1994).

Continentality refers to the proximity of a particular region to an ocean. The diurnal and annual ranges of temperatures in a maritime climate are markedly less than in regions far removed from the oceans; this is essentially due to the large thermal capacity of the sea, which warms and cools far less rapidly than land. Because the ocean represents a large source of moisture, there is also more precipitation in a maritime climate than in a continental one, provided the dominant wind direction is onshore. Examples of maritime mountain climates include the Cascade Ranges in Oregon and Washington States of the United States, the Alaskan coastal mountains, the New Zealand Alps, the Norwegian Alps, and the southern Chilean Andes. Mountains under the dominant influence of continental-type climates include the Tibetan Plateau, the mountains of Central Asia (Pamir, Tien Shan, Urals), and the Rocky Mountains in the western states of the United States. However, many other mountain regions often define and separate climatic regions; for example, the European Alps act as a boundary between Mediterranean-type, Atlantic, and continental climates.

While mountains in continental regions experience more sunshine, less precipitation, and a larger range of temperatures than maritime mountains, they are not necessarily harsher environments. For certain ecosystems, the larger amounts of sunshine compensate for lower mean temperatures in continental regions. Increased cloudiness and precipitation (both rain and snow) in coastal mountain ranges such as the Cascades or the New Zealand Alps limit the growth of certain species despite the milder overall temperatures. The timberline in continental regions is often located at higher elevations than in maritime zones, which confirms the importance of these compensating factors for regional ecological systems (WARDLE, 1973; WADE and McVEAN, 1969).

Latitude determines to a large extent the amplitude of the annual cycle of temperature and, to a lesser extent, the amount of precipitation that a region experiences. Mountains tend to amplify some of the characteristics of tropical, mid-latitude and boreal climates for reasons related to topography. Altitude, however, is certainly the most distinguishing and fundamental characteristic of mountain climates. Atmospheric density, pressure and temperature decrease with height in the troposphere. At high elevations, thermal conditions are often extreme; the only source of energy is the direct solar radiation that is absorbed by the surface. Mountain soils can heat up rapidly during the day, and exchange some of this heat with the air in contact with the surface, but because of the lack of air and water vapor molecules, only a very thin envelope of air is likely to be warmed. This leads to steep vertical temperature gradients in the first few meters above the ground. Mountains serve in this sense as elevated heat sources, whereby diurnal temperatures are higher than at similar altitudes in the free atmosphere (FLOHN, 1968). The diurnal and annual range of temperature tends to decrease with altitude because of the lower heat capacity of the atmosphere at higher elevations. The altitudinal controls on mountain climates also exert a significant influence on the distribution of ecosystems. Indeed, there is such a close link between mountain vegetation and climate that vegetation belt typology has been extensively used to define climatic zones and their altitudinal and latitudinal transitions (for example KLÖTZLI, 1994; OZENDA, 1985; QUEZEL and BARBERO, 1990; RAMEAU et al., 1993).

Mountain systems generate their own climates (EKHART, 1948), as a function of the size of the land mass at a particular elevation. Topographic features also play a key role in determining local climates, in particular due to the slope, aspect, and exposure to climatic elements. These factors tend to govern the redistribution of solar energy as it is intercepted at the surface, as well as precipitation that is highly sensitive to local site characteristics. In many low and mid-latitude regions, precipitation is observed to increase with height; even modest topographic elements can exert an often disproportionate influence on precipitation amount. Precipitation mechanisms are linked to atmospheric dynamics and thermodynamics. When a mass of moist air is forced to rise above the condensation level, the excess vapor is converted to fine liquid water particles that become visible in the form of mist, fog or clouds. If uplift of air continues, then precipitation processes will be triggered if sufficient liquid water or ice crystals are available.

Precipitation in a mountain region will generally fall out on the windward-facing slopes of the mountains because of the dynamics associated with uplift. Because most of the moisture is extracted from the clouds on the windward slopes, the air that crosses over to the lee side of the mountains is essentially dry. Clouds may form and appear to adhere to the mountain crests, but little precipitation is likely to fall on the leeward sides of the mountains. Death Valley in California, one of the most arid desert regions in the world, is located on the leeward side of the highest part of the Californian Sierra Nevada range. Indeed, most of the desert regions of the western United States lie to the east of the coastal ranges that are located close to the Pacific coasts of North America. Precipitation gradients are steep from one side of the ranges to the other; for example, San Francisco, California receives an average of 475 mm of precipitation annually (double or triple that amount falls in the upper reaches of the Sierra Nevadas), while Las Vegas, Nevada receives only 99 mm. On more local scales, the windward and leeward influences on precipitation are particularly marked on islands which have a significant orography; the main city of Big Island, Hawaii (Hilo) experiences close to 3,500 mm annually, while the resort areas of Kailua-Kona, situated in the lee of the 4,000 m Mauna Laua and Mauna Kea volcanoes, receive only a sparse annual 65 mm.

As in many other cases of environmental stress, exogenous factors contribute to the overall environmental impacts in mountains and uplands. Air pollution is one such stress factor that has a number of ecological consequences, because it is so readily transported by atmospheric circulations from the emission sources to the receptor areas. The distances covered by chemical compounds in the atmosphere can cover large distances, according to the type of pollutant, its chemical reactivity and the dominant airflows in which they are embedded. As a result, particulate matter or acidic compounds can be detected in remote mountain regions which are far removed from their source regions.

NO_2 is one of the precursor gases of ozone (O_3), which is a highly corrosive and toxic gas that damages plants and leads to respiratory and ocular problems at sufficiently high concentrations. Ground-level ozone is not emitted directly into the atmosphere, but is a secondary pollutant produced by reaction between nitrogen dioxide (NO_2), hydrocarbons and sunlight. Whereas nitrogen dioxide (NO_2) participates in the formation of ozone, nitrogen oxide (NO) destroys ozone to form oxygen (O_2) and nitrogen dioxide (NO_2). For this reason, ozone levels are not as high in urban zones (where high levels of NO are emitted from vehicles) as in non-urban areas. As the nitrogen oxides and hydrocarbons are transported out of urban regions, the ozone-destroying NO is oxidized to NO_2, which participates in ozone formation. Ozone can therefore be present in mountain areas located at considerable distances from the source regions of the ozone precursors. Ozone is possibly one of the factors responsible for forest dieback in the European Alps for example, where damage to trees is believed to be linked to heavy industrial sites at the boundaries of the mountains, as in northern Italy. As always with environmental threats, the large time lag between load and full reaction, where ecosystems and soils are involved, has probably still not reached the peak impact.

GRASSL (1994) has shown that there has been an upward trend in tropospheric ozone concentrations at Alpine sites, such as the Zugspitze, at 2950 m altitude in Germany, since measurements began in the late 1960s. In contrast, ozone levels at the nearby urban site of Garmisch-Partenkirchen have not risen (SCHNEIDER, 1992), typically due to the ozone-destroying chemical reactions taking place in an urban environment with a high density of motorized traffic. Since sunlight is a prerequisite for the physico-chemical transformation of NO and NO_2 into O_3, high levels of ozone are generally observed during periods of persistently warm weather, and in locations where poor ventilation allows concentrations of hydrocarbons and nitrogen oxides to reach critical levels. Because of the time required for chemical processing, ozone formation tends to take place downwind of pollution centers. The resulting ozone pollution or "summertime smog" may persist for several days and be transported over long distances.

Sulfur contained in coal, which is used in many industrial processes worldwide for energy transformations, smelting, etc., is released into the atmosphere in the form of sulfur dioxide (SO_2). This can then be dissolved in liquid water droplets in clouds and fog to form sulfuric acid (H_2SO_4), which is then transported in atmospheric flows and ultimately falls to the surface in a process known as acid deposition or acid rain, with a number of adverse environmental consequences. Acidification of soils damages vegetation and releases heavy metals such as aluminium, copper or mercury into hydrological systems; these toxic metals are absorbed by fish and other aquatic species, with potentially serious consequences for the food chain. The interactions between living organisms and the chemistry of their aquatic habitats are extremely complex. If the number of one species or group of species changes in response to acidification, then the ecosystem of the entire water body is likely to be affected through predator-prey relationships. Terrestrial animals dependent on aquatic ecosystems for their food supply, such as birds, are also affected.

In the European Alps, the consequences of acid precipitation may be exacerbated in many regions by the fact that precipitation increases with altitude. SMIDT (1991) has shown that for two Austrian alpine sites contamination of precipitation with ions resulting from pollution decreases with altitude, but deposition increases because of enhanced precipitation with height. Since the concentration of basic anions and cations in precipitation is rather uniform over central Europe, the Alps are subject to as much acid deposition as other areas because of the orographic controls on precipitation, despite the fact the Alps are not in themselves a major source of sulfate-based pollutants.

Future trends in pollution and its long-distance transport are difficult to assess. On the one hand, increasing industrialization has the potential of generating high levels of pollution. On the other hand, the awareness that pollution is detrimental to human health and the environment has led to policies aimed at controlling emissions, particularly in western Europe, and thus the additional impacts on the Alps may be limited.

3. Shifts in Alpine Climates during the 20th Century

The climate of the Alpine region is characterized by a high degree of complexity, due to the interactions between the mountains and the general circulation of the atmosphere, which result in features such as gravity wave breaking, blocking highs, and föhn winds. A further cause of complexity inherent to the Alps results from the competing influences of a number of different climatological regimes in the region, namely Mediterranean, Continental, Atlantic, and Polar.

Figure 1 shows the changes in yearly mean surface temperature anomalies in the course of the 20th century from the 1961–1990 climatological mean, averaged for eight sites in Switzerland (namely Altdorf, Basel, Bern, Davos, Lugano, Neuchâtel, Säntis and Zürich). The global data discussed in JONES and MOBERG (2003) have been superimposed here to illustrate the fact that the interannual variability in the Alps is higher than on a global or hemispheric scale; the warming experienced since the early 1980s, while synchronous with global warming, is of far greater amplitude and exceeds 1.5°C for this ensemble average. This represents roughly a three-fold amplification of the global climate signal in the Alps (DIAZ and BRADLEY, 1997).

Closer investigation reveals that climatic change in the alpine region during the 20th century has been characterized by increases in minimum temperatures of up to 2°C, a more modest increase in maximum temperatures, little trend in the precipitation data, and a general decrease of sunshine duration through to about the mid-1980s (BENISTON, 2000). Several periods of warming can be observed during the instrumental record, with the 1940s exhibiting a particularly strong warming and

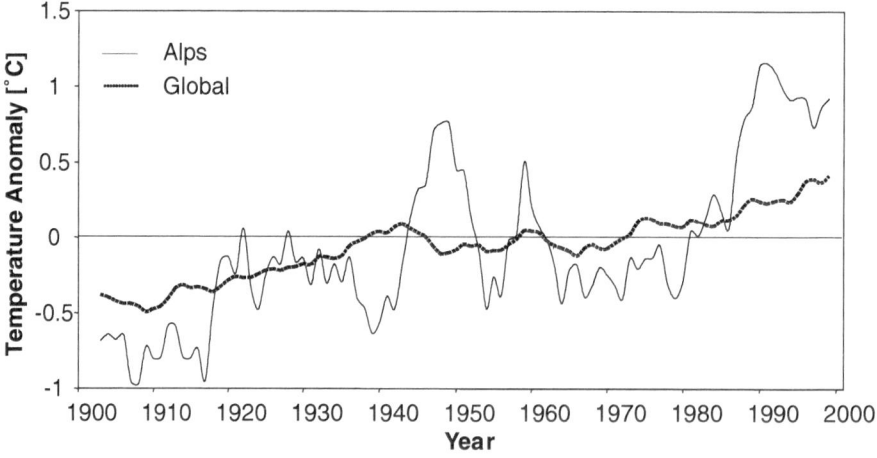

Figure 1
Temperature departures from the 1961–1990 climatological mean in the Swiss Alps compared to global temperature anomalies, for the period 1901–2000. A five-year filter has been applied to remove interannual noise (BENISTON et al., 1997).

then a cooling into the 1950s. The most intense warming occurs in the 1990s, however (JUNGO and BENISTON, 2001), which can be explained in part by the behavior of the North Atlantic Oscillation (NAO; BENISTON and JUNGO, 2002). This phenomenon is a large-scale alternation of atmospheric pressure fields (i.e., atmospheric mass), whose centers of action are near the Icelandic Low and the Azores High. One mode of the NAO is when sea-level pressure is lower than normal in the Icelandic low pressure center, it is higher than normal near the Azores, and *viceversa*, hence the notion of an oscillatory behavior of the system. Another possible mode occurs when pressures also rise or fall simultaneously in both centers of meteorological activity, but in this case the NAO signal is not quite as strong. The NAO index is a normalized pressure difference between the Azores (or Lisbon, Portugal) and Iceland; it is a measure of the intensity of zonal flow across the North Atlantic and the associated position of storm tracks and regions of strongest storm intensity. This flow is itself driven by the temperature (and hence pressure) contrasts between polar and tropical latitudes.

The NAO represents one of the most important modes of decadal-scale variability of the climate system after ENSO (El Niño/Southern Oscillation), and accounts for up to 50% of sea-level pressure variability on both sides of the Atlantic (HURREL, 1995). It has been observed to strongly influence precipitation and temperature patterns on both the eastern third of North America and western half of Europe; the influence of the NAO is particularly conspicuous during winter months. It was shown in recent years (BENISTON *et al.*, 1994; HURREL, 1995; ROGERS, 1997;

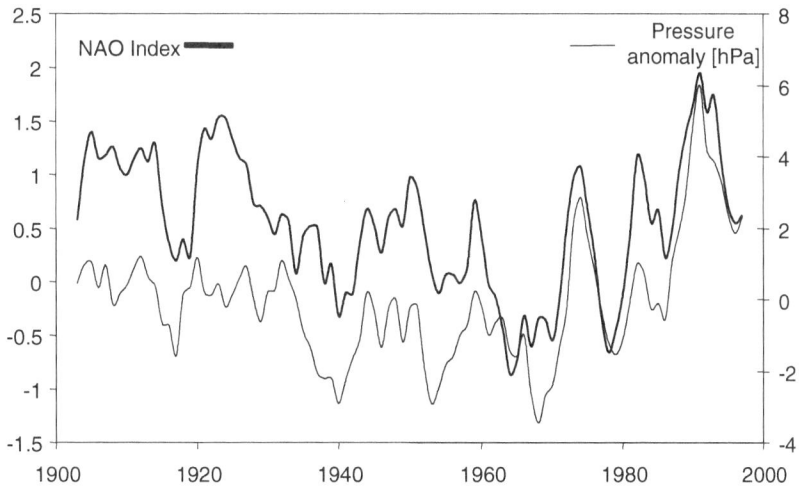

Figure 2

20th Century time series of the wintertime (DJF) index of the North Atlantic Oscillation (NAO) and surface pressure anomalies at Säntis (2,500 m above sea-level). A 5-point filter is used to eliminate high-frequency oscillations in the series (BENISTON and JUNGO, 2002).

SERREZE *et al.*, 1997) that a significant fraction of climatic anomalies observed on either side of the Atlantic are driven by the behavior of the NAO.

BENISTON (2000) has shown that temperature, moisture and pressure trends and anomalies at high elevations stand out more clearly than at lower levels, where boundary-layer processes, local site characteristics and urban effects combine to damp the large-scale climate signals. Climatic processes at high elevation sites can thus in many instances be considered to be the reflection of large-scale forcings, such as the NAO. These findings have been confirmed through numerical experimentation by GIORGI *et al.* (1997), who have underlined the altitudinal dependency of the regional atmospheric response to large-scale climatic forcings.

Figure 2 depicts the time series of the wintertime NAO index during the 20th century, and the associated surface wintertime (December-January-February; DJF) pressure anomalies in Zürich, based on the 30-year climatological average period 1961–1990; a 5-point filter has been applied to both curves in order to remove the higher-frequency fluctuations for the purposes of clarity. Average pressure values, even at a single site, can be considered to be a measure of synoptic-scale conditions influencing the Alpine region, as discussed in BENISTON *et al.* (1994). The pressure measured at Zürich or elsewhere, when averaged on a seasonal or longer time span, is therefore representative of the large-scale pressure field over Switzerland. The very close relationship between the two curves in Figure 2 highlights the subtle linkages between the large-scale NAO forcings and the regional-scale pressure response over Switzerland. When computed for 1901–1999, 56% of the observed pressure variance in Switzerland can be explained by the behavior of the NAO. From 1961–1999, this

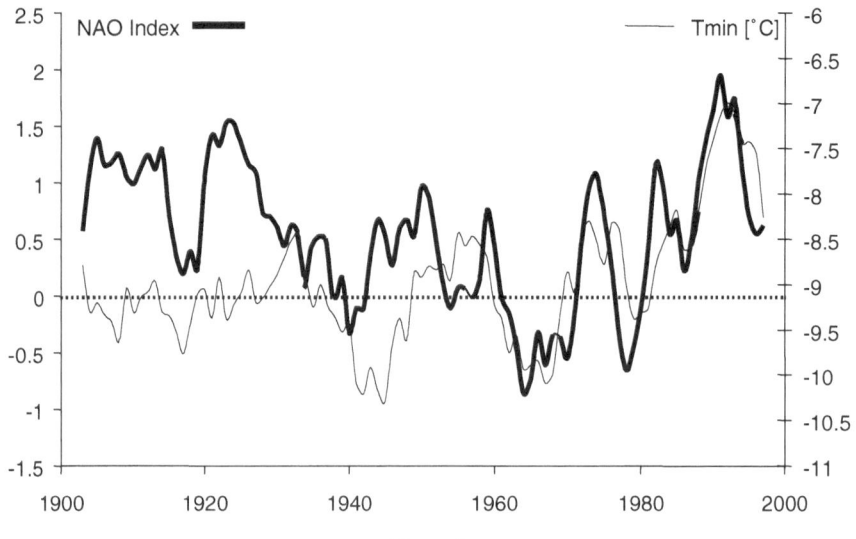

Figure 3

As Figure 2, except for DJF minimum temperature time series (BENISTON and JUNGO, 2002).

figure rises to 83%, which is considerable bearing in mind the numerous factors that can also determine regional pressure fields. WANNER et al. (1997) speculate that the persistent Alpine high pressure observed in the 1980s and early 1990s is linked to rising NAO index values through a northern shift of the upper-level jet stream associated with the polar front. When this occurs, the Alps lie to the right exit zone of the diverging jet streamlines, and are thus subject to mass influx and hence positive pressure tendencies.

Figure 3 illustrates the relation between the wintertime NAO index and the DJF temperature time series for Zürich, where both curves are smoothed as in Figure 2. As for pressure trends, the synchronous behavior between temperature and the NAO is striking, particularly in the second half of the 20th Century, where the minimum temperature variance which can be accounted for by the NAO fluctuations from 1961–1999 exceeds 72%.

A particular feature of the positive phase of the NAO index is that it is invariably coupled to anomalously low precipitation and milder than average temperatures, particularly from late fall to early spring, in southern and central Europe (including the Alps and the Carpathians), while the reverse is true for periods when the NAO index is negative. As an illustration of the impacts of NAO behavior on the seasonality and quantity of snow in the Alps, BENISTON (1997) has shown that periods with relatively low snow amounts are closely linked to the presence of persistent high surface pressure fields over the Alpine region during late Fall and through to early Spring. In his 1997 paper, Beniston discussed a number of recurring links between the NAO and winter temperature and precipitation characteristics; in general, strongly positive NAO leads to persistent high pressure over the alpine region, resulting in warmer than average temperatures and lower than average precipitation, because of the fact that Atlantic storms track much further to the north than when the NAO is less strongly positive. Since the mid-1980s and until 1996, the length of the snow season and the general quantities of snow amount decreased substantially in the Alps, as a result of pressure fields which were far higher and more persistent than at any other time during the 20th century, as shown in Figure 4, from BENISTON, (1997). Here, the pressure data is from Zürich which, though not in the same location as the two alpine sites, is nevertheless representative of a large-scale weather anomaly over the entire alpine region.

In order to highlight the possible relationships of high or low NAO index values with shifts in the frequency distributions of climate variables such as pressure, temperature and moisture, two thresholds for the wintertime NAO index have been chosen, namely the lower and upper 10 percentiles of the NAO index distribution during the 20th century (i.e., the 10% and 90% thresholds, which correspond roughly to index values around −1.5 and +2.0, respectively). These thresholds are representative of two highly contrasting synoptic regimes affecting the Alps, namely above-average pressure and associated positive temperature and negative moisture anomalies when the threshold is above the 90% level, and lower than average

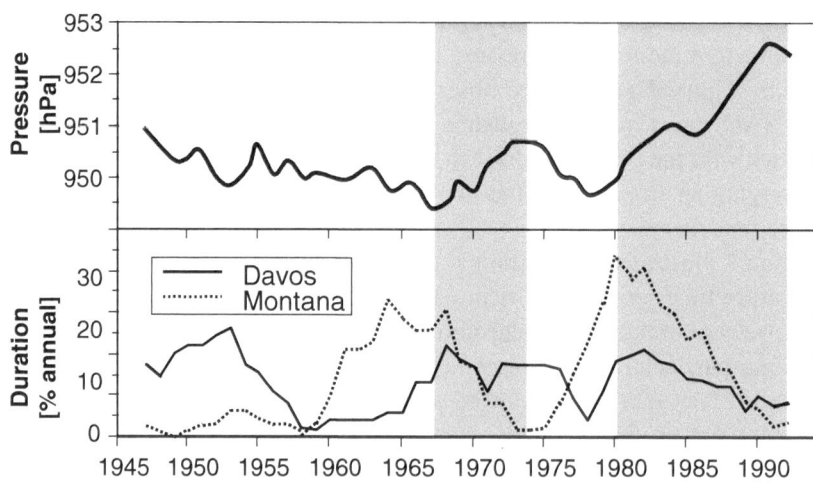

Figure 4

50-year time series of mean winter pressure (upper) and duration of winter snow cover at two alpine sites,
for a snow-depth threshold of 1 m (BENISTON, 1997).

pressure and its controls on temperature and humidity when the index is lower than the 10% level. The probability density functions (PDF) of pressure, maximum and minimum temperatures, and relative humidity, have been computed for periods where the NAO index is greater than the 90% threshold, and the temperature PDFs for winters where the NAO anomaly index is less than the 10% threshold.

Figure 5 illustrates the behavior of these pressure PDFs at Säntis (a high elevation site in NE Switzerland, located at 2,500 m above sea-level), computed for periods of the 20th century where the NAO index exceeds the upper threshold (90%), or is inferior to the lower threshold (10%). A substantial shift towards significantly higher pressures is observed for periods when the NAO index exceeds the 90% threshold; the PDF curve is translated to the right of the diagram, with a change in both the skewness and kurtosis of the distribution. Similar to the changes in pressure PDF, minimum temperatures exhibit a shift from the extreme low to the extreme high tails of the distribution (results not shown here). The changes are conspicuous at most observation sites in Switzerland. At Säntis, for example, the extreme low tails of the minimum temperature distribution disappear during periods of high NAO index, in favor of much warmer temperatures. Temperatures below −15°C at Säntis, which account for roughly 30% of the winters where the NAO index is below the 10% level, occur only 15% of the time in winter months that experience high NAO values. This implies that the periods with extreme cold conditions are reduced by 50%. Conversely, temperatures at the upper end of the distribution, for example above −5°C represent 12% of the winter days for low NAO values and 23% for high NAO values. The duration of milder temperatures at Säntis thus doubles when the NAO shifts from its low negative to its high positive phase.

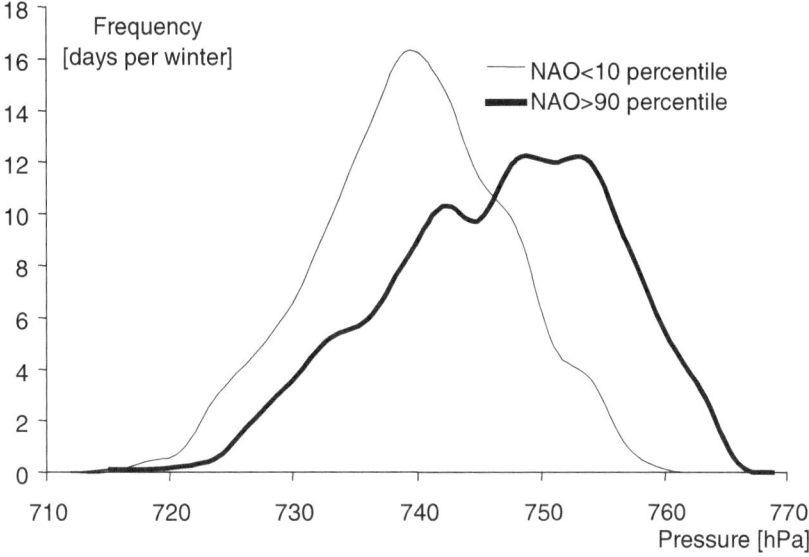

Figure 5
Probability density function of pressure at Säntis (2,500 m above sea-level) for periods when the negative and positive NAO index thresholds are exceeded (BENISTON and JUNGO, 2002).

Similar conclusions can be reached for the distribution of T_{max} (Fig. 6). This particular figure emphasizes the fact that the number of days in winter in which

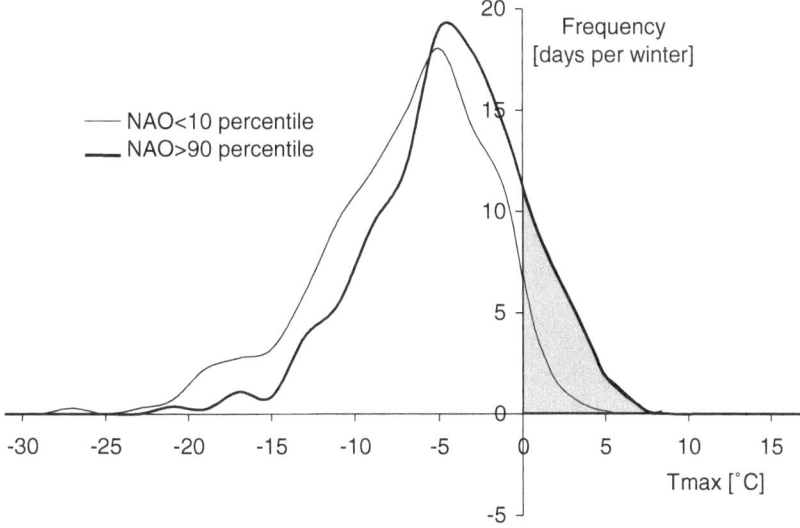

Figure 6
As Figure 5, except for maximum temperature PDF. Gray shading emphasizes PDF domain where temperatures exceed the freezing point (BENISTON and JUNGO, 2002).

maximum temperatures exceed the freezing point range from 10 days for the lowermost NAO 10-percentile to 25 days for the uppermost NAO 10-percentile. This obviously has implications for physical variables such as snow amount and duration, and biologically-relevant factors such as the start of the vegetation period.

Moisture and precipitation in the alpine region are also influenced by the behavior of the NAO. In the case of the negative index threshold, over 50% of the values recorded in winter exceed 90% relative humidity, while in the case of the positive threshold this level of relative humidity is exceeded only 35% during the winter months (results not shown here). There is thus a clear reduction in ambient moisture at high elevations.

In the last decade, considerable interest has focused on the North Atlantic Oscillation, and it is sometimes used as an empirical predictor for precipitation and temperature in regions where climatic variables are well correlated with the NAO index. This is clearly seen to be the case in Northern and Southern Europe, while Central Europe and the Alps are generally a "pivot" around which the forcing of the NAO is amplified with distance north or south. While at low elevations, the NAO signal may be weak or absent in the Alps, higher elevation sites are on the contrary sensitive to changes in NAO patterns (BENISTON et al., 1994; BENISTON and REBETEZ, 1996; HURRELL, 1995; HURREL and VAN LOON, 1997; GIORGI et al., 1997). The processes associated with periods when the NAO index exceeds the 90% level include frequent blocking episodes, where pressure fields over the Alpine area are high, and vertical circulations induce subsiding air with associated compression warming. Such circulations invariably generate positive temperature anomalies, and reductions in moisture and precipitation. In addition, diurnal warming at high elevations is enhanced by above-average sunshine, since there is a lowering of cloud amount and duration during periods of blocking high pressures. The reverse is generally true for periods when the NAO index is below the 10-percentile of its distribution. It could be argued that nocturnal cooling should also be stronger in a cloud-free atmosphere, thus leading to lower minimum temperatures; however, in complex terrain, radiative cooling at night will lead to down slope flow and accumulation of cold air in the valleys. The nocturnal cooling effect is, proportionally, not as strong at mountain summits such as Säntis as further down in the valleys.

Such anomalies are not only reflected in the means of the analyzed climatic variables, but also — and perhaps especially — in their extremes. The previous section has shown that there are clear links between strongly positive or negative modes of the NAO, and extremes of pressure, temperature, and moisture; high NAO values systematically shift the distributions from the lower extremes to the upper extremes.

Since the early 1970s, and until 1996, the wintertime NAO index has been increasingly positive, indicative of enhanced westerly flow over the North Atlantic. This has led to synoptic situations in recent decades which have been associated with abundant precipitation over Norway, as cyclonic tracks enter Europe relatively far to the north of the continent (HURRELL, 1995). Over the Alpine region, on the other hand, positive NAO indices have resulted in surface pressure fields that have been higher than at any time this century. Investigations by BENISTON *et al.* (1994) concluded that close to 25% of the pressure episodes exceeding the 965 hPa threshold recorded this century in Zürich (approximately 1030 hPa reduced sea-level pressure) occured in the period from 1980–1992, with the four successive years from 1989–1992 accounting for 16% of this century's persistent high pressure in the region.

Table 1 shows an analysis of mean wintertime values for minimum and maximum temperatures, relative humidity and precipitation at Säntis for four distinct periods of the 20th century, namely 1901–1999, 1950–1999, 1975–1999, and 1989–1999. In each case, the observed DJF mean is given, followed by the mean which would have occurred without the influence of highly-positive NAO index (beyond the 90-percentile threshold). The third column for each variable represents the bias which NAO index exceeding the 90-percentile has imposed on temperature and moisture variables. It is seen in this Table that the bias is relatively small when considering the entire 20th century (1901–1999), but then increases as one approaches the end of the 20th century. In the last decade, from 1989–1999, the bias for minimum temperatures exceeds 1°C. In the absence of the large forcings imposed by NAO index values greatrer than the 90-percentile, this decade would in fact have been slightly cooler in terms of minimum temperature than the average conditions observed between 1975 and 1999. Indeed, had there not been such a strong positive NAO forcing in the latter years of the 20th century, minimum temperatures would not have risen by almost 1.5°C (decadal mean for the 1990s minus century mean from 1901–1999) but by less than 0.5°C, as seen in the second column of the minimum temperature analyses. The bias imposed by strongly-positive NAO thresholds on maximum temperatures follows the same trends, but is not as high as for minimum temperatures; even in the absence of the NAO forcing, maximum temperatures would have risen substantially in the latter part of the 20th century.

In terms of moisture, relative humidity has decreased in winter, with a bias of close to 10% in the period 1989–1999, resulting from the NAO forcing; mean DJF relative humidity would have otherwise remained relatively constant throughout the century. Precipitation is also seen to be considerably marked by the NAO forcing in the last decade of the 20th century, with a substantial drop of 20% of winter precipitation linked to the high and persistent NAO index recorded during this period.

Because over half of the NAO index values exceeding the 90-percentile have occurred since 1985, it may be concluded that the NAO is a significant driving factor for the climatic anomalies observed in recent years in the Alps. In particular, the highly anomalous nature of temperatures and their extremes that have been observed and discussed *inter alia* by JUNGO and BENISTON (2001) are largely explained by the large-scale influence on regional climate generated by the recent trends of the NAO. Removal of the biases imposed by high NAO episodes would have resulted in relatively modest increases in minimum temperatures and reduced rates of maximum

Table 1

Mean winter (December-January-February, or DJF, averaged) values for minimum temperature, maximum temperature, relative humidity and precipitation for different periods of the 20th Century. The first column represents the mean recorded during each period; the second column represents the mean that would have been observed in the absence of NAO forcing beyond the 90-percentile threshold; the third column represents the bias imposed by NAO index values beyond this threshold (difference between the first two columns).

Period	Tmin (°C)			Tmax (°C)		
	Observed Mean	Mean without NAO≥90% level	Bias	Observed Mean	Mean without NAO≥90% level	Bias
1901-1999	-8.95	-9.03	0.08	-4.33	-4.42	0.10
1950-1999	-8.66	-8.74	0.13	-3.36	-3.55	0.19
1975-1999	-8.20	-8.38	0.18	-2.74	-2.95	0.21
1989-1999	-7.50	-8.56	1.06	-2.15	-2.50	0.34

Period	Relative Humidity (%)			Precipitation (mm/day)		
	Observed Mean	Mean without NAO≥90% level	Bias	Observed Mean	Mean without NAO≥90% level	Bias
1901-1999	75.70	76.06	-0.36	7.04	7.10	-0.06 (-0.79%)
1950-1999	73.44	66.76	6.69	6.42	5.81	0.60 (+9.41%)
1975-1999	69.14	69.71	-0.56	6.73	6.90	-0.17 (-2.54%)
1989-1999	65.82	75.34	-9.52	8.12	9.76	-1.63 (-20.1%)

temperature warming, thus leading to Alpine-scale warming comparable to global-average warming (JONES and MOBERG, 2003).

4. Future Climatic Change in the Alps

General Circulation Model (GCM) simulations of the response of global climate to enhanced GHG concentrations have been a key focus of the IPCC reports (IPCC, 1996; 2001). Model simulations point to a warming which on a global average is in the range of 1.5–5.8°C in 2100, depending on the scenario used. Scenarios of greenhouse-gas trajectories into the future are a complex mix of social, economic, political, and technological forecasts for the 21st century. However, the complexity and mutual interdependency of mountain environmental and socio-economic systems pose significant problems for climate impacts studies (BENISTON *et al.*, 1997). The assessment of current and future trends in regional climate is limited by the current spatial resolution of General Circulation Models (GCM) which is generally too crude to adequately represent the orographic detail of most mountain regions. On the other hand, most impacts research requires information at fine spatial resolution where the regional detail of topography or land-cover is an important determinant in the response of natural and managed systems to change. Since the mid-1990s, the scaling problem related to complex topography has been addressed through regional modeling techniques, pioneered by GIORGI and MEARNS (1991), and through statistical-dynamical downscaling techniques (e.g., ZORITA and VON STORCH, 1999).

So-called "nested" approaches to regional climate simulations, whereby large-scale data or GCM outputs are used as boundary and initial conditions for regional climate model (RCM) simulations, have been applied to future climatic change in the course of the 21st century (GIORGI and MEARNS, 1999) over a given geographical area. The technique is applied to specific periods in time ("time slices" or "time windows"). GCM results for a given time window include the long-term evolution of climate prior to the chosen time window, based on an incremental increase of greenhouse gases over time. The nested modeling approach represents a trade-off between decadal- or century-scale, high resolution simulations that are today unattainable, even with the most sophisticated computational resources and relying only on coarse resolution results provided by long-term GCM integrations. When driven by analyses and observations, RCMs generally simulate a realistic structure and evolution of synoptic events. Model biases with respect to observations are in the range of a few tenths to a few degrees for temperature and 10–40% for precipitation. These biases tend to decrease with increasing resolution or decreasing RCM domain size.

Although the method has a number of drawbacks, in particular the fact that the nesting is "one-way" (i.e., the climatic forcing occurs only from the larger to the finer scales and not *vice versa*), RCMs may in some instances improve regional detail of climate processes, particularly when they are driven by measured boundary conditions. This can be an advantage in areas of mountainous terrain, where for example topographically-enhanced precipitation may represent a significant fraction of annual or seasonal rainfall in a particular mountain region. Such improvements are related to the fact that RCM simulations capture the regional detail of forcing elements like topography or large lakes, and the local forcings of such features on regional climate processes, in a more realistic manner than GCMs (BENISTON, 2000). GCM-driven RCMs tend to produce less convincing results, however, which is to be expected since GCMs can generate unreliable results in terms of storm tracks and precipitation belts, for example. As a result, the errors of a GCM will tend to propagate into an RCM. In the example illustrated in Figure 7, results for precipitation and temperature have been compiled by GIORGI *et al.* (1994) for an RCM operating at 50 km resolution nested into a GCM with a 300 km resolution.

The salient differences observed in this illustration are directly related to the enhanced RCM resolution and to the improved regional detail (Fig. 8). In this example, predictions of precipitation changes in Europe show a decrease in a warmer climate according to the GCM simulations and, conversely, an increase in the RCM. This can be explained by the fact that the GCM grid poorly resolves European mountain ranges such as the Alps, the Pyrenees, or the Carpathians, while the RCM is capable of better representing the mountains and thus the enhancement of precipitation which they generate. In other regions, discrepancies between the two climate models are not so important, in part because the orographic effect is either

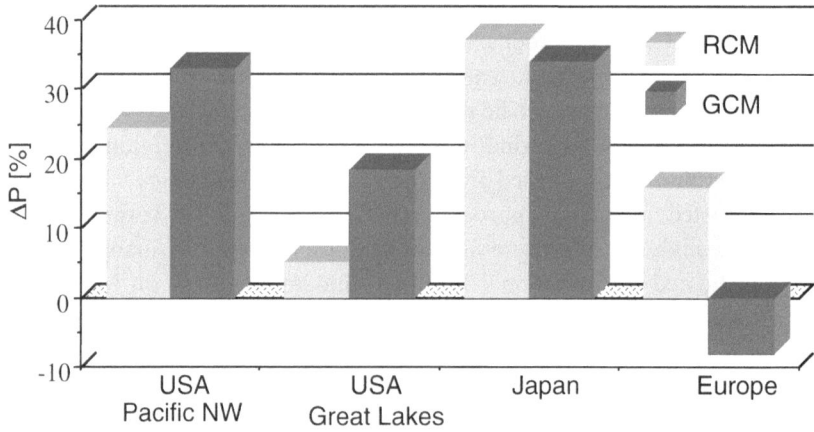

Figure 7

Global (GCM) and regional (RCM) climate model simulations of precipitation change between current climatic conditions and future climate (doubled atmospheric CO_2 concentrations) for four regions in the world (BENISTON, 2000).

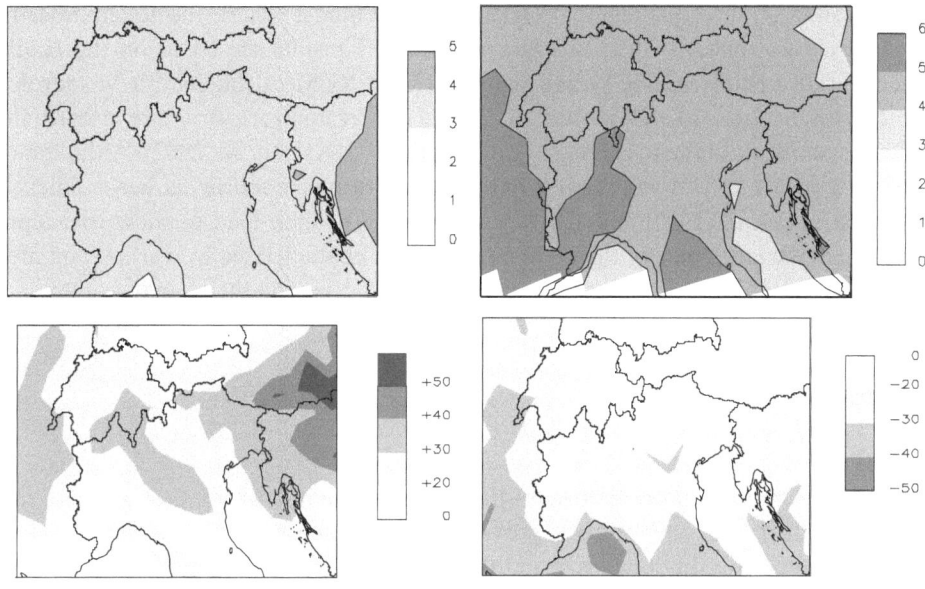

Figure 8

RCM results for climatic change in the Alpine region in the period 2071–2100 compared to the climatological average period 1961–1990. Upper left: Change in mean winter (DJF) temperatures (°C); Upper Right: Change in mean summer (JJA) temperatures (°C). Lower left: Change in mean winter (DJF) precipitation (%); Lower right: Change in mean summer (JJA) precipitation (%).

not the dominant factor for precipitation, or because the mountain ranges involved are sufficiently extensive to be adequately resolved in the GCM.

Over time, the increase in spatial resolution of RCMs, that has been enabled by the rapid evolution of computational resources, has enhanced the understanding of regional climate processes and the assessment of the future evolution of regional weather patterns influenced by a changing global climate. MARINUCCI et al. (1995) tested the nested GCM-RCM technique at a 20 km resolution to assess its adequacy in reproducing the salient features of contemporary climate in the European Alps, while ROTACH et al. (1997) repeated the numerical experiments for a scenario of enhanced greenhouse-gas forcing. Over the past five years, RCM spatial resolution has continually increased, partially as a response to the needs of the impacts community. Currently, detailed simulations with 5 km or even 1 km grids are used to investigate the details of precipitation in relation to surface runoff, infiltration, and evaporation (e.g., ARNELL, 1999; BERGSTRÖM et al., 2001), extreme events such as precipitation (FREI et al., 1998), and damaging wind storms (GOYETTE et al., 2001). In its Third Assessment Report, the IPCC (2001) has extensively used the nested modeling technique in an attempt to improve regional climate information.

As an example of the regionalization of global model results, the nested modeling technique is applied to the European Alps. These results are based on the Hadley Centre (UK) HadCM3 GCM and the HIRHAM RCM of the Danish Meteorological Institute, used in the context of an EU 5th Framework Program project that focuses on future climatic change in Europe (CHRISTENSEN et al., 2002). Although the RCM grid is a relatively coarse 50 km, the results confirm earlier studies by MARINUCCI et al. (1995) and ROTACH et al. (1997) which tend to show that alpine climate in the latter part of the 21st century will be characterized by warmer and more humid conditions in winter, and much warmer and drier conditions in the summer, as illustrated in Figure 8.

5. Conclusions

Mountains are unique features of the Earth system in terms of their scenery, their climates, their ecosystems; they provide key resources for human activities well beyond their natural boundaries; and they harbor extremely diverse cultures in both the developing and the industrialized world. The protection of mountain environments against the adverse effects of economic development should be a priority for both today's generation and the generations to come (BENISTON, 2000).

The assessment of climatic change and of its related impacts in mountain regions has been shown to be particularly difficult because of the complexity of a number of interrelated factors in regions where topography is a dominant feature of the environment. However, despite numerous uncertainties and issues that still need to be addressed, there is today a large consensus concerning the very real threat which abrupt global warming may pose to a wide range of environmental, social and economic systems both globally and regionally such as in the Alps. The IPCC (1996; 2001) has been instrumental in providing the state-of-the-art information on climatic change and its environmental and economic consequences, so that while science can continue to refine its predictions for the future, our current understanding of the system and its evolution is sufficient to justify international action for reducing the risks related to climatic change and to define strategies for rapidly adapting to change.

In facing up to climatic change, it will be necessary to plan in terms of decades and centuries. Many of the impacts that can be expected on mountain environments related to the amplitude and speed of change may not become unambiguously apparent for several generations. The policies and decisions related to pollution abatement, climatic change, deforestation or desertification would provide opportunities and challenges for the private and public sectors. A carefully selected set of national and international responses aimed at mitigation, adaptation and improvement of knowledge can reduce the risks posed by environmental change to water resources and natural hazards.

REFERENCES

ARNELL, N. (1999), *The Effect of Climate Change on Hydrological Regimes in Europe* Global Environ. Change *9*, 5–23

BARRY, R.G., *Past and potential future changes in mountain environments; A review*. In (Beniston, M., ed. Routledge Publishing Company, London and New York 1994), *Mountain Environments in Changing Climates* pp. 3–33.

BECKER, A. and BUGMANN, H. (1997), *Predicting Global Change Impacts on Mountain Hydrology and Ecology: Integrated Catchment Hydrology/Altitudinal Gradient Studies* IGBP Report *43*, Stockholm.

BENISTON, M. (1997), *Variations of Snow Depth and Duration in the Swiss Alps over the last 50 Years: Links to Changes in Large-scale Climatic Forcings*, Climatic Change *36*, 281–300.

BENISTON, M. (2000), *Environmental Change in Mountains and Uplands*. (Arnold/Hodder Publishers, London, UK, and Oxford University Press, New York, USA), 172 pp.

BENISTON, M. and JUNGO, P. (2002), *Shifts in the Distributions of Pressure, Temperature and Moisture in the Alpine Region in Response to the Behavior of the North Atlantic Oscillation*, Theor. Appl. Clim. *71*, 29–42.

BENISTON, M. and REBETEZ, M. (1996), *Regional Behavior of Minimum Temperatures in Switzerland for the period 1979–1993*, Theor. Appl. Clim. *53*, 231–243.

BENISTON, M., DIAZ, H. F., and BRADLEY, R. S. (1997), *Climatic Change at High Elevation Sites; A Review*. Climatic Change *36*, 233–251.

BENISTON, M., REBETEZ, M., GIORGI, F, and MARINUCCI, M.R. (1994), *An Analysis of Regional Climate Change in Switzerland*, Theor. Appl. Clim. *49*, 135–159.

BERGSTRÖM, S., CARLSSON, B., GARDELIN, M., LINDSTRÖM, G., PETTERSSON, A., and RUMMUKAINEN, M. (2001), *Climate Change Impacts on Runoff in Sweden—Assessments by Global Climate Models, Dynamical Downscaling and Hydrological Modelling*, Climate Research *16*, 101–112.

CHRISTENSEN, J.H., CARTER, T.R., and GIORGI, F. (2002), *PRUDENCE Employs New Methods to Assess European Climate Change*, EOS, Trans. Am. Geophy. Union *83*, 147.

DIAZ, H. F. and BRADLEY, R. S. (1997), *Temperature Variations during the Last Century at High Elevation Sites*, Climatic Change *36*, 253–279.

EKHART, E. (1948), De *la structure de l'atmosphère dans la montagne*, La Météorologie *3*, 3–26.

FLOHN, H. (1968), *Contributions to a Meteorology of the Tibetan Highlands*, Atmos. Phys. Paper, *130*, Dept.of Atmospheric Sciences, Colorado State University, Fort Collins. 120 pp.

FREI, C., SCHÄR, C., LÜTHI, D., and DAVIES, H.C. (1998), , Geophys. Res. Lett. *25*, 1431–1434.

GIORGI, F. and MEARNS, L.O. (1991), *Approaches to the Simulation of Regional Climate Change: A Review*. Rev. Geophy *29*, 191–216.

GIORGI, F. and MEARNS, L. O. (1999), *Regional Climate Modeling Revisited*. J. Geophys. Res. *104*, 6335–6352.

GIORGI, F., BRODEUR, C.S., and BATES, G.T. (1994), *Regional Climate Change Scenarios over the United States Produced with a Nested Regional Climate Model*. J. Clim. *7*, 375–399.

GIORGI, F., HURRELL, J., MARINUCCI, M., and BENISTON, M. (1997), *Height Dependency of the North Atlantic Oscillation Index. Observational and Model Studies*, J. Clim. *10*, 288–296.

GOYETTE, S., BENISTON, M., JUNGO, P., CAYA, D., and LAPRISE, R. (2001), *Numerical Investigation of an Extreme Storm with the Canadian Regional Climate Model: The Case Study of Windstorm Vivian, Switzerland, February 27, 1990*, Climate Dynamics *18*, 145–168.

GRASSL, H., *The Alps under local, regional and global pressures*. In *Mountain Environments in Changing Climates* (Beniston, M., ed. Routledge Publishing Company, London and New York 1994), pp. 34–41.

HURRELL, J. W. (1995), *Decadal Trends in the North Atlantic Oscillation Regional Temperatures and Precipitation*, Science *269*, 676–679.

HURRELL, J. W., and VAN LOON, H. (1997), *Decadal Variations in Climate Associated with the North Atlantic Oscillation*, Climatic Change *36*, 301–326.

IPCC (1996), Clim*ate Change. The IPCC Second Assessment Report*. Cambridge University Press, Cambridge and New York. Volumes I (Science), II (Impacts) and III (Socio-economic implications).

IPCC (2001), *Climate Change. The IPCC Third Assessment Report*. Cambridge University Press, Cambridge and New York.

JONES, P.D. and MOBERG, A. (2003), *Hemispheric and Large-scale Surface Air Temperature Variations: An Extensive Revision and an Update to 2001,* J. Climate *16,* 206–223.

JUNGO, P., and BENISTON, M. (2001), *Changes in 20th Century Extreme Temperature Anomalies at Swiss Climatological Stations Located at Different Latitudes and Altitudes,* Theor. Appl. Clim. *69,* 1–12.

KAPOS, V., RHIND, J., EDWARDS, M., RAVILIOUS, C., and PRICE, M. (2000), *Developing a map of the world's mountain forests.* In *Forests in a Sustainable Mountain Environment.* (Price, M.F. and Butt, N., eds.),(CAB International, Wallingford 2000)

KLÖTZLI, F. (1994), *Vegetation als Spielball naturgegebener Bauherren,* Phytocoenologia *24,* 667–675.

MARINUCCI, M. R., GIORGI, F., BENISTON, M., WILD, M., TSCHUCK, P., and BERNASCONI, A. (1995), *High Resolution Simulations of January and July Climate over the Western Alpine Region with a Nested Regional Modeling System,* Theor. Appl. Clim. *51,* 119–138.

MEYBECK, M., GREEN, P., and VÖRÖSMARTY, C. (2001), *A New Typology for Mountains and other Relief Classes: An Application to Global Continental Water Resources and Population Distribution,* Mountain Res. Develop *21,* 34–45.

OZENDA, P. (1985), *La végétation de la chaîne alpine dans l'espace montagnard européen,* Masson, Paris, 344 pp.

QUEZEL, P. and BARBERO, M. (1990), *Les forêts méditerranéennes: problèmes posés par leur signification historique, écologique et leur conservation,* Acta Botanica Malacitana *15,* 145–178.

RAMEAU, J.C., MANSION, D., DUMÉ, G., LECOINTE, A., TIMBAL, J., DUPONT, P.,and KELLER, R. (1993), *Flore Forestière Française, Guide Ecologique Illustré.* Lavoisier TEC and DOC Diffusion, Paris, 2419 pp.

ROGERS, J. C. (1997), *North Atlantic Storm Track Variability and its Association to the North Atlantic Oscillation and Climate Variability of Northern Europe,* J. Climate *10,* 1635–1647.

ROTACH, M., WILD, M., TSCHUCK, P., BENISTON, M., and MARINUCCI, M. R. (1997), *A Double CO$_2$ Experiment over the Alpine Region with a Nested GCM-LAM Modeling Approach,* Theor.Appl. Clim. *57,* 209–227.

SCHNEIDER, U. (1992), *Die Verteilung des troposphärischen Ozons in Bayrischen Nordalpenraum,* Ph.D. Dissertation, University of Mainz, Germany.

SERREZE, M. C., CARSE, F., BARRY, R. G., and ROGERS, J. C. (1997), *Icelandic Low Cyclone Activity: Climatological Features, Linkages with the North Atlantic Oscillation, and Relationships with Recent Changes in the Northern Hemisphere Circulation,* J. Climate *10,* 453–464.

SMIDT, S. (1991), *Messungen nasser Freilanddepositionen der Förstlichen Bundesversuchsanstalt,* FBVA-Berichte, ISSN 1013-0713 50, Nasse Deposition, Austria.

WADE, L. K. and MCVEAN, D. N. , *Mt. Wilhelm Studies; The Alpine and Subalpine Vegetation* (Australian National University, Canberra 1969), 225 pp.

WANNER, H., RICKLI, R., SALVISBERG, E., SCHMUTZ, C., and SCHUEPP, M. (1997), *Global Climate Change and Variability and its Influence on Alpine Climate—Concepts and Observations,* Theor. Appl. Climatology *58,* 221–243.

WARDLE, P. (1973), *New Zealand Timberlines,* Arctic and Alpine Res. *5,* 127–136.

ZORITA, E. and VON STORCH H. (1999), *The Analog Method—A Simple Statistical Downscaling Technique: Comparison with more Complicated Methods,* J. Climate *12,* 2474–2489.

(Received May 27, 2003, accepted January 21, 2004)
Published Online First: May 25 2005

To access this journal online:
http://www.birkhauser.ch

B. Weather

Pure appl. geophys. 162 (2005) 1609–1626
0033–4553/05/091609–18
DOI 10.1007/s00024-005-2685-8

❙ Pure and Applied Geophysics

Persistent, Widespread, and Strongly Absorbing Haze Over the Himalayan Foothills and the Indo-Gangetic Plains

V. RAMANATHAN,[1] and M.V. RAMANA[1]

Abstract—We examine the impact of the Atmospheric Brown Clouds on the direct radiative forcing of the Himalayan foothills and the Indo-Gangetic Plains (IGP) regions, home for over 500 million S. Asians. The NASA-Terra MODIS satellite data reveal an extensive layer of aerosols covering the entire IGP and Himalayan foothills region with seasonal mean AODs of about 0.4 to 0.5 in the visible wavelengths (0.55 micron), which fall among the largest seasonal mean dry season AODs for the tropics. We show new surface data which reveal the presence of strongly absorbing aerosols that lead to a large reduction in solar radiation fluxes at the surface during the October to May period. The three-year mean (2001 to 2003) October to May seasonal and diurnal average reduction in surface solar radiation for the IGP region is about 32 (± 5) W m^{-2} (about 10% of TOA insolation or 20% of surface insolation). The forcing efficiency (forcing per unit optical depth) is as large as -27% (note that the forcing is negative) of top-of-atmosphere (TOA) solar insolation, and exceeds the forcing efficiency that has been observed for other polluted regions in America, Africa, East Asia, and Europe. General circulation model sensitivity studies suggest that both the local and remote influence of the aerosol induced radiative forcing is to strengthen the lower atmosphere inversion, stabilize the boundary layer, amplify the climatological tendency for a drier troposphere, and decrease evaporation. These aerosol-induced changes could potentially increase the life times of aerosols, make them more persistent, and decrease their single scattering albedos, thus potentially leading to a detrimental positive feedback between aerosol concentrations, aerosol forcing, and aerosol persistence. In addition, both the model studies and observations of pan evaporation suggest that the reduction in surface solar radiation may have led to a reduction in surface evaporation of moisture. These results suggest the vulnerability of this vital region to air pollution related direct and indirect (through climate changes) impacts on agricultural productivity of the region.

Key words: Aerosol and particles, anthropogenic effects, transmission and scattering of radiation, aerosol radiative forcing efficiency, dimming of the surface, pan evaporation.

1. Introduction

The Indo-Gangetic Plains (IGP) is a vast stretch of land in S. Asia extending about 1600 km in length and 400 km in breadth, cutting across Pakistan, India, Nepal, and Bangladesh (ABROL *et al.*, 2002), and houses about one third of the population of S. Asia. IGP is a major region of food production for S. Asia since it is

[1] Center for Atmospheric Sciences, Scripps Institute of Oceanography (SIO), University of California in San Diego, La Jolla, 92037, U.S.A. E-mail: vram@fiji.ucsd.edu

remarkably flat and includes the largest fluvial plains of the world, fed by the Indus and the Ganges rivers. The northern boundary of IGP is the Himalayan foothills and to the south is the Indian peninsula. Between 1901 to 1991, the population of the Indian sector of IGP increased three-fold, from 105 million to 339 million (SINGH et al., 2002). The region has witnessed impressive growth rates in agricultural productivity since the dawn of the industrial era, accompanied by significant land-use modification and changes. However, the agricultural productivity has been decreasing since the glory days of the green revolution in the 1970s. For example, growth rates for rice in the Indian sector of IGP decreased from about 5%–9% in the 1970s to as low as 1% in the 1990s (ABROL et al., 2002). There is now a real concern about the sustainability of the region's ability to support the population due to loss of biodiversity, soil degradation, and air pollution (ABROL et al., 2002), which brings us to the present study.

IGP is also one of the Asian regions with a thick layer of anthropogenic absorbing aerosols (Fig. 1), alternately referred to as Atmospheric Brown Clouds (RAMANATHAN and CRUTZEN, 2003). The Indian Ocean Experiment (INDOEX) provided well-documented evidence for the long-range transport of manmade aerosols and how it transforms the so-called urban haze into regional and continental scale brown clouds (RAJEEV et al., 2000; RAMANATHAN et al., 2001a; LELIEVELD et al., 2001; RAMANATHAN and RAMANA, 2003). Characteristics of a brown cloud are

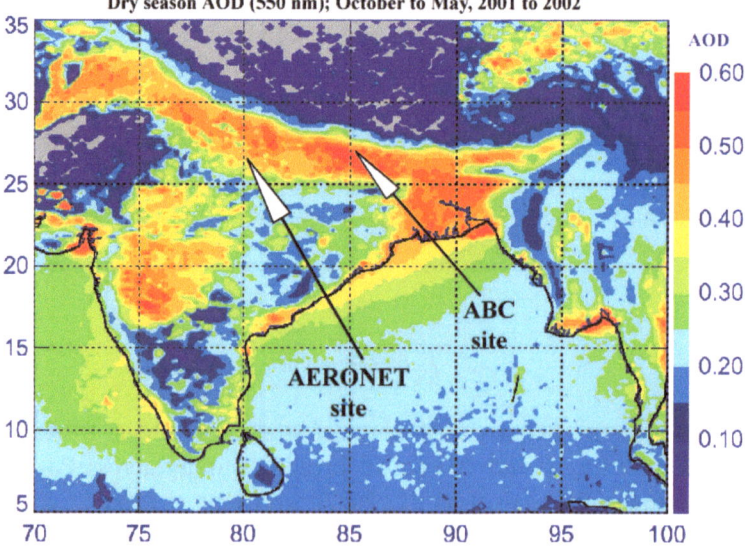

Figure 1

Regional distribution of aerosol optical depth (AOD) at 550 nm, averaged over October to May for 2001 and 2002. The data were obtained from MODIS instrument onboard NASA's Terra satellite. The figure also shows the location of surface sites, namely; ABC site (Kathmandu) and AERONET site (Kanpur).

the presence of carbonaceous (black and organic carbon) aerosols, large reductions in the flux of solar radiation reaching the earth's surface over large geographic areas and a correspondingly large increase in atmospheric absorption of solar radiation due to the black carbon in the brown cloud. The climatic impacts of the surface cooling and atmospheric heating are multifaceted, as shown by CHUNG et al. (2002), CHUNG and RAMANATHAN (2003) and KOREN et al. (2004). The calculated effects for the S. Asian dry season (October to May) include a large surface cooling over regions directly underneath the aerosol layer and a surface warming elsewhere due to large-scale horizontal advection effects (KRISHNAN and RAMANATHAN, 2002).

The recently launched NASA-TERRA satellite not only confirmed INDOEX findings, but also revealed the presence of a widespread pollution haze layer downwind of many other continents, including Africa, S. America, N. America, and Europe (RAMANATHAN et al., 2001b; KAUFMAN et al., 2002). The focus in this study is the S. Asian region. In particular we emphasize the IGP region and the Himalayan foothills because AODs in this region are among the largest found in S. Asia and as we will shortly show, are persistent throughout the winter and spring season until May. Figure 1 shows the regional distribution of aerosol optical depth averaged over 2001 and 2002 for the October to May period (the so-called dry season) which shows very high AODs over the IGP region and the Himalayan foothills (e.g., see over Nepal). It also indicates the locations of surface observations used in this study.

The major objective of this paper is to examine the impact of the brown clouds on the solar radiation budget of the IGP and the Himalayan foothill region where we have made surface-based observations from December 2002 to April 2003. We will compare the direct surface forcing of aerosols over the Himalayan foothill region with surface forcing observations from other regions of the world. In addition we will estimate the IGP forcing from the available aerosol data over this region. This is followed by a description of general circulation model results that reveal how the aerosol forcing over the IGP and rest of S. Asia impact the IGP climate.

The study uses surface data described by RAMANA et al. (2004) over the Himalayan foothills from January to February 2003, and satellite AOD data for two years from 2001 to 2002 as a part of the project Atmospheric Brown Clouds (ABC). The ABC is built upon the Indian Ocean Experiment (INDOEX) completed during 1996–1999. The ABC is an international research program initiated by the United Nations Environment Programme (UNEP) with financial support from the National Oceanic and Atmospheric Administration (NOAA), the National Science Foundation (NSF), and participating member nations to ascertain the major environmental challenges facing the Indo-Asia-Pacific region, specifically the environmental consequences of rising air pollution levels due to rapid industrialization and population growth (RAMANATHAN and CRUTZEN, 2003). The project includes the construction of 12–15 aerosol-climate observatories in the Indo-Asian-Pacific region to perform the long-term measurements of solar radiation and aerosols (http://www-abc-asia.ucsd.edu). Towards this goal, the first direct surface observations of aerosol (a): Time series of spectral AOD at KCO during INDOEX (Satheesh et al., 2002)

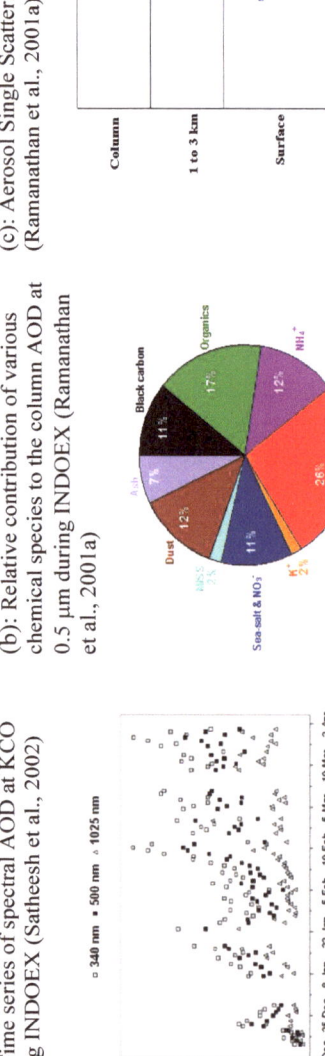

(b): Relative contribution of various chemical species to the column AOD at 0.5 μm during INDOEX (Ramanathan et al., 2001a)

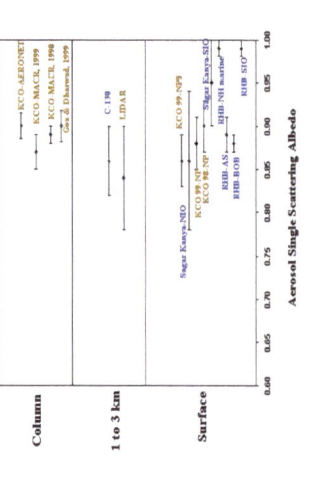

(c): Aerosol Single Scattering Albedo during INDOEX (Ramanathan et al., 2001a)

(d): Typical aerosol profiles during INDOEX (Ramanathan et al., 2001a)

(e): Comparison of clear-sky global measured fluxes with MACR fluxes in the 0.4 - 0.7 μm spectral region at the surface (Ramanathan et al., 2001a)

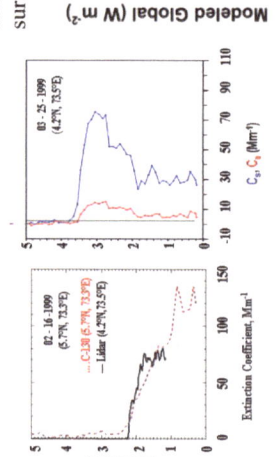

(f): Direct Aerosol forcing from radiometers during INDOEX (Satheesh and Ramanathan, 2000)

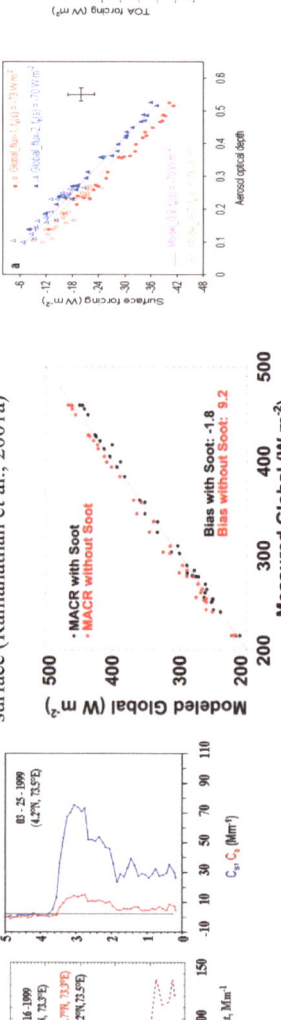

radiative forcing have been carried out during the winter of 2003 in the Nepalese Himalayan region.

The approach for integrating surface observations with satellite observations is similar to those developed under the INDOEX project. INDOEX was conducted during the Northern Hemispheric winter monsoon season over the tropical Indian Ocean and a detailed description of the experiment was given by RAMANATHAN et al. (1995). The purpose of INDOEX was to study the effect that South Asian aerosols and trace gases have on the radiative and chemical processes occurring over the North Indian Ocean during the northeast Asian monsoon.

2. Estimation of the Aerosol Radiative Forcing: The INDOEX Approach

The Indian Ocean Experiment (INDOEX) was conducted over the tropical Indian Ocean for a period of 4 years (1995 to 1999) with an intensive field phase in Feb.–March 1999. Simultaneous measurements were made during INDOEX from ships, aircraft, satellites, and surface stations. The data obtained from the INDOEX campaign were integrated to yield the aerosol forcing from observations as illustrated in Figure 2, which shows how the coupling of aerosol optical depth, chemical and physical properties of aerosols, and the data from the radiometric instruments yield information on aerosol radiative forcing. Time series of AOD (Fig. 2a) at the surface (The Kasshidhoo Climate Observatory in Maldives; see SATHEESH and RAMANA-THAN, 2000) reveals the variability of column aerosol loading and reveals how aerosol amounts build with the progression of the dry season. Chemical analyses of filters collected at KCO as well as from aircraft are used to infer the contribution of various species to the column AOD (Fig. 2b). The magnitude of solar absorption of an aerosol particle can be inferred from the data in Figure 2b and this is validated by direct measurements of single scattering albedos (SSA) shown in Figure 2c. The quantity (1-SSA) is the fraction of solar radiation absorbed by the aerosol. For the INDOEX region, SSA varied from about 0.85 to 0.95, which indicates a highly absorbing aerosol. The next important information is the vertical distribution of the aerosols shown in Figure 2d. The direct aerosol forcing is very sensitive to the vertical distribution, particularly when the aerosol concentrations peak aloft as shown in Figure 2d. During INDOEX, roughly 30% of the profiles had peak

◄

Figure 2

Figure showing how data from the INDOEX integrated to yield direct radiative forcing: (a) time series of aerosol optical depth (AOD), (b) chemical composition of aerosols to the column AOD, (c) surface and columnar measurements of aerosol single scattering albedo, (d) vertical distribution of extinction coefficient, scattering coefficient (Cs in blue curve) and absorption coefficient (Ca in red curve), (e) comparison of measured fluxes with the MACR predicted fluxes, and (f) direct aerosol radiative forcing at the surface and at the TOA during INDOEX (RAMANATHAN et al., 2001a).

concentrations near the surface, while 70% had peak concentrations around 3 km (Fig. 2d). In the presence of low clouds (between 1 km and 3 km), the elevated profiles (right hand panel of Fig. 2d) led to column (surface-atmosphere) heating because the absorbing aerosols absorbed the solar radiation reflected by clouds, whereas, when the aerosols are confined below low clouds, they led to net cooling. The data shown in Figures 2a to 2d were incorporated in a Monte Carlo Aerosol-Cloud Radiation (MACR) model (PODGORNY *et al.*, 2000) and the simulated solar radiation fluxes were validated with observed fluxes (using Pyranometers and Biospherical instruments) as in Figure 2e. In addition, the observed surface and TOA forcing (reduction of absorbed solar radiation at the surface and at the top-of-the atmosphere) by aerosols were compared with computed forcing as in Figure 2f. The excellent agreement with observed fluxes and forcing validated the MACR model, which was then used to estimate direct forcing for clear and cloudy skies (see RAMANATHAN *et al.*, 2001a for details).

The aerosol forcing efficiency (forcing per unit optical depth at 0.5 micron) at the surface ranged from -70 to -75 W m^{-2} and about -22 to -23 W m^{-2} at the TOA

Figure 3
Haze over the Himalayan region: (a) at Nagarkot on Dec. 19, 2002 (looking towards west); (b) looking at the Himalayas on Dec. 19, 2002 taken from Nagarkot; (c) haze over Kathmandu on Jan. 17, 2003; and (d) looking at the Himalayas on Feb. 3, 2003, taken from Kathmandu (just after rains).

(SATHEESH and RAMANATHAN, 2000; PODGORNY *et al.*, 2000; CONANT, 2000; VALERO *et al.*, 1999; MEYWERK and RAMANATHAN, 1999, 2002).

3. Aerosol Radiative Forcing over the Himalayan Foothills

We initiated surface measurements at Nepal during the winter of 2003 as part of the project ABC. The measurements were made at two sites, one at Kathmandu valley (27.67°N 85.31°E) at an elevation of about 1.3 km and another at Nagarkot (27.71°N 85.52°E) located at an altitude of 1.97 km above sea level and 32 km east of Kathmandu. The measurements at Kathmandu and Nagarkot included AODs at multiple wavelengths, broadband (280–2800 nm) global radiative fluxes at the surface, total column water, and column ozone. Aerosol samples, aerosol absorption and scattering measurements were made only at Kathmandu valley. In addition, Micro Pulse Lidar (at 523 nm) was used at the Kathmandu site for vertical profiles of aerosols. Figure 3 illustrates through photographs the buildup of the brownish haze over Nepal during the observation period. The details of the measurements and the campaign have been reported in RAMANA *et al.* (2004).

The aerosol optical depths during this period were in the 0.3 to 0.5 range. There were also some relatively clean cases when AOD was about 0.1–0.15 as well as some

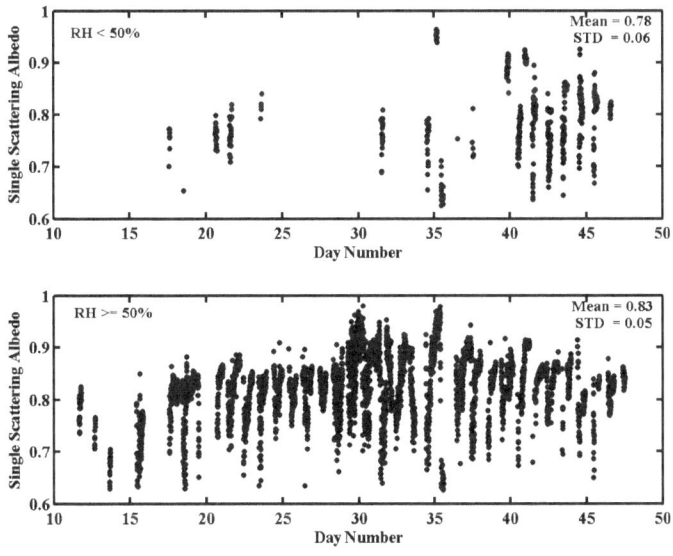

Figure 4
Scatter plot of Single Scattering Albedo (SSA) versus day number when (a) relative humidity (RH) < 50% and (b) RH ≥ 50%. Day numbers 10 and 45 correspond to Jan. 10, 2003 and Feb. 14, 2003, respectively.

high aerosol events in which AOD was over 0.6. SSA was derived from the simultaneous measurements of absorption and scattering coefficients. For practical

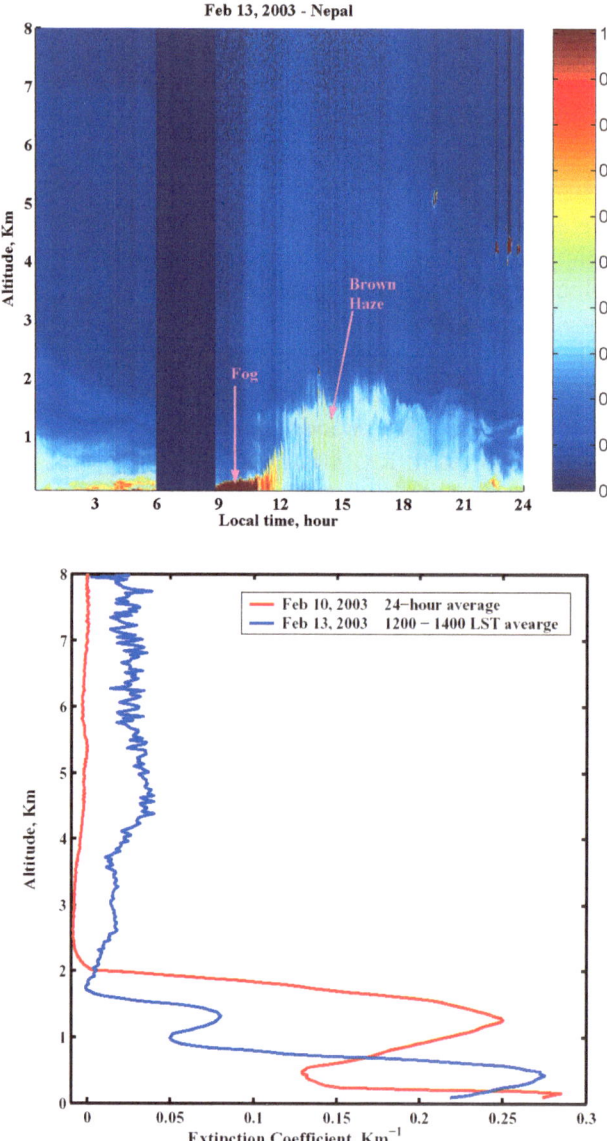

Figure 5

(a) MPL normalized relative backscatter (counts. (km)^2 / (μJ. μs)) time verses altitude plot at Kathmandu on Feb. 13, 2003 (blue area between 6 to 9 LST represents no data), (b) vertical profiles of aerosol extinction on Feb. 10, 2003 (in red color) (RAMANA et al., 2004) and Feb. 13, 2003 (in blue color) over Kathmandu. The profiles are calculated using 30-min MPL signal averages.

purposes, SSA data were segregated based on relative humidity (RH) and are plotted as a function of day in Figures 4a and 4b. In general, SSA values varied between 0.6 and 0.9, indicating the presence of a significant amount of absorbing aerosols. During rainy events (day number 30, 35), SSA values increased to 0.96. Figure 5a displays Micro Pulse Lidar (MPL) near-range corrected backscatter returns on Feb. 13, 2003, which show the layered structures (see RAMANA *et al.*, 2004 for details of Lidar data reduction procedures). Early morning fog is a common phenomenon over Kathmandu and on Feb. 13, 2003 it started dissipating at ~1100 hrs. The extremely large backscatter returns (near surface red color) in the lowermost 200 m result from light scattering by fog and moist aerosols. Figure 5b plots vertical aerosol extinction profiles on Feb. 10, 2003 (RAMANA *et al.*, 2004) and Feb. 13, 2003. Both profiles exhibit a secondary maximum at ~1.3 km above the surface (i.e., about 2.6 km above mean sea level), and these lofted layers are observed on most days and

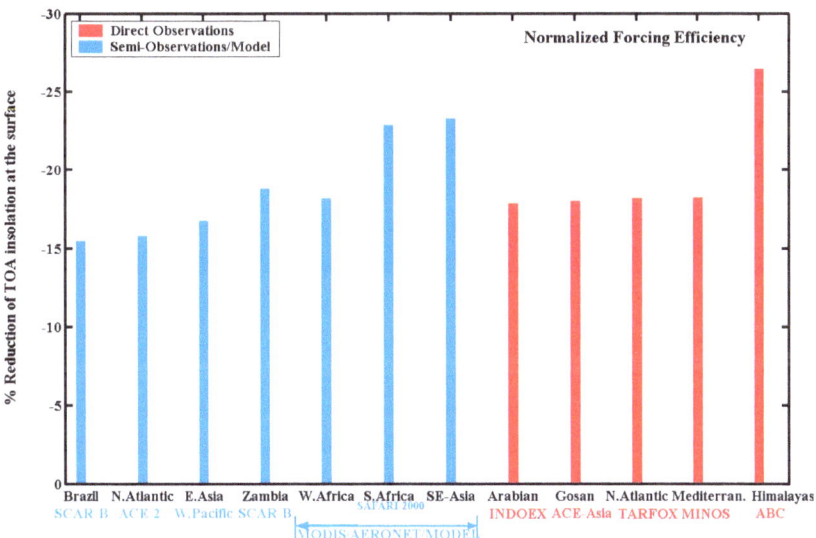

Figure 6

Diurnal clear-sky Radiative Forcing Efficiency normalized with diurnal averaged TOA solar insolation (S) and converted into percent of TOA insolation. Data source and seasonal details: i) Brazil, SCAR-B: Spring, $S = 444$ Wm^{-2}, AOD $= 0.8$ (KINNE and PUESCHEL, 2001; CHRISTOPHER *et al.*, 2000); ii) N. Atlantic, ACE-2: Summer, $S = 443$ Wm^{-2}, AOD $= 0.40$ (KINNE and PUESCHEL, 2001); iii) E. Asia, W. Pacific: March, $S = 335$ Wm^{-2}, AOD $= 0.25$ (KINNE and PUESCHEL, 2001); iv) Zambia, SCA-B: Fall, $S = 444$ Wm^{-2}, AOD $= 0.6$ (KINNE and PUESCHEL, 2001); v) W. Africa: September, TOA $= 422$ Wm^{-2}, AOD $= 0.30$ (KAUFMANN *et al.*, 2002); vi) S. Africa, SAFARI-2000: September, $S = 423$ Wm^{-2}, AOD $= 0.31$ (KAUFMANN *et al.*, 2002; ICHOKU *et al.*, 2003); vii) SE-Asia: September, $S = 412$, AOD $= 0.24$, (KAUFMANN *et al.*, 2002); viii) Arabian, INDOEX: Feb. & March, $S = 420$ Wm^{-2} AOD $= 0.41$ (JAYARAMAN *et al.*, 1998, RAMANATHAN *et al.*, 2001a), ix) Gosan, ACE-Asia: July & August, $S = 468$, AOD $= 0.36$, x) N. Atlantic, TARFOX: July, $S = 477$, AOD $= 0.3$ (HIGNETT *et al.*, 1999), xi) Mediterranean, MINOS: July & August, TOA $= 468$ Wm^{-2}, AOD $= 0.21$ (MARKOWICZ *et al.*, 2002), xii) Himalayas, ABC: Jan. & Feb., TOA $= 272$, AOD $= 0.34$ (RAMANA *et al.*, 2004).

throughout the day. In general, pollutants transported above the boundary layer by turbulent mixing, first due to nocturnal collapse of the boundary layer, and second due to the orographically forced motions, could have caused the lofted layers. Complex topography in this region produces a variety of valley circulation including thermally and dynamically driven wind systems (see RAMAN et al., 1990). In addition, the dry convective rise of pollutants at distant sources and horizontal upper air transport could have brought about these elevated concentrations.

The diurnal mean aerosol radiative forcing efficiency is deduced directly from the observations following SATHEESH and RAMANATHAN (2000). The diurnally averaged aerosol forcing efficiency for the Kathmandu valley was about -73 W m^{-2} per AOD (RAMANA et al., 2004). The diurnally averaged surface aerosol forcing at Kathmandu is about -25 (± 5) W m^{-2} (for a seasonal mean AOD$_{500\ nm}$ of 0.34). The TOA forcing efficiency (estimated from MACR model; MACR is referred to in section 2) is about zero, because of the low single scattering albedo aerosols (i.e., highly absorbing). Such absorbing aerosols absorb the solar radiation reflected by the surface, which nearly balances the scattering of solar radiation back to space by the aerosols, yielding a zero forcing efficiency. Since the difference between the TOA forcing and the surface forcing denotes atmospheric solar absorption, the aerosols in the brown clouds lead to an increase in solar absorption of about 25 (± 5) W m^{-2}, equivalent to an increase in solar heating rate as large as 1 K/day between the surface and 2 km.

4. Comparison of Worldwide Data

In order to compare the Himalayan foothills forcing with available data from other regions of the world, we have to normalize the data to facilitate the comparison. We cannot compare the forcing in Wm^{-2} since it depends not only on the aerosol properties but also on the TOA insolation which varies strongly with season and latitude. Hence we normalized the forcing efficiency with TOA solar insolation and compare the forcing efficiency as a percent of TOA insolation in Figure 6. The data from other regions of the world include the MINOS (the Mediterranean Intensive Oxidant Study, see LELIEVELD et al., 2002) campaign radiation data from MARKOWICZ et al. (2002), Arabian Sea during INDOEX (JAYARAMAN et al., 1998; RAMANATHAN et al., 2001a), western Atlantic off the USA during the TARFOX campaign (HIGNETT et al., 1999), the western Pacific off E. Asia during the ACE-Asia campaign (BUSH and VALERO, 2003), W. Africa and SE Asia (MODIS from KAUFMAN et al., 2002), S. Africa during the SAFARI campaign (KAUFMAN et al., 2002; ICHOKU et al., 2003), Brazil during the SCAR-B campaign (KINNE and PUESCHEL, 2001; CHRISTOPHER et al., 2000), and the polluted W. Pacific (KINNE and PUESCHEL, 2001). The pollution sources include Asian outflow (INDOEX, ACE-Asia), biomass burning (SCAR-B, SAFARI-2000), urban-industrial pollutants (TARFOX, MINOS), and continental polluted out flow from E.

Asia, W. Africa, S. Africa, and SE Asia. Thus the data include a variety of pollution sources from around the world. Anthropogenic aerosols in general lead to a large reduction in surface solar radiation (Fig. 6). The forcing efficiency varies from −15% to −26% of the TOA insolation (Fig. 6). Both backscattering and absorption of solar radiation contribute to these large reductions. In terms of the monthly and seasonal monthly mean forcing, it ranges from about −5% to −10% of TOA insolation. The forcing efficiency normalized with the surface insolation (not shown) ranges from −30% to 50%. As shown in Figure 6, the measured forcing over the Himalayan foothills exceeds those estimated for other regions. The reasons for the minima in the forcing (i.e., maximum negative values) over the Himalayas are largely due to the strongly absorbing nature of the aerosols.

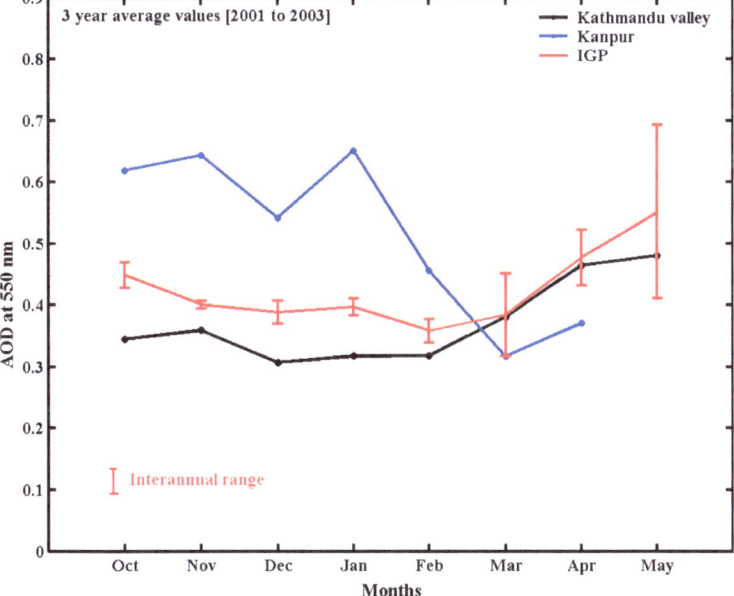

Figure 7

Three-year average aerosol optical depth (AOD) at 550 nm over Kathmandu, Kanpur, and Indo-Ganges plains (IGP) for dry season (October – May) averaged over 2001–2003. The data over Kathmandu and IGP were obtained from MODIS instrument onboard NASA's Terra satellite and data over Kanpur were obtained from AERONET instrument (HOLBEN et al., 1998).

5. Conclusions

a. Implications to the Indo-Gangetic Plains

We first note from Figure 1 that the high AODs over the Nepalese Himalayas (location of our data presented thus far) during the dry season are just part of a larger scale high aerosol regime over the entire IGP. Figure 7 shows a three-year (2001, 2002 and 2003) average monthly variation of dry season AODs averaged over the entire IGP. The AODs typically are 0.4 or larger. The high aerosol regime in the IGP is a persistent feature of all three years. The Kathmandu valley values (also from MODIS) are shown for comparison and are similar to the IGP values, which is another confirmation that the two are part of the same large-scale brown clouds. The October to May average AOD (from Fig. 7) is about 0.43, while for the Kathmandu valley the three-year average for October to May is about 0.37. In addition, we will now compare these values with ground-based AOD observations within the IGP. At Kanpur (26.85°N, 80.34°E) which is within the IGP region (see the arrow for the "AERONET" site), there is a sun and sky-tracking radiometer as part of the worldwide AERONET (Aerosol Robotic Network maintained by NASA; see HOLBEN et al., 1998) ground stations. The AODs for Kathmandu and Kanpur are indicative of the spatial variance of the AOD values. The 2001 to 2003 average AOD for 2001 to 2003 is 0.52, slightly higher than the IGP mean value of 0.43, and furthermore, the AERONET instrument also yields the single scattering albedos, which is 0.88 for Kanpur, again indicating a highly absorbing aerosol. This is clearly an indication that the forcing observed for the Kathmandu valley reflects forcing over the larger scale IGP features seen in Figures 1 and 7. Thus, assuming that the Kathmandu valley forcing values are applicable for IGP, we infer that the dry season average (October to May) surface forcing over the IGP is about -32 (± 6) W m^{-2} {-73 W m^{-2} (forcing efficiency) * 0.43(AOD)}. Since the TOA forcing is about zero (again relying on TOA forcing of Kathmandu valley), the atmospheric forcing is about 32 W m^{-2}. We should note that this forcing includes both natural and anthropogenic aerosols. We have analyzed the NCAR (National Center for Atmospheric Research) aerosol assimilation model developed by COLLINS et al. (2001) and the so-called GOCART (NASA-Goddard aerosol assimilation model) model of CHIN et al. (2002) which simulate anthropogenic as well as natural aerosols, and both of these suggest that about 70 to 80% of the total aerosol is due to anthropogenic activities. In conclusion, the IGP region is subject to a large reduction in solar radiation at the surface by as much as 32 (± 6) W m^{-2} (20% reduction of the surface solar insolation or about 10% of the TOA insolation) and a very large increase in solar heating of the atmosphere (32 W m^{-2}) which is as much as 40% of the total tropospheric absorption or alternately a 100% increase in the absorption within the first three kilometers.

b. Implications to Boundary Layer Stability, Atmospheric Humidity and Aerosol Life Times

The reduction in surface solar absorption accompanied by a large increase in solar heating should strengthen the atmospheric inversion during the dry season. In order to explore the magnitude of this effect, we analyzed the model sensitivity studies conducted by CHUNG et al. (2002) and CHUNG and RAMANATHAN (2003). These two studies employed the regional aerosol forcing (both direct and indirect) estimated by RAMANATHAN et al. (2001a) in the NCAR general circulation model (GCM) with fixed sea-surface temperatures. The aerosol forcing was employed as a perturbation heating to the model thermodynamic equation. The forcing was begun on October 1 and continued until May 30 (see Fig. 2 of CHUNG et al., 2002).

The perturbation forcing field introduced in the GCM was spatially smoothed (see Figs 1 to 3 of CHUNG et al., 2002) to correspond to the coarse spatial resolution of the GCM (about 450 km by 450 km) and hence did not employ the sharp buildup of AODs seen in Figure 1 over the IGP. The real effect of the smoothing was that the GCM aerosol forcing field over IGP was smaller by about 30% to 50% when compared with the forcing estimated in this study. The GCM with the perturbation was run for 60 model years and used the average of the last 50 model years to

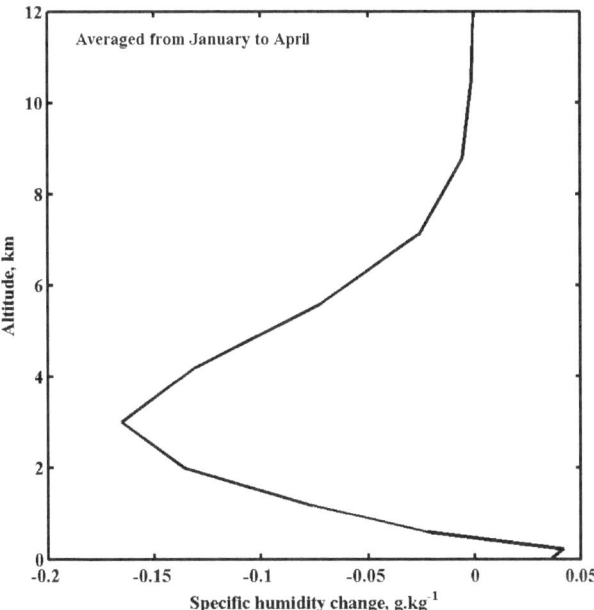

Figure 8
GCM simulated effects of haze forcing on the vertical humidity profile for the Indo-Gangetic Plains region. The results show a 25-year average difference of model humidity values for the IGP with and without the aerosol direct forcing.

compare with the 50-year average of the "Control" run without the aerosol heating perturbation, to estimate the impact of aerosol forcing on the model climate. We estimated the interannual variability of the model fields (see Fig. 3 of CHUNG et al., 2002) for control to estimate the statistical significance of the differences and interpret only those differences that are statistically significant (greater than 2 sigma). The results were thoroughly described in two papers (CHUNG et al., 2002; CHUNG and RAMANATHAN, 2003). However, the boundary layer stability for the IGP was not examined in that study. For the purposes of this paper, the IGP is the near continuous red and yellow shaded regions (AOD > 0.4) extending from Pakistan to Bangladesh, i.e, the red and yellow shaded regions between north of 25°N and 35°N and between 70°E and 85°E; and north of 20°N and between 85°E and 92°E.

In response to the aerosol forcing, the IGP surface cooled by about 0.5 K and the lower atmosphere (1 to 3 km) warmed by about 0.5 K, thus strengthening the low-level inversion. The boundary layer humidity increased by about 5 to 10% (Fig. 8) and this increase in humidity near the surface is due to the strengthening of inversion which inhibited the vertical transport of moisture from the boundary layer. The surface cooling, the strengthening of the inversion together with the increase in the near surface humidity decreased the latent heat flux due to evaporation by about 3 W m^{-2} (10% decrease) and sensible heat flux by about 10 W m^{-2} (25% decrease), which compensated for about 70% of the decreased solar radiation. The remaining 30% of the decrease in solar flux was compensated by enhanced back radiation from the warmer atmosphere. The strengthening of the inversion can lead to an increase in the lifetime of aerosols, thus positively feeding back on the radiative forcing and boundary layer response. These near surface changes are mostly due to the local response to the imposed aerosol forcing. There were also significant changes above the boundary layer, most of which are caused by the imposed aerosol forcing over the entire S. Asia and N. Indian Ocean (see the extent of the aerosols in Fig. 1). These changes due to remote forcing are described next.

The middle and upper tropospheric (3 to 12 km) humidity decreased substantially by about 10% to 20% as shown in Figure 8. The peak decrease of about 0.17 g/kg near 3 km corresponds to a humidity decrease of about 12% and corresponding decrease in the relative humidity (RH) of 6% (from 36% to 30% RH). The decrease in humidity can significantly impact the greenhouse effect of the atmosphere. In addition, the troposphere (up to about 12 km) warmed by about 1 K in response to the imposed aerosol heating over the rest of S. Asia and the N. Indian Ocean (e.g., see the extent of the aerosol into the low latitude Indian Ocean in Fig. 1). These mid-tropospheric changes, which were rather uniform over most of S. Asia, were driven by the response of the monsoonal circulation (northeast monsoon) to the lower tropospheric solar heating. As explained by a detailed analysis in CHUNG et al., the ITCZ in the equatorial region was strengthened and the low-level solar heating was mixed vertically throughout the troposphere (by moist convection and large-scale motions) leading to a uniform warming over most of S. Asia. The moisture decreased because the enhanced return flow (to the ITCZ) advected dry air from the north due to the northeasterly flow at the low levels and from the west due to the westerly flow aloft. The drying will inhibit the hygroscopic growth of particles, which will tend to enhance their lifetimes and in addition will decrease the single scattering albedos because the scattering optical depth is proportional to the square of the particle radius, while the absorption by the black carbon is not greatly affected by humidity.

The meteorology during the dry season (also referred to as the northeast monsoon or the winter monsoon) over S. Asia consists of northeasterlies, anticyclonic circulation, subsidence, and low-level inversion (KRISHNAMURTI et al., 1997), all of which leads to a dry and hazy atmosphere. Both the local and remote response to the aerosol solar heating profile, as estimated by this GCM, is to strengthen the inversion, decrease tropospheric humidity, and to increase the solar absorption, i.e., amplify the influence of the northeast monsoon on aerosol buildup. Thus the GCM results suggest a positive feedback loop between aerosols, their radiative forcing, and their dynamical influence on the thermal and moisture structure. Simultaneously, the decrease in humidity will have a negative feedback effect on the greenhouse warming. The fundamental concern raised by this study is whether this positive feedback has lead to the observed persistence and widespread nature of the aerosols over the IGP (Fig. 1).

We note however the NCAR GCM used in this study is a coarse resolution model whose treatment of orographic effects such as the Himalayas, and treatment of boundary layers and clouds are very approximate. Furthermore the model ignores the feedback effects between aerosols, their radiative forcing and the tropical dynamics. We are at the very early stage of modeling such phenomena. The present study, however, provides the motivation for further studies with higher spatial resolution models of climate and aerosol chemistry and physics.

c. Impact on the Surface Evaporation

STANHILL and COHEN (2001) have documented the so-called "Global dimming," i.e., a reduction in global radiation reaching the Earth's surface. STANHILL and COHEN analyzed the global (largely land-based) data set from 1958 to 1992, which showed statistically significant annual reductions averaging between 0.34 and 0.60 Wm^{-2} per year (or, relatively, between 0.23 and 0.32% per year). RODERICK and FARQUHAR (2002) provided indirect evidence for the Global dimming by examining trends in pan evaporation over the last 50 years. They illustrated that the decrease in pan evaporation is consistent with the observed widespread decrease in sunlight. In general, pan evaporation is more sensitive to variations in net irradiance and average vapor pressure deficit (difference in saturation vapor pressure at ambient temperature and at dew point temperature of the air) than to variations in wind speed (RODERICK and FARQUHAR, 2002). RODERICK and FARQUHAR (2002) assert that declining solar radiation is responsible for the decline in pan evaporation. Most, if not all, of the stations used in these studies are land-based observations.

The present data (Fig. 7) suggest a record reduction of sunlight in the foothills of the Himalayas and we examine the resulting impact on pan evaporation. We obtained the pan evaporation data from the Kathmandu Meteorological Department. The data were available for only two periods some 20 years apart, one for 1976 to 1986 and the other for 1996 to 2000. It is during this period that India witnessed a large growth in aerosol emissions. For example sulphur-dioxide (precursor to sulfate aerosols) increased from about 1 Tg/year to over 3 Tg/year in 2000 and black carbon emissions also increased by more than two-fold during this period (NOVAKOV et al., 2003). Thus we can expect that a major fraction (50% to 70%) of the forcing shown in Figure 6 should have been added during the 1975 to 2000 period. The pan evaporation between the two periods had decreased (from 1976–1986 to 1996–2000) by about 9 W m^{-2}, which can account for 30% of the estimated reduction in the solar flux. As shown by the model results, the remaining 70% of the aerosol induced solar reduction is accounted for by the decrease in sensible heat flux and reduced IR cooling. Consequently, the reduction in pan evaporation could have been caused by the aerosol induced reduction of the solar radiation reaching the surface ("dimming of the surface").

A reduction in land average evaporation, by itself, need not lead to a reduction in rainfall since the land average rainfall is also governed by evaporation from the ocean and its subsequent transport to the land areas. For example, observations of soil moisture show increasing tendencies of more than 1 cm per decade in large regions of Eurasia (ROBOCK et al., 2000) where the studies mentioned earlier observed reductions in sunlight and in pan evaporation.

The findings reported here clearly call for a comprehensive and interdisciplinary approach to assessing the environmental and climate effects of greenhouse gases, the atmospheric brown clouds, and land-use modifications (PIELKE, 2001) in the Indo-Gangetic Plains region.

Acknowledgements

We acknowledge support from the National Atmospheric and Oceanic Administration #NA17RJ1231 and the National Science Foundation #0201946. We also thank Professors Sethu Raman and Paul Crutzen for their comments on the paper.

REFERENCES

ABROL, Y.P., SANGWAN, S., and TIWARI, M.K. *Land Use—Historical Perspecteives, Focus on Indo-Gangetic Plains* (Allied Publishers PVT. LTD., New Delhi, 2002) 655 pp.

BUSH, B.C. and VALERO, F.P.J. (2003), *Surface Aerosol Radiaitve Forcing at Gosan during ACE-Asia Campaign*, J. Geophys. Res. *108*, 8660, doi:10.1029/2002JD003233.

CHIN, M., GINOUX, P., KINNE, S., TORRES, O., HOLBEN, B.N., DUNCAN, B.N., MARTIN, R.V., LOGAN, J.A., HINGURASHI, A., and NAKAJIMA, T. (2002), *Tropospheric Aerosol Optical Thickness from the GOCART Model and Comparison with Satellite and Sun Photometer Measurements*, J. Atmos. Sci. *59*, 461–483.

CHRISTOPHER, S.A., LI, X., WELCH, R.M., REDI, J.S., HOBBS, P.V., ECK, T.F., and HOLBEN, B. (2000), *Estimation of Surface and Top-of-atmosphere Shortwave Irradiance in Biomass-burning Regions during SCAR-B*, J. Appl. Meteor. *39*, 1742–1753.

CHUNG, C.E. and RAMANATHAN, V. (2003), *South Asian Haze Forcing: Remote Impacts with Implications to ENSO and AO*, J. Climate *16*, 1791–1806.

CHUNG, C.E., RAMANATHAN, V., and KIEHL, J.T. (2002), *Effects of the South Asian Absorbing Haze on the Northeast Monsoon and Surface-air Heat Exchange*, J. Climate, 2462–2467.

COLLINS, W. D., RASCH, P.J., EATON, B.E., KHATTATOV, B.V., LAMARQUE, J.F., and ZENDER, C.S. (2001), *Simulating Aerosols Using a Chemical Transport Model with Assimilation of Satellite Aerosol Retrievals: Methodology for INDOEX*, J. Geophys. Res. *106*, 7313–7336.

CONANT, W.C. (2000), *An Observational Approach for Determining Aerosol Surface Radiative Forcing: Results from the First Filed Phase of INDOEX*, J. Geophys. Res. *105*, 15,347–15,360.

HIGNETT, P., TAYLOR, J.P., FRANCIS, P.N., and GLEW, M.D. (1999), *Comparison of Observed and Modeled Direct Aerosol Forcing During TARFOX*, J. Geophys. Res. *104*, 2279–2287.

HOLBEN, B.N., ECK, T.F., SLUTSKER, I., TANRE, D., BUIS, J.P., SETZER, A., VERMOTE, E., REAGAN, J.A., KAUFMAN, Y.J., NAKAJIMA, T., LAVENU, F., JANKOWIAK, I., and SMIRNOV, A. (1998), *AERONET–A Federated Instrument Network and Data Archive for Aerosol Characterization*, Rem. Sens. Environ. *66*, 1–16.

ICHOKU, C., REMER, L.A., KAUFMAN, Y.J., LEVY, R., CHU, D.A., TANRE, D., and HOLBEN, B.N. (2003), *MODIS Observations of Aerosols and Estimation of Aerosol Radiaitve Forcing over Southern Africa during SAFARI 2000*, J. Geophys, Res. *108*, 8499, doi:10.1029/2002JD002366.

JAYARAMAN, A., LUBIN, D., RAMACHANDRAN, S., RAMANATHAN, V., WOODBRIDGE, E., COLLINS, W., and ZALPURI, K.S. (1998), *Direct Observations of Aerosol Radiative Forcing over the Tropical Indian Ocean during the Jan. Feb. 1996 pre-INDOEX Cruises*, J. Geophys. Res. *103*, 13,827–13,836.

KAUFMAN, Y., TANRE, D., and BOUCHER, O. (2002), *A Satellite View of Aerosols in the Climate System*, Nature *419*, 215–223.

KINNE, S. and PUESCHEL, R. (2001), *Aerosol Radiative Forcing for Asian Continental Outflow*, Atmos. Environ. *25*, 5019–5028.

KOREN, I., KAUFMAN, Y.J., REMER, L.A., and MARTINS, J.V. (2004), *Measurements of the Effect of Amazon Smoke on Inhibition of Cloud Formation*, Science *303*, 1342–1345.

KRISHNAN, R., and RAMANATHAN, V. (2002), *Evidence of Surface Cooling from Absorbing Aerosols*, Geo. Phys. Res. *29*, doi:10.1029/2002GL014687.

KRISHNAMURTI, T. N., JHA, B., RASCH, P.J., and RAMANATHAN, V. (1997), *A High Resolution Global Reanalysis Highlighting the Winter Monsoon. Part I, Reanalysis Field*, Meteorol. Atmos. Phys. *64*, 123–150.

LELIEVELD, J., CRUTZEN, P.J., RAMANATHAN, V., et al. (2001), *The Indian Ocean Experiment: Widespread Air Pollution from South and Southeast Asia*, Science *291*, 1031–1036.

LELIEVELD, J., et al. (2002), *Global Air Pollution Crossroads over the Mediterranean*, Science, 2002.

MARKOWICZ, K., FLATAU, P.J., RAMANA, M.V., CRUTZEN, PJ., and RAMANATHAN, V. (2002), *Absorbing Mediterranean Aerosols Lead to a Large Reduction in the Solar Radiation at the Surface*, Geophys. Res. Lett. *29*, 1968, doi:10.1029/2002GL015767.

MEYWERK, J. and RAMANATHAN, V. (2002), *Influence of Anthropogenic Aerosols on the Total and Spectral Irradiance at the Sea Surface during the Indian Ocean Experiment (INDOEX 1999)*, J. Geophys. Res. *107*, 8018, doi:10.1029/2000JD000022.

MEYWERK, J. and RAMANATHAN, V. (1999), *Observations of the Spectral Clear-sky Aerosol Forcing over the Tropical Indian Ocean*, J. Geophys. Res. *104*, 24,359–24,370.

NOVAKOV, T.V., RAMANATHAN, V., HANSEN, J.E., KIRCHSTETTER, T.W., SATO, M., SINTO, J.E., and SATHAYE, J.A. (2003), *Large Historical Changes of Fossil-fuel Black Carbon Aerosols*, Geophys. Res. Lett. *30(6)*, 1324, doi:10.1029/2002GL016345.

PIELKE, R.A. Sr. (2001), *Influence of the Spatial Distribution of Vegetation and Soils on the Prediction of Cumulus Convective Rainfall*, Rev. Geophys. *39*, 151–177.

PODGORNY, I.A., CONANT, W.C., RAMANATHAN, V., and SATHEESH, S.K. (2000), *Aerosol Modulation of Atmospheric and Surface Solar Heating Rates over the Tropical Indian Ocean*, Tellus *52B*, 947–958.

RAJEEV, K., RAMANATHAN, V., and MEYWERK, J. (2000), *Regional Aerosol Distribution and its Long-range Transport over the Indian Ocean*, J. Geophys. Res. *105*, 2029–2043.

RAMAN, S., TEMPLEMAN, B., TEMPLEMAN, S., MURTHY, A.B., SINGH, M.P., AGARWAAL, P., NIGAM, S., PRABHU A., and AMEENULLAH, S. (1990), *Observation of Mean Boundary Layer Structure and Turbulence during Pre-monsoon and Monsoon Periods in India*, Atmos. Environ. *24A*, 723–734.

RAMANA, M.V., RAMANATHAN, V., PODGORNY, I.A., PRADHAN, B.B., and SHRESTHA, B. (2004), *The Direct Observations of Large Aerosol Radiative Forcing in the Himalayan Region*, Geophys. Res. Lett. *31*, L05111, doi:10.1029/2003GL018824.

RAMANATHAN, V., and CRUTZEN, P. J. (2003), *New Directions: Atmospheric Brown Clouds*, Atmos. Environ. *37*, 4033–4035.

RAMANATHAN, V. and Ramana, M.V. (2003), *Atmospheric Brown Clouds: Long-range Transport and Climate Impacts*, EM *12*, 28–33.

RAMANATHAN, V. et al. (2001a), *The Indian Ocean Experiment: An Integrated Analysis of the Climate Forcing and Effects of the Great Indo-Asian Haze*, J. Geophys. Res. *106*, 22,371–22,398.

RAMANATHAN, V., CRUTZEN, P.J., KIEHL, J.T., and ROSENFELD, D. (2001b), *Atmospheric Aerosols, Climate, and the Hydrological Cycle*, Science *294*, 2119–2124.

RAMANATHAN, V. et al. (1995), *Indian Ocean Experiment (INDOEX) White Paper*, Center for Clouds, Chemistry, and Climate (c4)#143, Scripps Inst of Oceanogr., La Jolla, Calif., URL = http://www-indoex.ucsd.edu/publications/white_paper/.

ROBOCK, A., VINNIKOV, K.Y., SRINIVASAN, G., ENTIN, J.K., HOLLINGER, S.E., SPERANSKAYA, N.A., LIU, S., and NAMKHAI, A. (2000), *The Global Soil Moisture Data Bank*, Bull. Amer. Meteor. Soc. *81*, 1281–1299.

RODERICK, M.L., and FARQUHAR, G.D. (2002), *The Cause of Decreased Pan Evaporation over the past 50 Years*, Science *298*, 1410–1411.

SATHEESH, S.K., RAMANATHAN, V., HOLBEN, B.N., MOORTHY, K.K., LOEB, N.G., MARING, H., PROSPERO, J.M., and SAVOIE, D. (2002), *Chemical, Microphysical, and Radiative Effects of Indian Ocean Aerosols*, J. Geophys. Res. 107, 4725, doi:10.1029/2002JD002463.

SATHEESH, S.K., and RAMANATHAN, V. (2000), *Large Differences in the Tropical Aerosol Forcing at the Top of the Atmosphere and Earth's Surface*, Nature *405*, 60–63.

SINGH, N., SONTAKKE, A., and PATWARDHAN, *Hydroclimatic and environmental changes of the Indo-Gangetic Plains region: A historical perspective. In Land Use historical Perspectives, Focus on Indo-Gangetic Plains* (Allied Publishers PVT. LTD., New Delhi 2002), pp. 71–103.

STANHILL, G. and COHEN, S. (2001), *Global Dimming: A Review of the Evidence for a Wide spread and Significant Reduction in Global Radiation with Discussion of its Probable Causes and Possible Agricultural Consequences*, Agric. for. Meteorol. *107*, 255–278.

VALERO, F. P., Bucholtz, A., BUSH, B. C., and Pope, S. K. (1999), *Climate Forcing by Anthropogenic Aerosols over the Indian Ocean during INDOEX (abstract)*, EOS Trans. AGU *80(46)*, Fall Meet. Suppl., 162.

(Received October 12, 2003, accepted May 1, 2004)
Published Online First: May 25 2005

 To access this journal online:
http://www.birkhauser.ch

Pure appl. geophys. 162 (2005) 1627–1641
0033–4553/05/091627–15
DOI 10.1007/s00024-005-2686-7

© Birkhäuser Verlag, Basel, 2005

❘Pure and Applied Geophysics

Structures of Mesocirculations Producing Tornadoes Associated with Tropical Cyclone Frances (1998)

Gandikota V. Rao,[1,*] Joshua W. Scheck,[1] Roger Edwards,[2] and
Joseph T. Schaefer[2]

Abstract—Radar structures of one mesocyclone and one mesocirculation (the term mesocirculation refers to a class of rotating updrafts, which may or may not be as spatially and temporally large as a typical mesocyclone) that developed a total of four tornadoes in association with Tropical Cyclone (TC) Frances 1998 are presented. One tornado developed within an inner rainband near the time of landfall while three of the other tornadoes developed within an outer rainband nearly 24 hours after the landfall. Radar reflectivities of the tornadic circulations averaged upwards of 40 dBZ while Doppler radar wind components directed toward the radar averaged 11m s−1. It is realized that although TC Frances was a minimal hurricane it spawned several tornadoes (four of which were studied) causing damage exceeding $2 million. These tornadoes were not all located close to the TC center, serving as a caution to forecasters and emergency personnel that the immediate landfalling area is not the only place to watch.

While it is difficult to accurately predict the TC tornado location and time of occurrence, the degree of low-level baroclinicity seems to play an important role in tornadogenesis. Another significant finding is that the tornadoes were produced on the inward side of an inner rainband, as well as the inward side of an outer rainband. Consistent with climatology, the forward right quadrant of the TC developed the four tornadoes studied here.

Key words: Doppler radar, tornado, tropical cyclone, baroclinic boundary, supercell, mesocyclone.

1. Introduction

Since the 1940s, radar has become an increasingly valuable tool to meteorologists. Radars provide significant insights into the dynamics of mesoscale atmospheric motion and cloud microphysics. Radar data from reconnaissance research aircraft give excellent information about hurricanes and hurricane rainbands (e.g., Jorgensen, 1984). Marks (1990) has reviewed the radar studies of tropical weather systems such as squall lines and tropical cyclones (TCs). Kulshrestha (1990) gave a historical account of radar meteorology in India, while Raghavan (1997) summa-

[*]The lead author, Professor G. V. Rao died 31 July 2004 at the age of 70. He fell victim to the waves while swimming in Mazatlan, Mexico. This is the last paper he published as lead author.

[1]Saint Louis University, Saint Louis, Missouri, U.S.A.
[2]Storm Prediction Center, National Weather Service, Norman, Oklahoma, U.S.A.

rized the (conventional or non-Doppler) radar studies of Indian TCs. For a review of TC mesoscale characteristics see RAO and BHASKAR-RAO (2003).

During the 1990s, Doppler radars were installed at many stations in the United States (U.S.), making available routine data of mesoscale weather systems. The coastal stations, in particular, are able to view the incoming subsynoptic or mesoscale (2–200 km) systems and are useful in warning the public and emergency personnel about the imminent danger. Other tropical countries (e.g., Japan, Taiwan and India) are able to view mesoscale systems using similar Doppler radar technology.

TC is a term used in this study regardless of the intensity of a tropical cyclone. TCs, in general, move from east to west in the Atlantic Ocean and Bay of Bengal, with frequent northwestward or northward path curvature, although they may move in any direction. As a TC experiences landfall, widespread damage is often expected and a potential disaster is prepared for.

The destruction due to heavy rain, floods and surges is well known throughout the world. Not as widely addressed is the damage caused by tornadoes associated with TCs. While the prevalence of tornadoes in many landfalling TCs has been known for a long time, low-level cloudiness, rain and intense winds make it difficult to directly observe this phenomenon in the case of landfalling TCs. SMITH (1965) noted that the right front quadrant of a moving TC is likely to develop tornadoes while PEARSON and SANDOWSKI (1965) observed that most of the TC tornadoes occurred outside of the area where hurricane force winds prevailed. HILL *et al.* (1966), in a seminal paper, noted that dry air intrusions into the TC circulation likely caused an environment that favored tornadogenesis. CURTIS (2004) studied this concept in more detail and confirmed the association with analysis of a dataset of observed upper air data from multiple landfalling TCs. BURPEE (1986) discussed the mesoscale structure of landfalling TCs and reviewed tornado studies. The structure of mesocyclones supporting the tornadoes and downbursts found in the middle latitudes was analyzed by FUJITA (1981) based on conventional observations, reflectivity radar and visual observations. Mesometeorological systems now are being understood better as a consequence of Doppler radar data. The structures of mesocyclones spawning TC tornadoes based on Doppler radar were reported by SPRATT *et al.* (1997), SUZUKI *et al.* (2000), RAO *et al.* (2000) and SCHECK (2001).

The purpose of this article is to present the structures of mesocirculations that generated tornadoes associated with TC Frances (1998) in Texas and Louisiana. The term mesocirculation is used to denote a class of rotating updrafts, which are not always as large in spatial and temporal dimensions, nor always as strong as a mesocyclone commonly observed in U.S. Great Plains severe weather outbreaks. Using this definition, a mesocyclone may be referred to as a mesocirculation. Illustrations of the mesocirculations, such as their horizontal and vertical extents and the accompanying reflectivity patterns are shown in this paper. It appears that three of the four tornadoes were generated in part because of the presence of surface baroclinicity associated with a warm front in Louisiana. A concept analogous to that for middle latitude tornadic supercells was proposed earlier by MARKOWSKI *et al.* (1998) who, on the basis of high-resolution observations in the U.S. Midwest, found tornadoes to occur when supercells interacted with surface thermal boundaries such as fronts. A similar observation was made by SUZUKI *et al.* (2000) in the context of a western Pacific TC (typhoon). SCHECK *et al.* (2004) also documented the presence of a low-level baroclinic zone in the vicinity of two tornadic supercells that formed north of Tampa, Florida, and were contained within the circulation of TC Earl (1998). Because the data were rather coarse in this case, the authors can offer no insight into the mechanism of tornadogenesis for a tornado in Texas during the landfall of TC Frances.

2. Data and Methodology

To present detailed structures of TC spawned tornadoes, TC Frances (1998) was chosen because it generated a number of verified tornadoes for which reliable data exist. The archive of tornadoes and other weather-related phenomena is called *Storm Data* (NCDC, 1998). *Storm Data* provides the counties, times, path width, F-scale (FUJITA, 1981), a damage summary, and track of confirmed tornadoes. Knowing the times and locations of confirmed tornadoes enables a researcher to request specific level II Doppler radar data (CRUM *et al.* 1993) from the National Climatic Data Center in Asheville, North Carolina. Radar data are collected both in the azimuthal and vertical directions every five or six minutes, corresponding to the completion time for volumetric scans. The data the authors utilized consisted of conventional surface and upper air maps and Doppler radar observations. The Doppler radar data are viewed using the WSR-88D Algorithm Testing and Display System (WATADS 9.0) software program supported by the National Severe Storms Laboratory (NATIONAL SEVERE STORMS LABORATORY, 2000). WATADS graphically displays reflectivity as well as radial velocity at about 5-minute intervals. Reflectivity values range from 0 to 65 dBZ and are color coded and displayed on the Plan Position Indicator (PPI) and vertical cross section views. Radial velocity values are also denoted by color (cool colors such as green represent winds blowing toward radar while warm colors such as red represent winds blowing away from radar).

3. TC Frances Evolution

A TC formed at 2100 UTC on 8 September, 1998 in the western Gulf of Mexico. The poorly organized depression located 400 km south of Galveston, Texas, had a large area of convection on the east side of its center and became TC Frances 24 hours later. By 10 September, environmental vertical wind shear had decreased enough to allow intensification. Maximum sustained winds increased to 26 ms^{-1}.

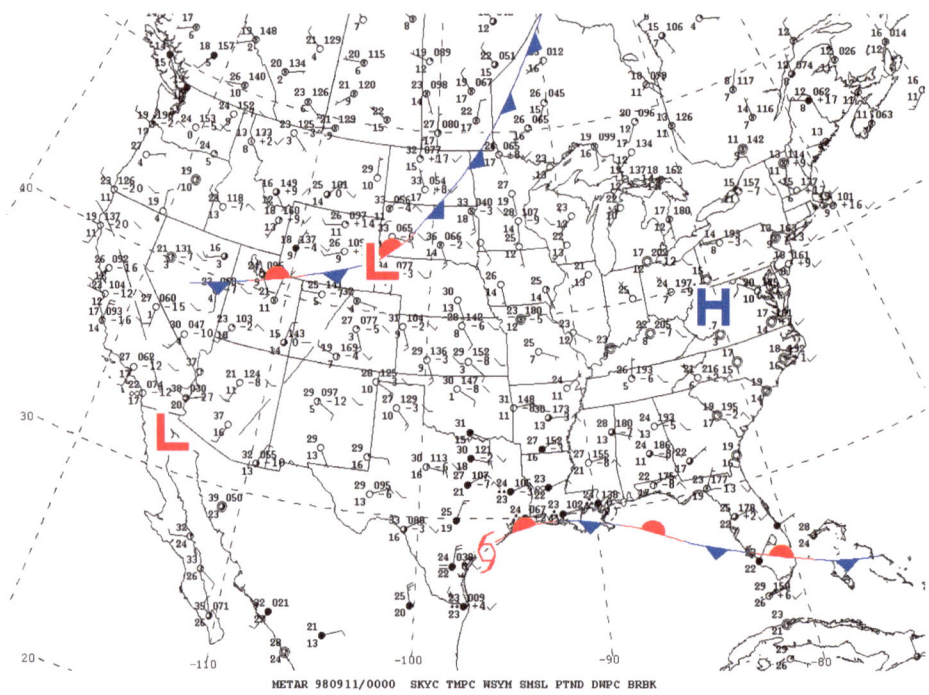

METAR 980911/0000 SKYC TMPC WSYM SMSL PTND DWPC BRBK

Figure 1
Surface analysis at 0000 UTC 11 September, 1998. Notice the quasi-stationary boundary that may have increased the magnitude of low-level vorticity available to convective cells, thereby increasing tornado potential.

Figure 1 shows the sea level pressure map (nearly 6 hours after the storm made landfall) at 1200 UTC 11 September with the TCs central pressure at 992 hPa (not shown). As the TC strengthened, the surface pressure gradient intensified between Frances and a large high pressure system over the southeastern U.S. A quasistationary front existed along the eastern Texas and Louisiana coasts stretching into Florida.

The temperature contrast between the stations along the coast and those inland in eastern Texas and Louisiana is noteworthy. As the storm-induced surface flow crossed the quasistationary frontal zone, it backed to a more easterly component, increasing the lower tropospheric vertical shear and, apparently resulted in increased helicity and vorticity magnitudes. MADDOX *et al.* (1980) documented the increased vertical wind shear in the vicinity, and on the cool side, of low-level horizontal thermal gradients (e.g., higher helicity values). See BROOKS *et al.* (1993) for a definition and discussion of helicity and evolution of low-level mesocyclones.

After making landfall near Victoria, Texas at about 0600 UTC 11 September, the storm moved west-northwest. By 0000 UTC 12 September TC Frances generated six tornadoes in Texas and Louisiana, as shown in Table 1. Frances became a depression

Table 1

Chronological list of tornadoes spawned by TC Frances (1998) according to Storm Data (NCDC 1998). The symbol "*" denotes a case in which WSR-88D radar data were available, but to the extended distance from the radar station, there was only a poor view of the circulation. The symbol "+" denotes cases in which no WSR-88D level II data were available for examination.

TC Frances (1998) Tornadoes					
Date	Time (UTC)	County	State	WSR-88D	F-Scale
09/11/98	0015	Matagorda*	TX	KHGX	F0
09/11/98	0932	Harris +	TX	KHGX	F0
09/11/98	1555	Jefferson Davis	LA	KPOE	F1
09/11/98	2218	Acadia	LA	KPOE	F0
09/11/98	2256	Evangeline	LA	KPOE	F1
09/11/98	2315	Allen	LA	KPOE	F0

at 0400 UTC 12 September and quickly lost its tropical character, based on best-track information provided by the National Hurricane Center (PASCH et al. 1998). Figure 2 shows a sea-level pressure map at 1200 UTC 12 September with a low in Texas with two troughs and a warm front undergoing frontolysis over eastern Texas and Louisiana. Figure 3 shows the various locations of TC Frances at different times. It may be noted that after the landfall the storm moved westwards and then moved east-northeastwards and finally northeastwards occupying the position indicated by the letter L. Most of the tornadoes in Louisiana occurred shortly after the storm reached this L position.

Figure 4 shows a skew-T log p thermodynamic diagram for Lake Charles, Louisiana (LCH) at 1800 UTC, 11 September. Although it was launched 18 hours after the landfall it constitutes the nearest rawinsonde sounding to the tornadoes, spatially and temporally. Four of the five tornadoes happened within a few hours (one occurring before and three after) of this sounding. The sounding is characterized by a nearly saturated profile with south-southeasterly flow in the low levels and minimal directional shear. The Convective Available Potential Energy (CAPE) was 240 J kg^{-1}. According to the examination of TC supercell and tornado environments performed by MCCAUL, (1993), mean climatological values of CAPE for TC tornadoes range from 300 to 1100 J kg^{-1} at a distance of 400 km from the TC center, suggesting that the unmodified LCH sounding depicted buoyancy is unfavorable for tornadic storms. Because of their near-moist adiabatic lapse rates, and without the presence of strong diabatic forcing of the thermodynamics, saturated soundings do not develop much CAPE. The storm relative helicity (SREH) through the layer 0–3 km above ground level (AGL) was 239 m^2 s^{-2}, high enough to suggest mesocyclone formation. The climatological total helicity value given by MCCAUL (1993) at 400 km from the center varies between 30 and 120 m^2 s^{-2}.

The Bulk Richardson Number (BRN, after WEISMAN and KLEMP, 1982) value obtained was 9 for the current situation. The MCCAUL (1993) climatological values

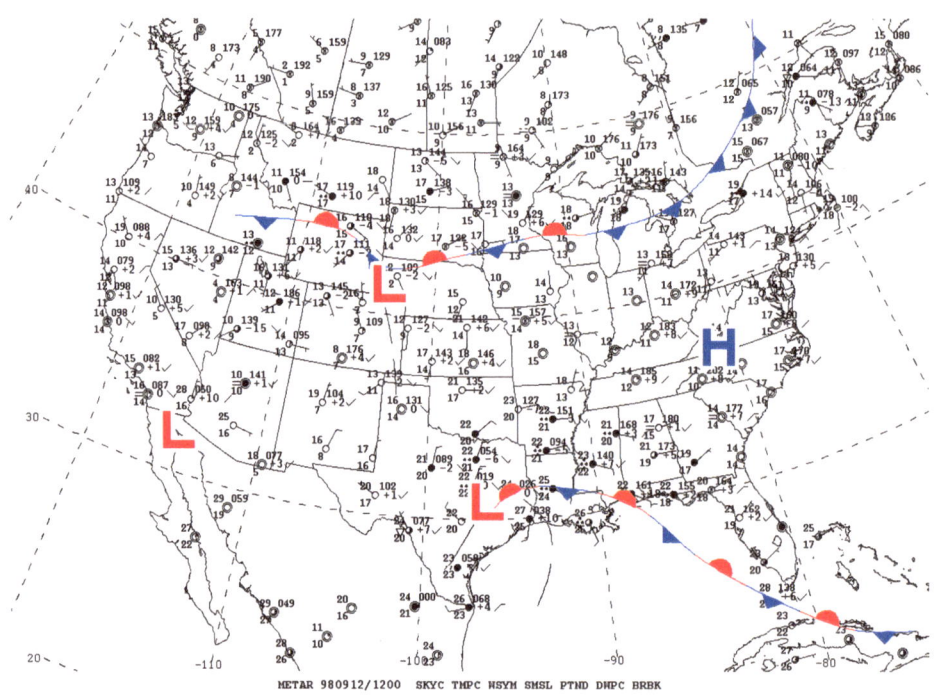

Figure 2
As in Figure. 1, except analysis valid 1200 UTC 12 September 1998. Note the baroclinic zone extending eastward from the TC in eastern Texas and central Louisiana.

range from 25 to 200, suggesting that the current value is unfavorable for the development of supercells. With a level of free convection (LFC) of 816 hPa and a convective inhibition of 42 J kg^{-1}, it appears that some low-level forcing was necessary to support supercells. Supercells are frequently characterized by deep and persistent mesocyclones (DOSWELL and BURGESS, 1993) and only some minority of mesocyclones bear tornadoes. Overall, the situation as represented by the LCH sounding was not climatologically favorable for supercellular tornadogenesis.

4. *Radar Characteristics*

Shortly before landfall, TC Frances generated a tornado near Sargent in Matagorda County, Texas at 2315 UTC 10 September. This tornado was rated F0 resulting in damage totaling $15,000. It appears that the small, short-lived mesocirculation responsible for the tornado was embedded in an inner rainband. This is in contrast to the rest of the TC Frances tornadoes, which occurred within outer rainbands.

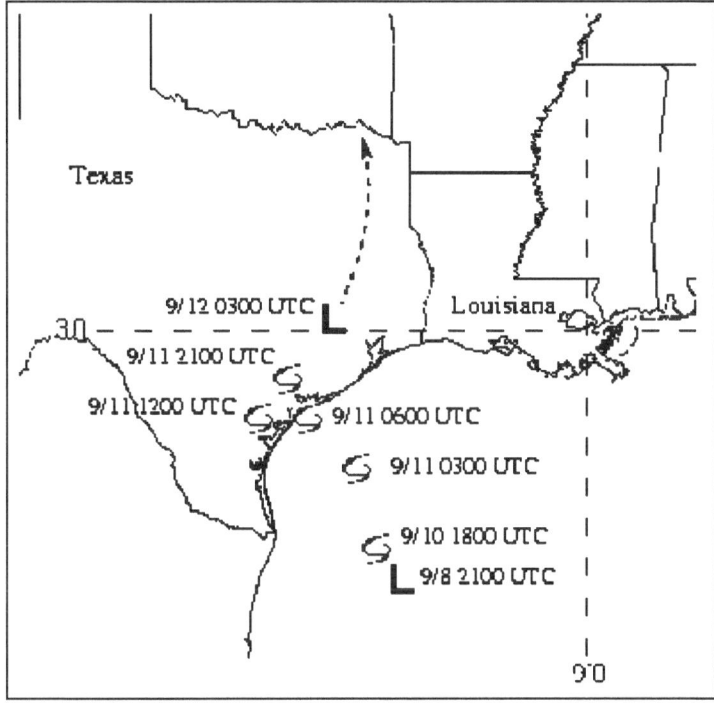

Figure 3
Track of Tropical Cyclone Frances (1998). Strength and location values are available on-line at http://
www.nhc.noaa.gov/1998frances.html.

Figure 5 displays four panels of radar (Houston, KHGX) imagery at the aforementioned time, with Matagorda County, Texas outlined in a thick white border. The town of Sargent is labeled, and is located in the southeast corner of the county. Shown in A is a PPI of reflectivity (shown in dBZ) at 1300 m elevation and in B are Doppler radar radial winds (shown in m s^{-1}; red, indicating the component away from radar and green, toward radar). BURGESS and RAY (1986) discussed the various means to identify mesocyclones through Doppler radar observations. The line AA' indicates a line along which vertical cross sections were constructed. In C, a vertical cross-section of reflectivity is shown while in D the corresponding Doppler radar winds are shown. The PPI in A shows a curved rainband with strong reflectivity values (> 34 dBZ). The left upper half of the B panel shows a wind component away from the radar while the rest is toward the radar. The tornado happened near the coastal town of Sargent and on the TC center side of the band. Doppler radial velocity values near Sargent show relatively low horizontal shears maximizing at 16 ms^{-1} over a distance of 5 km. In C, reflectivity values are moderately high between 15 and 21 km along the abscissa, and such values (> 40 dBZ) stretch vertically to 6 km. The horizontal shears in D are strong (24 ms^{-1} over

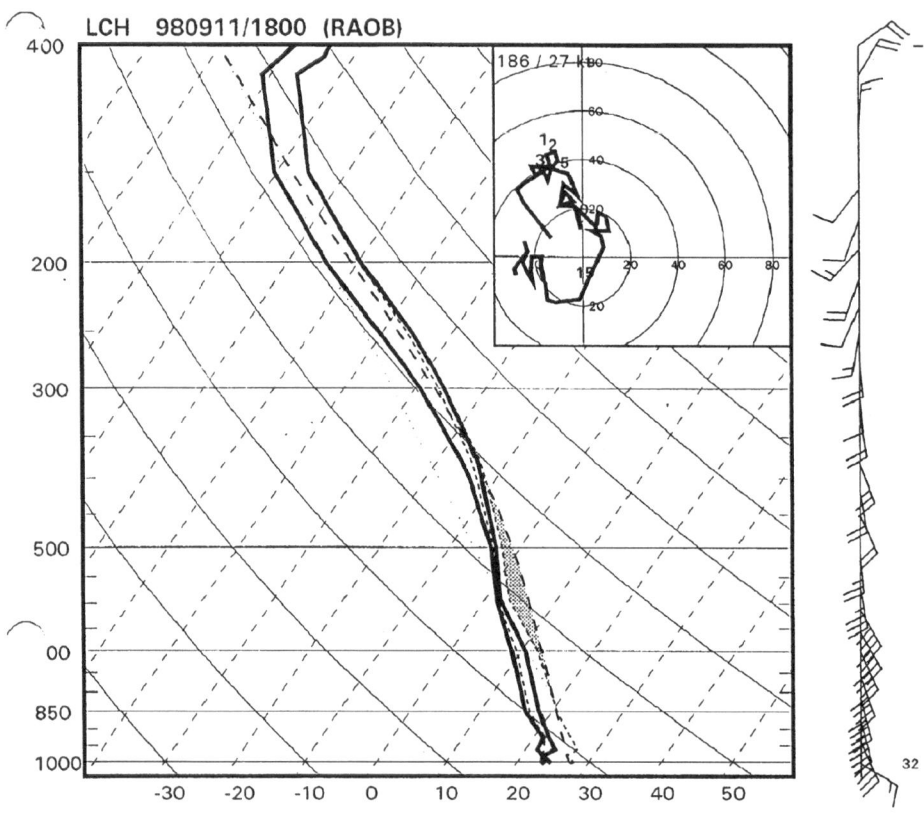

Figure 4
The thermodynamic diagram for Lake Charles, Louisiana at 1800 UTC 11 September 1998.

a distance of 3 km) and extend from 4 to 7 km vertically. Near-surface shears are not resolved because of the minimum beam elevation of 0.5° at the distance from the radar. The associated mesocirculation existed for several minutes over water prior to landfall and as it crossed onto land, a tornado was generated. It is possible that differential friction contributed to this tornadogenesis. No verification of this hypothesis is made in this article.

Tornadoes were first reported in Louisiana around 1552 UTC 11 September. At this time there was an outer rainband close to the Texas-Louisiana border. This band was approximately 350 km from the TC center, and one of the cells formed a tornado at 1555 UTC in Jefferson Davis Parish (not shown). Because of the meager size and strength of the mesocirculation, this tornado is not discussed further in this study. It is worthwhile to note that there appears to be no clear difference between the environment of smaller, short-lived, weak mesocirculations and the environment of larger, long-lived, strong mesocyclones. The authors suggest that more research on

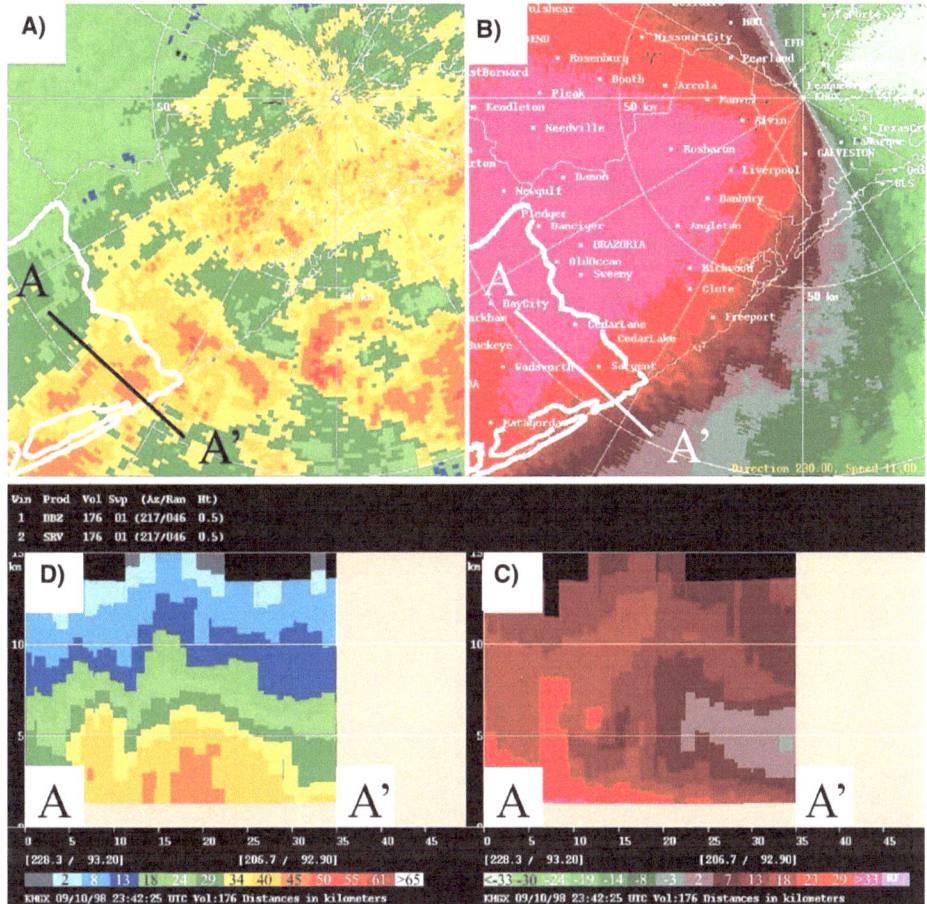

Figure 5
Four-panel image at 2342 UTC 10 September, 1998 from the Houston, Texas WSR-88D (KHGX) showing
the Plan Position Indicator (PPI) of reflectivity (a), storm relative radial velocity (b) and crosssections of
the two (c) and (d) for the convective cell responsible for a tornado in Matagorda County, Texas (outlined
in thick white line) near the town of Sargent, which is located in the southeast corner of the county.

this phenomenon is necessary so that meteorologists can better prepare to warn the
public when the possibility exists for tornadic mesocirculations.

Figure 6 is a four-panel presentation of the mesocyclone that produced a tornado
at 2221 UTC 11 September in Acadia Parish. The damage totaled $5 million and was
rated an F0 (NCDC 1998). The PPI at 1500 m elevation in A shows an intensifying
weak echo region (WER; MARWITZ, 1972) and a hook echo. This region is defined to
be the primary inflow and updraft region of the mesocyclone. The WER has
developed weaker reflectivity than the surrounding echoes because the ascent there is

too rapid to form sizable drops to produce higher reflectivities. Obviously a lag exists between cloud microphysics and kinematics of motion (DOVIAK and ZRNIC, 1993). The mesocyclone is distinct and has developed an inbound radial wind component (not particularly strong) of about 11 ms^{-1}. The vertical cross section in C starts at 1500 m and thus truncates some of the low level structure. High reflectivities (between 20 and 40 km in the horizontal) are building upward of 4 km. In panel D a discontinuity in color is seen near the 35 km mark along the abscissa, signifying the location of a mesocirculation. Like the Sargent tornado, this tornado also occurred on the TC center side of the rainband.

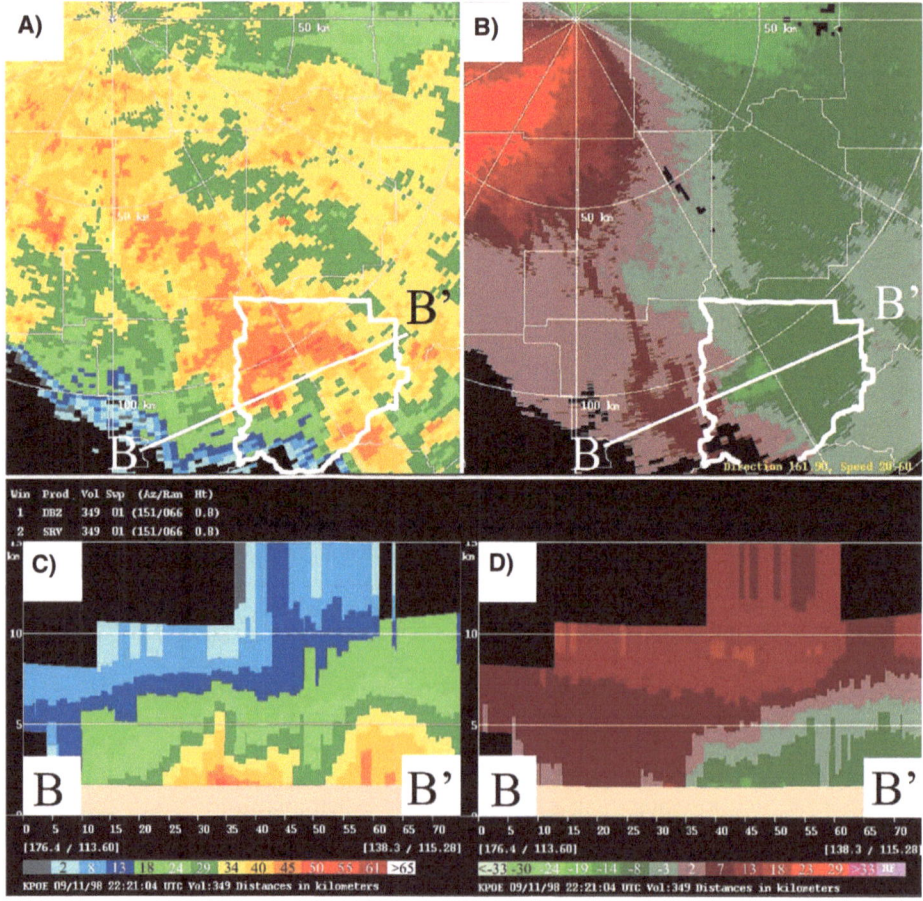

Figure 6
As in Figure. 5, but imagery from the Fort Polk, Louisiana WSR-88D (KPOE) at 2221 UTC 11 September, 1998 for the convective cell that caused a total of three tornadoes, this one in Acadia Parish, Louisiana (outlined in thick white line).

Figure 7 shows that the same mesocyclone traveled north-northwestward into Evangeline Parish and generated a tornado. This tornado caused F1 damage of $2 million. In A, a comma-shaped reflectivity with solid values upwards of 40 dBZ was developed. The reflectivities became stronger and covered a larger area at 2256 UTC 11 September, compared to 35 minutes earlier. Panel B shows that the mesocyclonic circulation widened. Cross section C shows that moderately intense (45 dBZ) convection extended to 5 km. The circulation (in D) prevailed up to 8 km. This mesocyclone traveled further north-northwestwards to Allen Parish and caused a third tornado causing F0 damage. The damage was limited to $10,000.

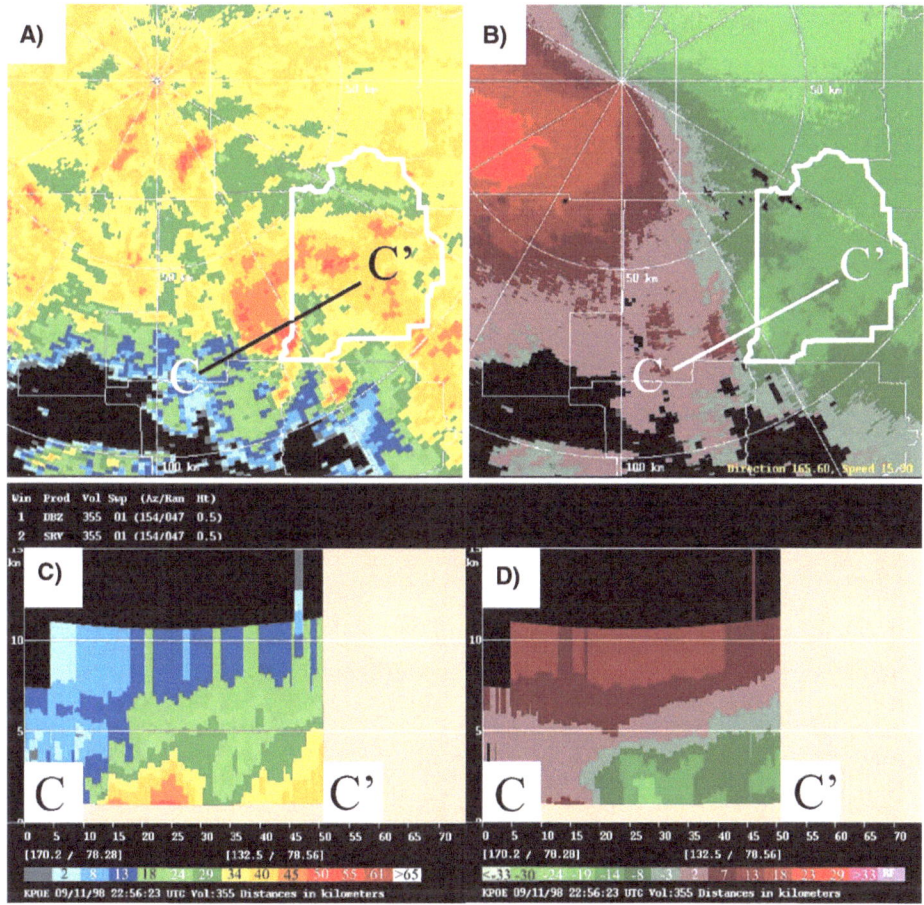

Figure 7

As in Figure. 6, but at 2256 UTC 11 September, 1998. At this time, the storm was in Evangeline Parish, Louisiana (outlined in thick white line). Note that the center of the mesocyclone is located to the county west of the tornado report. This is probably due to a slight time difference between the radar sample and the tornado occurrence.

Figure 8 deals with the Allen Parish tornado, the parent of which is the mesocyclone previously discussed. Panel A shows that the reflectivity pattern is reminiscent of two rain bands rotating around a TC center. Panel B shows that the circulation covered a larger but less intense area than Figure. 7. Vertical cross section of reflectivity (C) shows an upward growth of the convection. Panel D shows that the mesocyclonic circulation increased in depth, which is to be expected with the upward growth discussed above. This particular mesocyclone was observable for 75 minutes. This shows that some TC mesocyclones are long lasting and one mesocyclone can generate multiple tornadoes at different times.

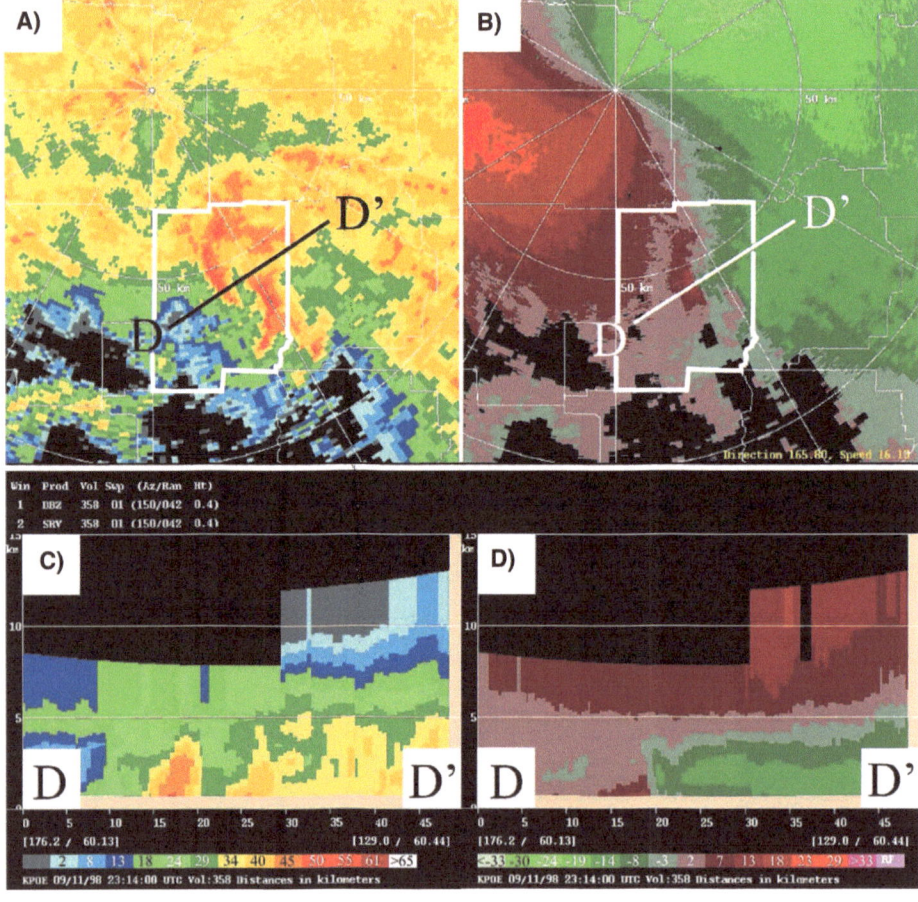

Figure 8
As in Figure. 7, except at 2314 UTC 11 September, 1998. This was during the last tornado generated by this storm, in Allen Parish, Louisiana (outlined in thick white line).

5. Discussion

It is to be cautioned that TC tornadoes are not simply a phenomenon to be concerned with in the immediate vicinity of the center of a landfalling TC. Tornadoes can strike even after landfall and can strike as far as 400 km from the TC center. Because of the routine availability of Doppler radar data it is possible to understand the structure of the parent mesocirculations. Using the Matagorda County, Texas tornado example near Sargent, the authors doubt that the current WSR-88D observing network can be expected to sample the "typical" mesocirculation that is responsible for short-lived, weak tornadoes. The supercell analyzed above, however, serves as an example of a larger, longer-lived mesocyclone that generated multiple tornadoes. It is that type of storm that must be scrutinized during the warning process. The baroclinic zone was likely a major contributor to the tornado production potential of the supercell. Because the radar imagery shown with this case resembles that which has been previously published, the authors cannot stress enough the importance of a low-level baroclinic zone in warning situations. Forecasters should know the locations and movements of these boundaries before the first convective storm forms. This should lead to more effective warning strategies that reduce false alarm rates.

The density of the Doppler radar network presently deployed in the U.S. may not be adequate to resolve, to optimal benefit of warning forecasters, low-level features of tornadic convection within TCs. This event emphasizes the value of the capability to assess tornadic convection within TCs at the closest possible proximity using densely spaced Doppler radars — both in the U.S. and abroad. Further, it is hoped that tropical countries with extensive coastlines will introduce Doppler radars and study this challenging problem, which deserves more attention both in research and operational forecasting.

Acknowledgements

This work was funded in part by a UCAR—COMET Partners Project grant between Saint Louis University and the Storm Prediction Center. Partial funding was provided by Dr. Brennan, Dean, Graduate School, Saint Louis University for travel. The authors are grateful to Saint Louis University for the award of a SLU 2000 research assistantship used in completing this research. Finally, a special thank you to Charles E. Graves for computer support.

REFERENCES

BROOKS, H.E., DOSWELL III, C.A., and DAVIES-JONES, R., *Environmental helicity and the maintenance and evolution of low-level mesocyclones.* In *The Tornado: Its Structure, Dynamics, Prediction and Hazards* (eds., C.R. Church) (Amer. Geophys. Union, Washington, D.C. 1993) pp. 97–104.

BURGESS, D. and RAY, P.S., *Principles of radar*. In *Mesoscale Meteorology and Forecasting* (ed. P.S. Ray) (Am. Meteor. Soc., Boston 1986) pp. 85–117.

BURPEE, R.W., *Mesoscale structure of hurricanes*. In *Mesoscale Meteorology and Forecasting*. (ed. P.S. Ray) (Am. Meteor. Soc., Boston 1986) pp. 311–330.

CRUM T.D., ALBERTY R.L., and BURGESS, D.W. (1993), *Recording, Archiving, and Using WSR-88D Data*. Bull. Am. Met. Soc. *74*, 645–653.

CURTIS, L. (2004), *Midlevel Dry Intrusions as a Factor in Tornado Outbreaks Associated with Landfalling Tropical Cyclones from the Atlantic and Gulf of Mexico*. Wea. Forecasting *19*, 411–427.

DOVIAK, R.J., DOSWELL and BURGESS and ZRNIC, D.S., *Doppler Radar and Weather Observation, second edition* (Academic Press Inc., New York 1993).

FUJITA, T.T. (1981), *Tornadoes and Downbursts in the Context of Generalized Planetary Scales*, J. Atmos. Sci. *38*, 1511–1534.

HILL, E.L., MALKIN, W., and SCHULZ, Jr., W.A. (1966), *Tornadoes Associated with Cyclones of Tropical Origin — Practical Features*, J. Appl. Meteor. *5*, 745–763.

JORGENSEN, D.P. (1984), *Mesoscale and Convective Scale Characteristics of Mature Hurricanes. Part 1: General Observations by Research Aircraft*, J. Atmos. Sci. *41*, 1268–1285.

KULSHRESTHA, S.M. (1990), Role and status of meteorological and hydrological services in WMO regions II and V in economic and social development. Proc. *Technical Conference, Economic and Social Benefits of Meteorological and Hydrological Services,* World Meteorological Organization No. 733, 366–369.

MADDOX, R.A., HOXIT, L.R., and CHAPPELL, C.F. (1980), *A Study of Tornadic Thunderstorm Interactions with Thermal Boundaries*, Mon. Wea. Rev. *108*, 322–336.

MARKOWSKI, P.M., RASMUSSEN, E.N., and STRAKA, J.M. (1998), *The Occurrence of Tornadoes in Supercells Interacting with Boundaries during VORTEX-95*, Wea. Forecasting *13*, 852–859.

MARKS, F.D., *Radar observations of tropical weather systems*. In *Radar in Meteorology*, (ed. D. Atlas) (Am. Meteor. Soc., Boston 1990) pp. 401–425.

MARWITZ, J.D. (1972), *The Structure and Motion of Severe Hailstorms*, J. Atmos. Sci. *11*, 166–179.

MCCAUL, Jr., E.W., *Observations and simulations of hurricane-spawned tornadic storms*. In *The Tornado: Its Structure, Dynamics, Prediction and Hazards* (ed. C.R. Church) (Am. Geophys. Union Press, Washington, D.C. 1993) 119–142.

NCDC, 1998: *Storm Data*. Vol. 40, No. 9, 165 pp. [Available from National Climatic Data Center (NCDC), Rm. 120, 151 Patton Ave., Ashville, North Carolina 28801–5001.]

NATIONAL SEVERE STORMS LABORATORY, 2000: WATADS (WSR-88D Algorithm Testing and Display System): Reference Guide for version 9.0. [Available from Storm Scale Applications Division, National Severe Storms Laboratory, 1313 Halley Circle, Norman, Oklahoma 73069].

PASCH, R.J., AVILA, L.A., and GUINEY, J.L. (2001), *The Atlantic Hurricane Season of 1998*, Mon. Wea. Rev. *129*, 3085–3123.

PEARSON, A.D., and SANDOWSKI, A.F. (1965), *Hurricane Induced Tornadoes and their Distribution*. Mon. Wea. Rev. *93*, 461–464.

RAGHAVAN, S. (1997), *Radar Observations of Tropical Cyclones over the Indian Seas*, Mausam *48*, 169–188.

RAO, G.V., and BHASKAR-RAO, D.V. (2003), *A Review of Some Observed Mesoscale Characteristics of Tropical Cyclones and Some Preliminary Numerical Simulations of their Kinematic Features*, Prof. Indian Nat. Sci. Acad., *69*(5), 523–541.

RAO, G.V., EDWARDS R., and SCHECK, J.W., *Case studies of tornadoes associated with tropical cyclones based on conventional and WSR-88D data*, In *Preprints, 24th Conf. on Hurricanes and Tropical Meteor.* (Am. Meteor. Soc., Fort Lauderdale, Florida 2000) pp. 306–307

SCHECK, J.W. (2001), *A WSR-88D Study of Tornadic Mesocyclones Embedded in Tropical Cyclones Danny 1997, Earl 1998 and Frances 1998*, Unpublished M.Sc. thesis in Meteorology, Saint Louis University, Saint Louis, Missouri, 91 pp.

SCHECK, J.W., RAO, G.V., SANTHANAM, K., EDWARDS, R., and SCHAEFER, J.T. (2004), *WSR-88D Investigation of Tornadic Mesocyclones Associated with Tropical Cyclone Earl (1998) and a Baroclinic Boundary in Central Florida*, [submitted to *Wea. Forecasting*, October 2004]

SMITH, J.S. (1965), *The Hurricane-tornado*, Mon. Wea. Rev. *93*, 453–459.

SPRATT, S.M., SHARP, D.W., WELSH, P., SANDRIK, A., ALSHEIMER, F., and PAXTON, C. (1997), *A WSR-88D Assessment of Tropical Cyclone outer Rainband Tornadoes*, Wea. Forecasting *12*, 479–501.

SUZUKI, O., NIINO, H., OHNO, H., NIRASAWA, H. (2000), *Tornado Producing Mini Supercells Associated with Typhoon 9019*, Mon. Wea. Rev. *128*, 1868–1881.

WEISMAN, M.L. and KLEMP, J.B. (1982), *The Dependence of Numerically Simulated Convective Storms on Wind Shear and Buoyancy*, Mon. Wea. Rev. *110*, 504–520.

(Received August 10, 2004, accepted October 12, 2004)

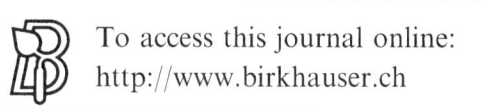

To access this journal online:
http://www.birkhauser.ch

Pure appl. geophys. 162 (2005) 1643–1672
0033–4553/05/091643–30
DOI 10.1007/s00024-005-2687-6

© Birkhäuser Verlag, Basel, 2005

❚ **Pure and Applied Geophysics**

Air-sea Coupling During the Tropical Cyclones in the Indian Ocean: A Case Study Using Satellite Observations

BULUSU SUBRAHMANYAM,[1] V.S.N. MURTY,[2] RYAN. J. SHARP,[1] and JAMES. J. O'BRIEN[1]

Abstract—In the years 1999 and 2001, three intense tropical cyclones formed over the northern Indian Ocean—two over the Bay of Bengal during 15–19 and 25–29 October, 1999 and one over the Arabian Sea during 21–28 May, 2001. We examined the thermal, salinity and circulation responses at the sea surface due to these severe cyclones in order to understand the air-sea coupling using data from satellite measurements and model simulations. It is found that the Sea Surface Temperature (SST) cooled by about 0.5 °–0.8 °C in the Bay of Bengal and 2 °C in the Arabian Sea. In the Bay of Bengal, this cooling took place beneath the cyclone center whereas in the Arabian Sea, the cooling occurred behind the cyclone only a few days later. This contrasting oceanic response resulted mainly from the salinity stratification in the Bay of Bengal and thermal stratification in the Arabian Sea and the associated mixing processes. In particular, the cyclones moved over the region of low salinity and smaller mixed layer depth with a distinct mixed layer deepening to the left side of the cyclone track. It is envisaged that daily satellite estimates of SST and Sea Surface Salinity (SSS) using Outgoing Longwave Radiation (OLR) and model simulated mixed layer depth would be useful for the study of tropical cyclones and prediction of their path over the northern Indian Ocean.

Key words: Tropical cyclones, Indian Ocean, EOL, OLR, sea-surface salinity, mixed layer depth, Remote Sensing.

1. Introduction

Tropical cyclones, the intense swirling motions of air moving cyclonically around a low-pressure center in the Northern Hemisphere and influencing vast horizontal distances in space, form over warmer oceanic regions where the Sea Surface Temperature (SST) is above 26 °C. In the tropical Indian Ocean, cyclones form over both the Bay of Bengal and the Arabian Sea and there are marked seasonal variations in their places of origin, tracks and intensities (ANONYMOUS, 1979). The

[1]Center for Ocean-Atmospheric Prediction Studies, The Florida State University, Tallahassee, FL 32306-2840, U.S.A.
[2]Physical Oceanography Division, National Institute of Oceanography, Goa, 403 004, India.

life span of a severe cyclonic storm over the Indian Ocean averages about 4 days from the time it forms until the time it makes landfall. On a yearly basis, the number of cyclones over the Bay of Bengal is 3–4 times more those over the Arabian Sea (ANONYMOUS, 1979; OBASI, 1997). Over the Bay of Bengal, tropical cyclones form during April-May over the southern and central Bay and move initially northwest striking the Indian coast and at times change gradually to a northward direction towards Bangladesh and at times recurve towards the Mynamar coast. Severe intensity tropical cyclones form during October-November mostly over the southern and central Bay and Andaman Sea and move westward towards the east coast of India and sometimes recurve between 15 ° and 18 °N affecting the Bangladesh coast. Weakened Pacific typhoons rejuvenate into low-pressure systems over the Andaman Sea and intensify further over the Bay during October-November. These cyclones give rise to copious rainfall and often cause major inland flooding and associated damage and loss of life in the eastern portion of the Indian subcontinent. In the Arabian Sea, cyclones generally form over southeast and central regions in May, October through December and in east central Arabian Sea in June. Some of the cyclones that originate in the Bay of Bengal travel across the peninsula, weaken and emerge into the Arabian Sea as low-pressure systems and may again intensify into cyclonic storms (ANONYMOUS, 1979). Knowledge of the ocean response to storm forcing is one of the key factors in tropical cyclone track prediction. Monitoring of surface weather parameters and the upper ocean thermal structure from moored buoys has become a key element in the studies of tropical cyclones and the ocean's response (PREMKUMAR *et al.*, 2000). Many studies were carried out to document the ocean's response to the tropical cyclones and to understand the associated air-sea interaction processes (for example, STRAMMA *et al.*, 1986; SHAY, 1994; SHAY *et al.*, 2000; GINIS, 2002). Tropical cyclones are driven by turbulent heat fluxes from the ocean and induce rapid changes in the heat and buoyancy fluxes and the currents in the oceanic mixed layer over a very short span of time (few days to weeks) and the magnitude of such processes are known only to some extent, particularly for the western Atlantic and Pacific. Because of this, one can conclude that the response of the upper ocean to the passing storm plays a major role in controlling the storm intensity through associated variability in the upper ocean conditions (such as surface temperature, surface salinity, mixed layer depth, currents in the mixed layer, etc.) and atmospheric conditions (such as winds, atmospheric pressure, and rainfall/ precipitation). In the case of the tropical Indian Ocean, earlier investigators (RAO, 1987; GOPALAKRISHNA *et al.*, 1993; MURTY, 1983; MURTY, *et al.*, 1996; SEE-TARAMAYYA *et al.*, 2001; CHINTHALU *et al.*, 2001) reported cooling of the sea-surface temperature (SST) due to the passage of cyclones in the Bay of Bengal and Arabian Sea. They noted that in regions of weaker upper ocean stratification (southern and western Bay of Bengal), magnitude of the SST decreases due to the cyclone was around 2 °C (RAO, 1987; GOPALAKRISHNA *et al.*, 1993) while in the northern Bay where the upper ocean stratification is strong due to low salinity waters, the SST decrease was between 0.3 °C and 0.9 °C (MURTY *et al.*, 1996; SEETARAMAYYA *et al.*, 2001; VINAYCHANDRAN *et al.*, 2002). This cyclone-induced SST decrease appears to lead to horizontal temperature gradients and in turn the pressure gradients in the atmosphere, thereby further affecting the intensity of the cyclone. However, the above studies have not indicated the variation of sea-surface salinity (SSS) during the storm period, as there was a lack of observations. During the Typhoon 90 Experiment in the western Pacific, the extensive observations indicated a significant decrease of SSS in the regions directly influenced by precipitation (SHAY, 1994).

Figures 1a–b show the paths of two tropical cyclones (TC) 04B and 05B that developed over the Bay of Bengal during 15–19 October and 25–30 October, 1999. TC 04B reached maximum intensity on October 17, 1999 before making landfall at the east coast of India. TC 05B was designated as a "Super Cyclone," it was the most

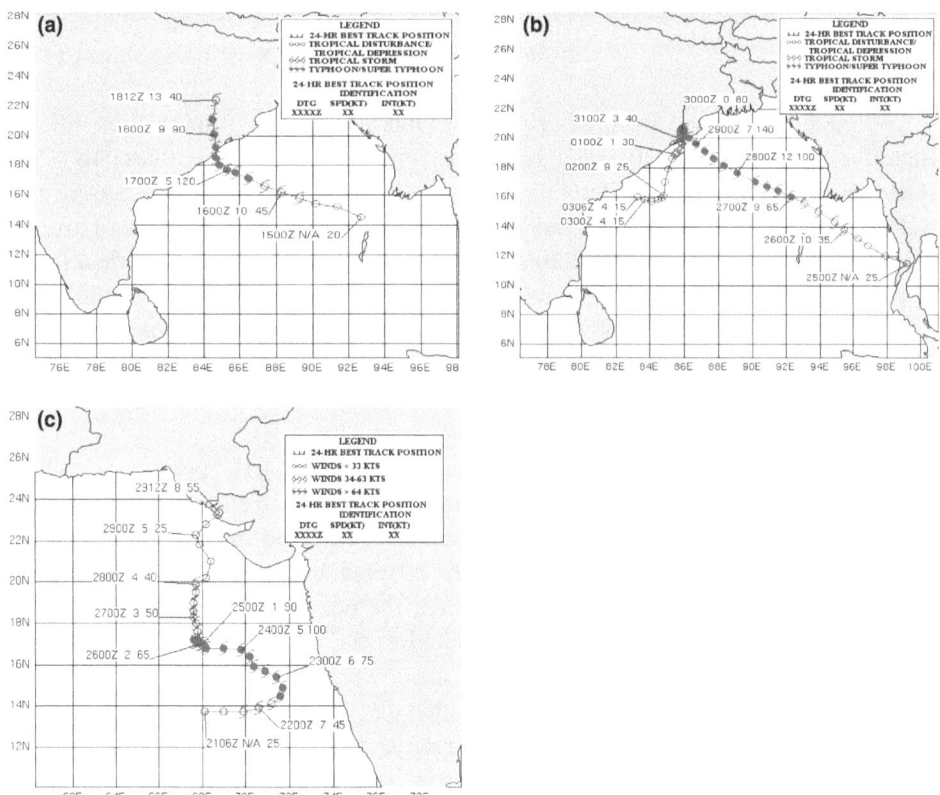

Figure 1
Tracks of tropical cyclones during (a) October 15–19, 1999 (TC 04B), (b) October 25–29, 1999 (TC 05B) in the Bay of Bengal and (c) May 21–28, 2001 (TC 01A) in the Arabian Sea (obtained from Joint Typhoon Warning Center, USA).

powerful tropical cyclone on record to affect coastal India. TC 05B developed from a disturbance that originated in the South China Sea and tracked through the Gulf of Thailand and across the Malay peninsula before strengthening over the Andaman Sea. It intensified at a greater rate than the climatological rate, peaking on October 28 at 1800Z as a Very Severe Cyclonic Storm (VSCS) and made landfall after 11 hours. LAL (2001) in his review on tropical cyclones had listed both TC 04B and 05B and discussed their consequences. Figure 1c shows the path of the first tropical cyclone in the year 2001 (TC 01A) which formed and developed over the eastern Arabian Sea during May 21–28. As the system intensified, it moved in a northwesterly direction and as it weakened, it moved in a northerly direction.

In this study, we describe the air-sea coupling through the ocean's thermal, salinity and circulation responses at sea surface due to these intense tropical cyclones in the northern Indian Ocean with a new observational method, based on satellite measurements and mixed layer model simulations.

2. Data and Methods

Wind stress derived from SeaWinds on QuikSCAT scatterometer was differentiated to derive the vorticity fields during the periods of tropical cyclones 04B (15–19 October, 1999), 05B (25–30 October, 1999) and 01A (21–28 May, 2001). SHARP *et al.* (2002) showed that vorticity could be used as a tool to identify areas of tropical cyclogenesis. Vorticity was calculated within the SeaWinds swath rather than from a regularly gridded product to obtain high resolution within the swath. SeaWinds data are freely available via the web http://www.coaps.fsu.edu from the Center for Ocean-Atmospheric Prediction Studies (COAPS) at Florida State University, U.S.A. Surface pressures were objectively calculated for the periods of tropical cyclones 04B, 05B and 01A using the method by HILBURN *et al.* (2003).

Daily OLR data on $2.5\,° \times 2.5\,°$ grids were obtained for the periods of cyclones (October 1999 and May 2001) from Climate Diagnostics Center (CDC), Boulder, Colorado, U.S.A. MURTY *et al.* (2000, 2002, 2003) reported that intense convection over the tropical Indian Ocean is closely coupled to the oceanic surface layer conditions (temperature and salinity) through the oceanic parameter, the Effective Oceanic Layer (EOL) which is linearly related to the OLR. The EOL is defined as the geopotential thickness (m^2/s^2) of the near-surface stratified layer, chosen as 30 m thick from sea-surface. This means that the EOL represents the sea-surface elevation in dynamic centimeters (dyn.cm) relative to the 30 db level. Climatological monthly temperature and salinity data from the World Ocean Atlas (WOA98; ANTONOV *et al.*, 1998 and BOYER *et al.*, 1998) were used to compute the monthly EOL. Regression relationships (algorithms) were then obtained between the EOL and the WOA98 SST for each month and the algorithms for October and May are mentioned in Sections 3.4 and 3.5. The 16 year-mean monthly OLR data and the computed monthly EOL data were used to develop relationships between OLR and EOL for each month for the Bay of Bengal and the Arabian Sea (MURTY et al., 2004). Similarly, algorithms were also obtained between the 16 year-mean monthly OLR and monthly WOA98 SSS for each month (MURTY et al., 2004). These monthly algorithms (OLR-EOL-SST and OLR-SSS) are applied to the synoptic data for the periods of the cyclones. Mention is made here that we noticed an improvement in the correlation coefficients for the OLR-EOL and OLR-SSS relationships for October, when compared to that reported in MURTY et al. (2002). This is attributed to the additional OLR data and the improved WOA98 climatology from that of LEVITUS et al. (1994a,b) climatology. The daily OLR values over the Bay of Bengal and Arabian Sea during the periods of cyclones were far lower than the respective 16 year-mean monthly OLR. Therefore, in order to estimate the SSS, a 2nd degree polynomial relationship between the 16 year-mean OLR and WOA98 SSS (specific for this study) was used for October only for the Bay of Bengal and a linear relationship for May for the Arabian Sea (as proposed by MURTY et al., 2004). The U.S. Navy's Modular Ocean Data Analysis System (MODAS) daily SSS data (FOX et al., 2002a,b), provided by Naval Research Laboratory (NRL) at Stennis Space Center, Mississipi, U.S.A., were also examined for the periods of cyclones 04B, 05B and 01A. In addition, daily values of precipitation and SST from the Tropical Rainfall Measuring Mission (TRMM) on $0.25° \times 0.25°$ grids were obtained from Remote Sensing Systems (http://www.remss.com/tmi). Finally, the Navy Layered Ocean Model (NLOM) simulated SST and mixed layer depth (MLD) data (KARA et al., 2000, 2003a,b), as provided by Naval Research Laboratory at Stennis Space Center, Mississipi, U.S.A. were also utilized. This NLOM has $1/8°$ resolution and forced with the six-hourly reanalysis winds obtained from the European Center for Medium-Range Weather Forecasts (ECMWF) Reanalysis; the SST simulation was available on a daily basis while the simulated mixed-layer-depth (MLD) was available as averages of 3.05 days to suppress the effect of inertial oscillations.

3. Results and Discussion

3.1 Surface Wind Field and Sea-level Pressure during the Tropical Cyclones

The cyclone 04B was located at 16.0 °N, 88.5 °E at 0300 Z and at 17.8 °N, 87 °E at 1200 Z on 16 October and moved across the Bay of the Bengal in three days while intensifying into VSCS on 17–18 October (Fig. 1a). This system weakened as Deep Depression (DD) on 19 October over the land. The cyclone 05B system moved in a northwesterly direction across the Bay of Bengal in 5 days from 25 October over the eastern Andaman Sea to 29 October over the northwestern Bay (Fig. 1b). This system also developed into VSCS stage on 28 October and further intensified over land into a 'Super Cyclone' on 30 October. The Indian Daily Weather Reports (IDWRs) issued by the India Meteorological Department (IMD) reported a central pressure of 996 hpa on 28 October at the VSCS stage.

The SeaWinds-derived voriticity fields and mean sea-level pressures associated with the cyclones 04B and 05B identified the regions with stronger wind speeds (30–35 knots), larger cyclonic (positive) vorticity (and associated convergence) and the rain band structures for each cyclone. On 16 October, the cyclonic vorticity was strong (20–25×10^{-5} s^{-1}) at the center of the cyclone (Fig. 2a) and extended over a $2°\times2°$ degree area with its axis oriented in a southwest-northeast direction,

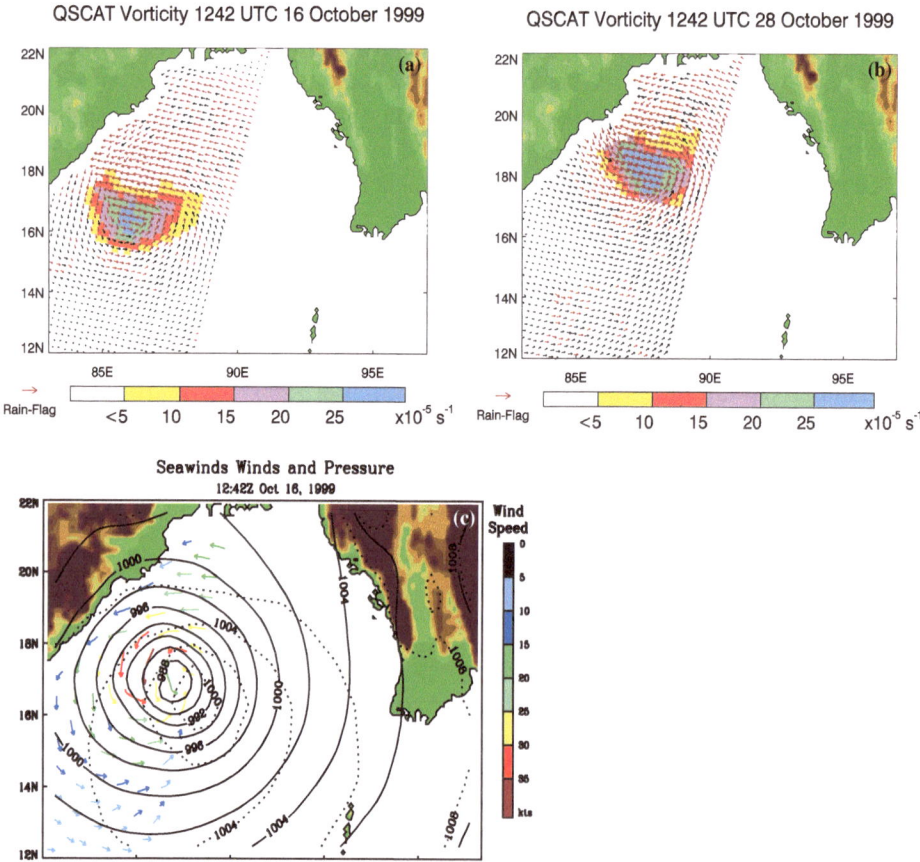

Figure 2

SeaWinds Scatterometer winds derived vorticity fields for (a) October 16, 1999, (b) October 28, 1999 during the intense stages of TC 04B and TC 05B respectively, with red arrows indicating the multi-dimensional histogram (MUDH) flagged rainfall, and (c) SeaWinds calculated mean sea-level pressure (thick black lines), with ECMWF background mean sea-level pressure (black dotted lines) and wind speed color vectors for October 16, 1999 during TC 04B.

identifying the path of the cyclone. The intense positive vorticity field at 1242 UTC (Fig. 2a) weakened towards the north by 2348 UTC on the same day while the rain band intensified in the western sector (figure not shown). The SeaWinds-derived vorticity fields during cyclone 05B was more intense (than that of 04B) reaching a vorticity of $25–30\times10^{-5}$ s^{-1} by 1242 UTC on 28 October (Fig. 2b). The SeaWinds-derived mean sea-level pressure (solid lines) and the ECMWF reanalysis mean sea level pressure (dotted lines) on 16 October (Fig. 2c) showed a difference of 12 hPa at the center of the cyclone. The SeaWinds calculated pressure was 988 hPa whereas the ECMWRF analyzed pressure was 1000 hPa. The IDWRs reported a central pressure of 1002 hPa at the center of the cyclone (CHINTHALU et al., 2001) which was close to the ECMWF pressure at the cyclone center. The isobars were closer in the northwestern sector of the cyclone and wide in its southeastern sector. The isobars of mean sea-level pressure nearly coincided with those reported by the IDWRs (CHINTHALU et al., 2001). The wind vectors showed a converging inward flow of speed up to 30–35 knots at the northwestern sector of the cyclone. The techniques developed by TURK and BANKS (1996) were used to plot the wind vectors superimposed on the surface pressure.

In the Arabian Sea, the SeaWinds-derived vorticity field for cyclone 01A clearly shows its location, intensity, extent and shift (movement) . The cyclone moved 4 ° in latitude from 23 May to 28 May, when it started weakening. The cyclone intensified on 24 May reaching a vorticity of $25–30\times10^{-5}$ s^{-1} at the cyclone center (17.5 °N, 68.5 °E)(Fig. 3a). The rain band was located on the southern edge of the cyclone. The SeaWinds-calculated mean sea-level pressure showed convergence of air towards the center of the cyclone and the central pressure reached as low as 988 hPa on 28 May (Fig. 3b). However, the ECMWF reanalysis mean sea-level pressure at the center of the cyclone was 996 hPa (black dotted contours in Fig. 3b).

3.2 OLR during the Tropical Cyclones

In association with the development of the weather systems 04B and 05B from depression to VSCS/Super cyclone stages, intense deep-convection took place over the Bay of Bengal. The zone of intense convection was identified with the zone of low OLR. The cyclone track was superimposed on the OLR distribution maps and the position of cyclone center at 1200Z on each day was marked (Fig. 4). On 15 October, the center of the 'depression' was characterized with a low OLR of 160 W/m^2 west of the Andaman Islands (Fig. 4a). On subsequent days, the low OLR zone followed the track of the weather system. The low OLR zone shifted to the northwestern Bay when the weather system intensified into the VSCS stage on 17 October (Fig. 4b) and the OLR was relatively high ($\sim$200 W/m^2) in the southeastern sector of the system. The weather system 05B was characterized by low OLR ranging from 120–140 W/m^2 (this low was much lower than that of cyclone 04B) at its center (Fig. 4c) and could be used to track the system during its movement across the Bay. On 28 October, the

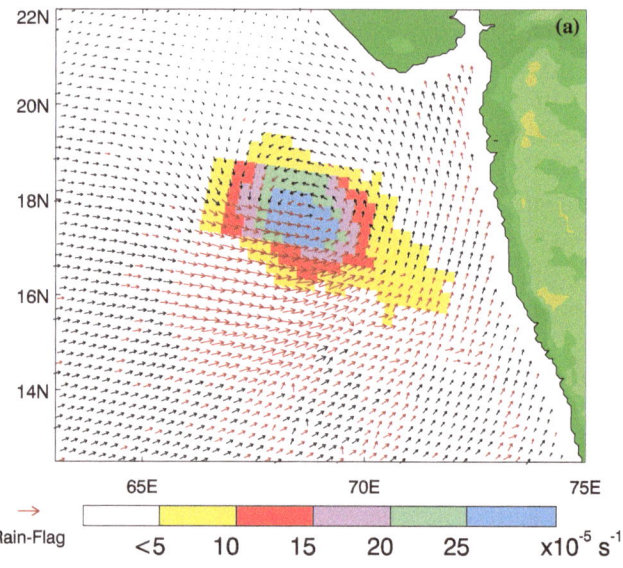

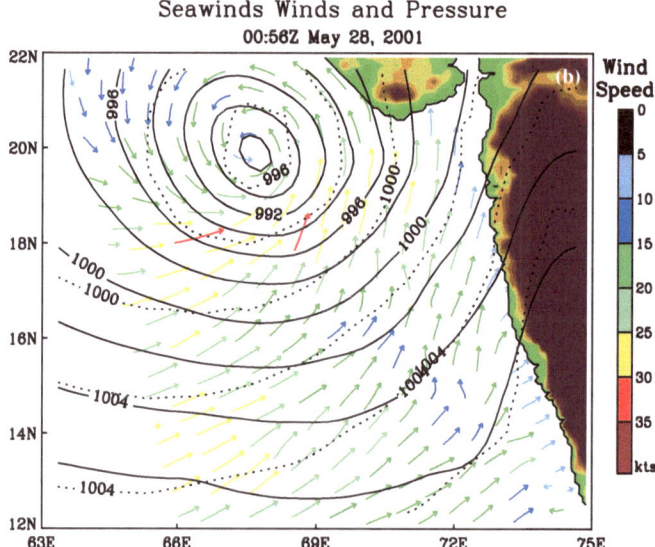

Figure 3

SeaWinds Scatterometer winds derived (a) vorticity field for May 24, 2001 during the intense stage of TC 01A, with red arrows indicating the multi-dimensional histogram (MUDH) flagged rainfall and (b) SeaWinds calculated mean sea-level pressure (thick black lines), with ECMWF background mean sea-level pressure (black dotted lines) and wind speed color vectors for May 28, 2001 during weakening stage of TC 01A.

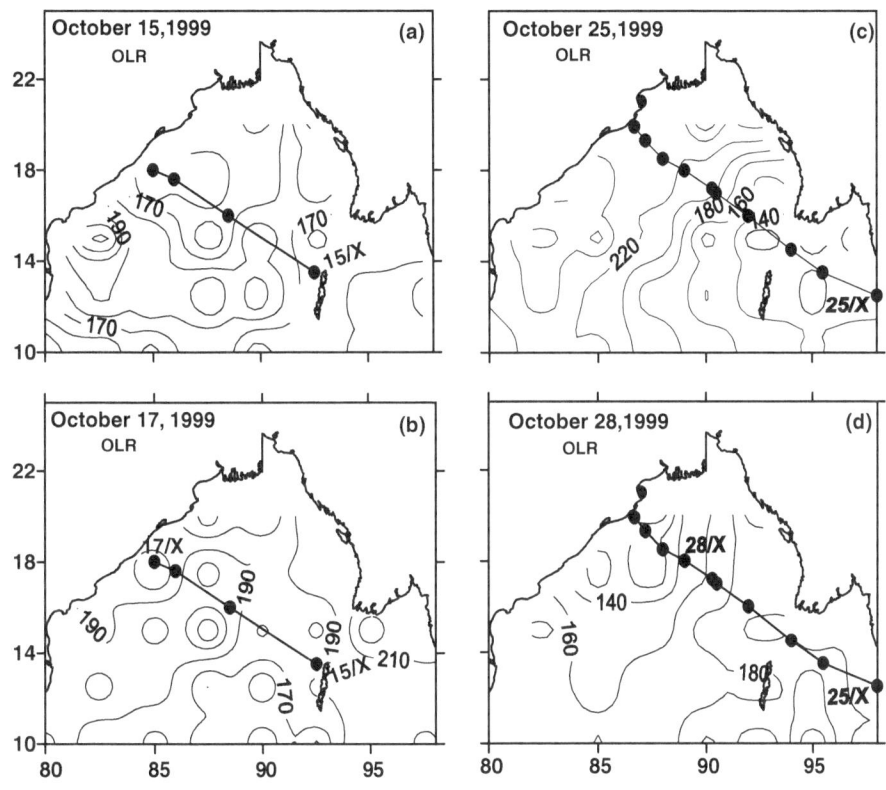

Figure 4

Variation of satellite measured Outgoing Longwave Radiation (OLR) during various stages (initial and intense) of (a–b) TC 04B (left panel) and (c–d) TC 05B (right panel).

system intensified into VSCS stage and the low OLR zone (< 140 W/m^2) shifted to the northwestern Bay (Fig. 4d).

In the Arabian Sea, one would see a change in the pattern of OLR during the TC 01A. Initially, on 21 May the OLR was minimum at 180 W/m^2 before the development of the weather system (Fig. 5a). On the day of its development into depression, i.e., on 24 May, a zone of low OLR (< 140 W/m^2) formed slightly to the left of the cyclone track (Fig. 5b). At the weakening stage of the cyclone on 28 May, the low OLR (< 240 W/m^2) zone shifted westwards while the cyclone moved northward towards the Indian peninsula (Fig. 5c).

3.3 EOL and the Inferred Surface Circulation during the Tropical Cyclones

The linear relation between the 16 year-mean OLR and climatological EOL had a correlation coefficient of $r^2 = 0.36$ in October in the Bay of Bengal (MURTY et al., 2004). MURTY et al. (2002) obtained a correlation coefficient of $r^2 = 0.16$ for

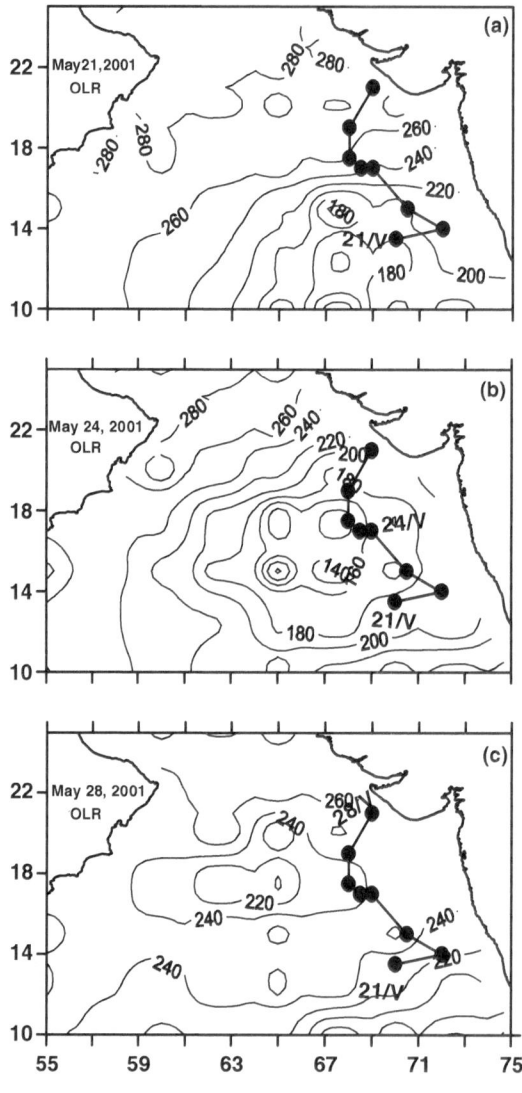

Figure 5
Daily variation of (a–c) satellite measured Outgoing Longwave Radiation (OLR) during various stages (initial, intense and weakening) of TC 01A.

October, with the 13 year-mean INSAT 1-D OLR and the climatology of LEVITUS *et al.* (1994a,b). Using this relationship (EOL = 0.0171*OLR − 1.511, r^2 = 0.36), the EOL was estimated (E_{eol}) at each grid point (2.5 ° × 2.5 °) from the daily OLR during the periods of TC 04B and TC 05B. The variation of E_{eol} facilitates to understanding the near-surface circulation under the influence of weather systems. Cyclonic (anticyclonic) circulation is inferred around the lower (higher) E_{eol}, with higher E_{eol} to the right of the flow direction (in the Northern Hemisphere).

Figures 6a–d show the variation of E_{eol} during the initial and intense stages of TC 04B (left panel) and TC 05B (right panel). The sea level height varied within a range of 8–10 dyn.cm from 1.2 to 2.0 m^2/s^2 on 15 October, and from 1.2 to 2.2 m^2/s^2 on 17 October (Figs. 6a–b). During the TC 05B, the sea-level height varied over a range of 20 dyn.cm from 0.6 to 2.6 m^2/s^2 on 25 October and a reduced range of 14 dyn.cm (0.6 and 2.0 m^2/s^2) during subsequent days (Figs. 6c–d). The inferred surface circulation is cyclonic (anticlockwise) around the center of the cyclone and anticyclonic (clockwise) at the rear end of the cyclone. On 15 October, the surface circulation was cyclonic north of 14 °N in the northwestern Bay and anticyclonic in the central Bay between 12 ° and 16 °N (Fig. 6a). The cyclonic circulation cell appeared to be coupled to the cyclonic vorticity zone associated with the winds around the center of cyclone (Fig. 2a). This is consistent with the dynamics associated with the response

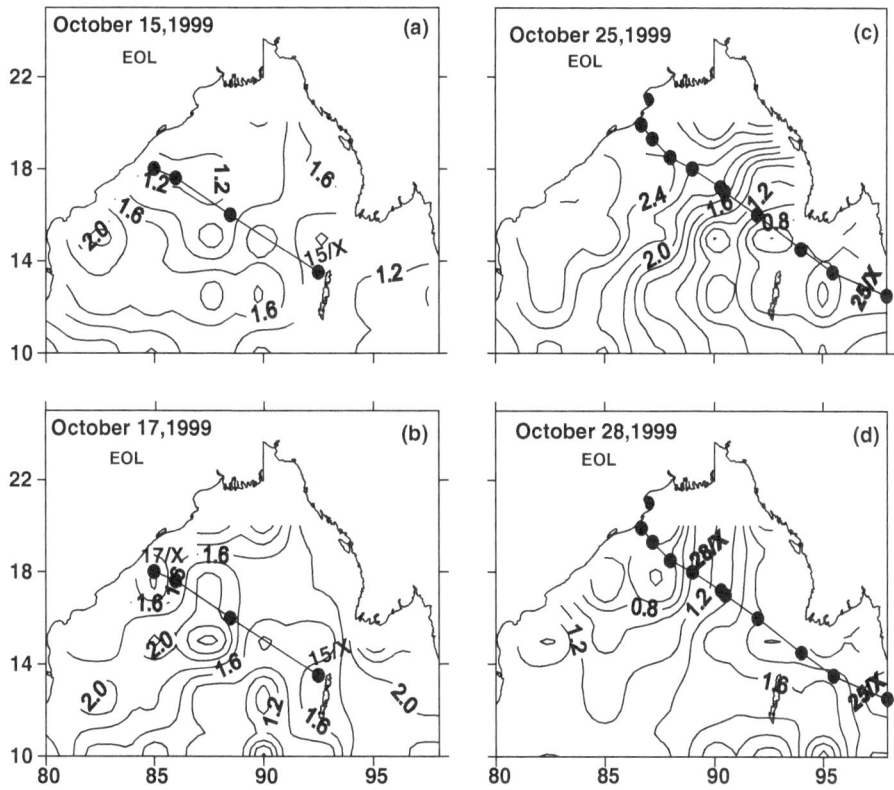

Figure 6
Variation of Effective Oceanic Layer (EOL) during various stages (initial and intense) of (a–b) TC 04B (left panel) and (c–d) TC 05B (right panel).

of the ocean to the wind field (p. 327; GILL, 1982). This coupling leads to the important observation of the presence of the surface Ekman divergence underneath the cyclone center (SHAY *et al.*, 2000) and geostrophic convergence at the outer edge of the cyclone.

As TC 05B intensified and moved across the Bay, the surface divergence zone also moved with it (Fig. 6b). The north/northwestward gradient in the estimated sea-surface height computed from the variation of E_{eol} during the period of cyclones resembles (qualitatively) that of the mean sea-level pressure, once again showing the response of ocean surface to the atmospheric forcing. The inferred surface eastward flow on 15 October (Fig. 6a) and southwestward flow on 17 October (Fig. 6b) in the central Bay agree with the measured currents at 3 m below the surface (qualitatively) at a moored buoy located at 13 °N, 87 °E as reported by CHINTHALU *et al.* (2001). On 25 October, the inferred surface circulation was cyclonic with its center (lower sea level) in the northern Andaman Sea and a south/southwesterly flow in the central Bay along the westward flank of the storm (Fig. 6c). Currents measured at 3 m below the sea surface by the moored data buoy at 13 °N, 87 °E confirm the inferred south/southwesterly flow in the central Bay (CHINTALU *et al.*, 2001). The cyclonic cell shifted northwestward as the weather system moved across the Bay and reached the northwestern Bay by 29 October. On the rear flank end of the system the surface circulation was anticyclonic. On 28 October the surface flow became northeasterly in the central Bay, opposite that seen on 25 October (Fig. 6d), which may be related to the inertial currents generated by the cyclone forcing. This identifies the quick response/change the surface flow adjusts/undergoes with the cyclonic wind forcing.

In the Arabian Sea, the 16 year-mean OLR and climatological EOL are negatively correlated in the month of May (EOL = −0.00072*OLR + 3.413 and r^2 = 0.9; MURTY *et al.*, 2004). Using this relation, the EOL was estimated (E_{eol}) from the daily OLR. During TC 01A the zone of intense convection (low OLR, 140–200 W/m^2), located to the left of the cyclone track (Figs. 5a–c), was characterized with higher E_{eol} (Figs. 7a–c). The inferred surface circulation associated with E_{eol} pattern is anticyclonic and the center of the weather system lay over the eastward flank of this anticyclonic cell. As the cyclone moved northward by 28 May, the intense convection zone also moved far northward. At the southeastern sector of the system a region of relatively higher OLR and lower EOL occurs, leading to cyclonic surface flow (Fig. 7c) and cooling at sea surface and deepening of MLD (see Sections 3.5.1 and 3.5.3).

3.4 Upper Ocean Response in the Bay of Bengal to the Tropical Cyclones

SHAY (1994) elaborated the thermal response and the momentum response, within the near-inertial dynamics, to the tropical cyclones on the upper ocean. RAO (2002) points out that there are two stages in the upper ocean response to the passage of cyclones; (1) local response which is mainly due to wind stress of the cyclone and

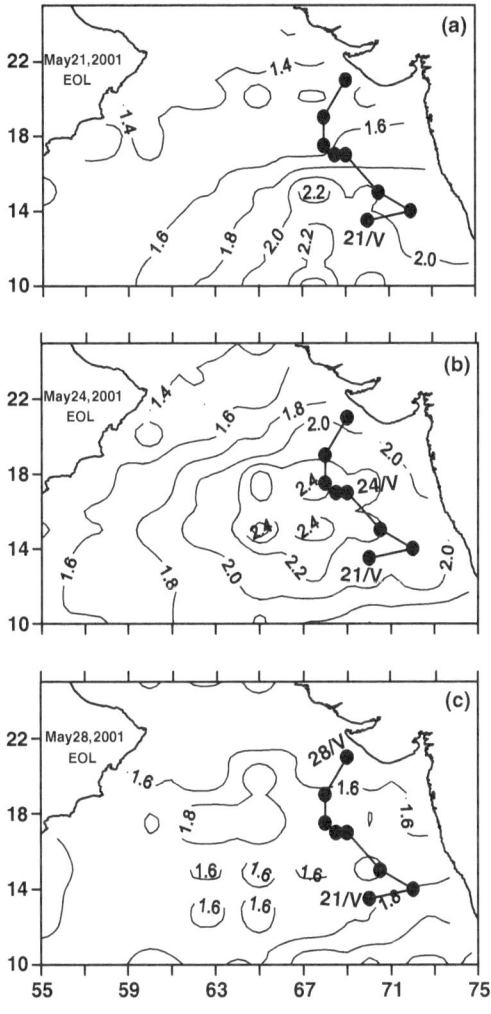

Figure 7
Variation of the Effective Oceanic Layer (EOL) during various stages (initial, intense and weakening) of
TC 01A.

includes substantial cooling of SST and mixed layer and strong currents in the mixed layer (baroclinc) and geostrophic currents associated with the trough in sea-surface height (barotropic) and (2) nonlocal baroclinic response due to the wind-stress curl. In Section 3.3, the barotropic response of the upper ocean (top 30 m layer) was interpreted as a change in the surface circulation during the cyclone periods. In this section, the local-baroclinic response due to the cyclones on the SST, SSS and mixed layer depth will be examined.

The thermal response (SST variation) to the cyclones was obtained in three ways — first by estimating the SST from the algorithm ($E_{sst} = 0.718*E_{eol} + 27.26$ wherein

E_{eol} is the estimated EOL; $r^2 = 0.34$) for October; second by using the TRMM measured SST; and third by the NLOM simulated SST. The buoyancy response (SSS variation) to the cyclones is examined by comparing the estimated SSS (E_{sss}) from OLR with the MODAS SSS during the cyclone period. The variation of MLD during the cyclones is obtained from the NLOM simulations.

3.4.1 Sea Surface Temperature

When the weather system 04B formed as a 'Low pressure' on 14 October, the estimated SST (E_{sst}) showed a variation of ~1.4 °C between 27.8 °C and 29.2 °C with SST increasing towards the northwestern Bay and a cell of low SST (27.8 °C) north of Andaman Islands (not shown). When the 'Low pressure' area developed into a 'Depression' on 15 October, the range of SST sharply reduced to ~0.6 °C between 28.2 ° and 28.8 °C, with relatively cold waters (~28.2 °C) on to the right of the cyclone center (identifying the rightward bias) and warm waters (> 28.4 °C) to its left (Fig. 8a). As the weather system further developed into a 'cyclonic storm' and moved northwestward on 16 October, SST decreased to 27.6 °C in the northwestern Bay (ahead of the system). Between 15 and 16 October, the SST was 0.6 °C lower within the cyclone center in the northwestern Bay (18 °N, 85 °E). As the weather system intensified into VSCS on 17 October (Fig. 8b), the northwestern Bay experienced persistent cooling and the central Bay experienced warming of ~0.4 °C (from 28.4 °C on 16 October to 28.8 °C on 17 October) as the sky became cloud free. As the VSCS weather system weakened into a 'Deep Depression' and moved further north, an increase in SST was noticed on 20 October (figure not shown) ranging from 28.8 °C to 29.0 °C, which was identical to conditions that prevailed before the formation of the weather disturbance on 14 October. This suggests that it took about 7 days for revival of SST to return to "normal" conditions after the weather system preduced its effect on the Bay. GOPALAKRISHNA *et al.* (1993) reported a 2-week return period for the SST to reach normal conditions after a severe cyclone storm passed.

At the 'Depression' stage of TC 05B, the daily variation in the E_{sst} revealed a range of 1.6 °C with the occurrence of low SST (27.6 °C) beneath the cyclone center and warmer SST (28.8 °C) on the left and ahead of the storm (Fig. 8c). As the weather system moved northwestward, the region of cooler SST, coupled to the cyclone center, moved along with it while warmer SSTs occurred behind it (Fig. 8d). Therefore the direction of the SST gradient reversed, compared to that on 25 October (Fig. 8c). Surface cooling beneath the cyclone center and warming on its rear end are the noteworthy features. Though the weather system intensified further into a 'Super Cyclone' over land, its impact on the E_{sst} was not strong.

CHINTHALU *et al.* (2001), from the analysis of moored buoy data collected 3 m below the surface at 13 °N, 87 °E, reported a decrease in SST on the order of 0.7 °–0.9 °C as a response to the TC 04B and 05B. This measured SST was closer to the computed E_{sst} at the same location. This supports the finding that the

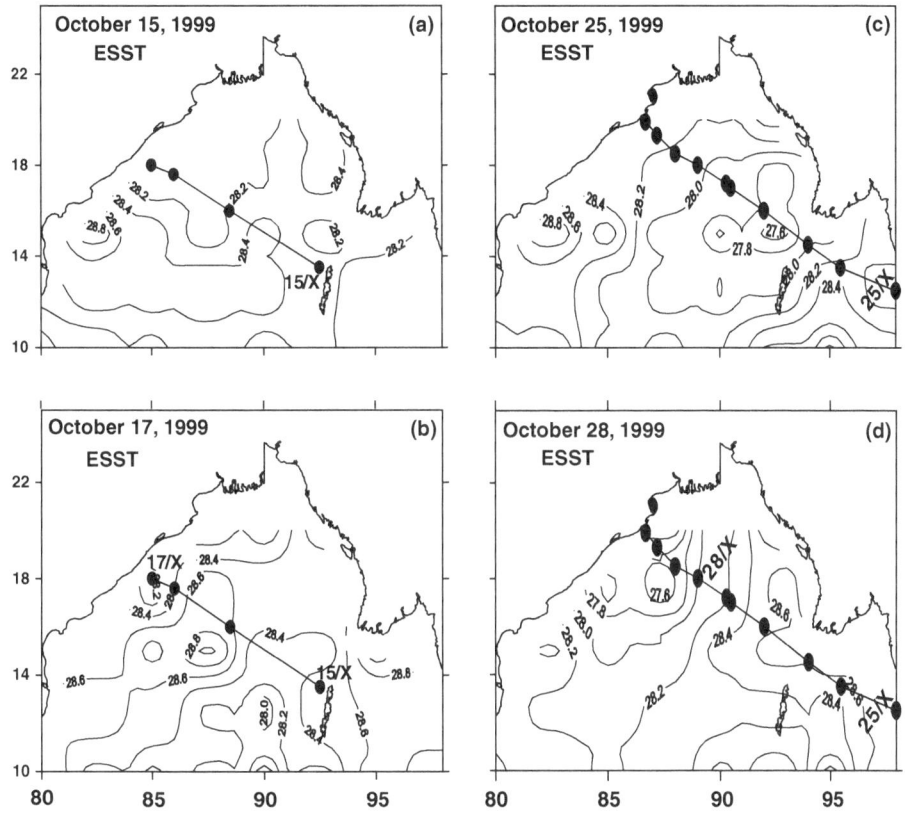

Figure 8
Variation of the estimated Sea Surface Temperature (E_{sst}) using the relationships between OLR, EOL and climatological SST during various stages (initial and intense) of (a-b) TC 04B (left panel) and (c-d) TC 05B (right panel).

estimation of SST using the daily OLR data through OLR-EOL-SST relationships could be a useful parameter to understand the thermal response processes and the prediction of the path of cyclones. Consideration of winds would improve the SST estimates.

The E_{sst} was also compared with TRMM SST (Figs. 9a–d) and NLOM SST (Figs. 10a–d) during the TC 04B and TC 05B. The TRMM SST was relatively higher compared to E_{sst}, with the TRMM SST on 17 October (weather system VSCS stage) warmer by 1.3 °C beneath the cyclone center and by 0.2 °–0.4 °C to the right of the cyclone center. Cooling in TRMM SST can also be seen along the tracks of TC 04 and TC 05B (Figs. 9b,d). While the variation of E_{sst} was ~0.6 °C, it was ~1.0 °C in TRMM SST and ~2.0 °C in NLOM SST along the cyclone track on 17 October (also on all the other days). On 28 October, the range of E_{sst} was 1 °C (27.6 °– 28.6 °C) while the range of TRMM SST was as high as 2.0 °C (27.5 °–29.5 °C) and

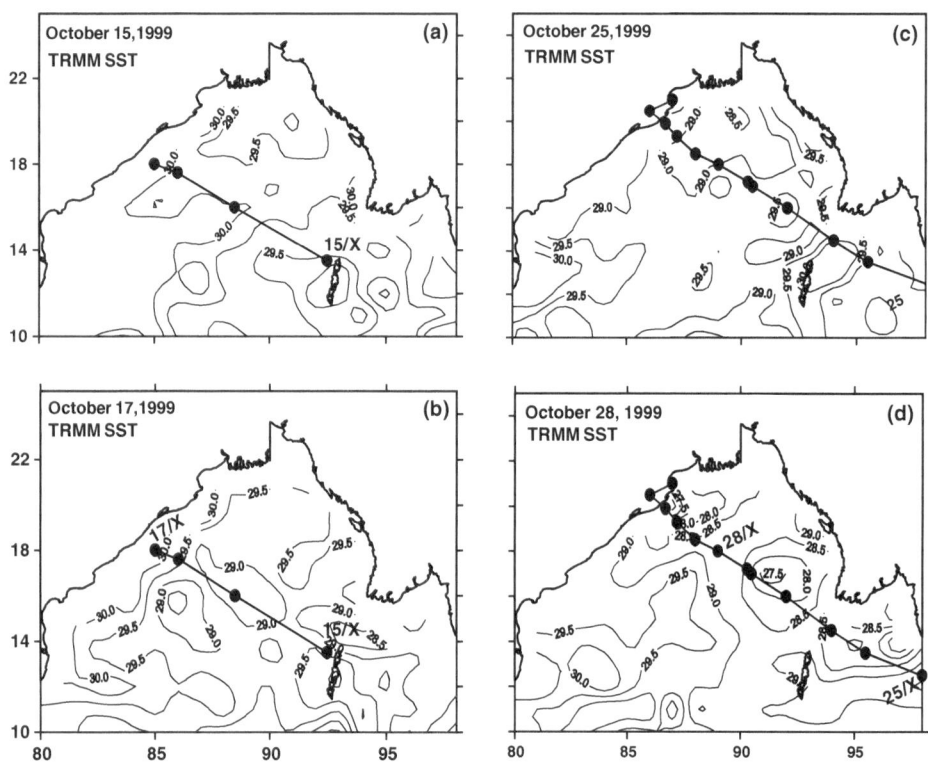

Figure 9
Variation of satellite measured TRMM SST during various stages (initial and intense) of (a–b) TC 04B (left panel) and (c–d) TC 05B (right panel).

the range of NLOM SST was 1 °C (27.5 °–28.5 °). The E_{sst} and the NLOM SST differed by 0.5 °C with cooling beneath the cyclone center and warming ahead of the cyclone center (Figs. 10a-d). The NLOM SST was always high in the direction of cyclone track (Figs. 10a-d), though it exhibited cooling beneath the cyclone center. The cooling of SST beneath the cyclone center arises from the upwelling due to the divergence of currents in the Ekman layer. It appears that the thermal response in SST cooling is evident in all the three types of SST variations.

3.4.2 Sea Surface Salinity

The estimated sea-surface salinity (E_{sss}) was computed from daily OLR using the 2nd degree polynomial relationship, $E_{sss} = -192.56 + 2.229*OLR - 0.0055*OLR*OLR$ and $r^2 = 0.41$. As well, we have limited the minimum estimated SSS to be 16.0. We find that the estimated SSS values from the 2nd degree polynomial are relatively better than those obtained from the linear relationship between OLR and SSS for the lower values of OLR associated with the tropical cyclones. The distribution of E_{sss}

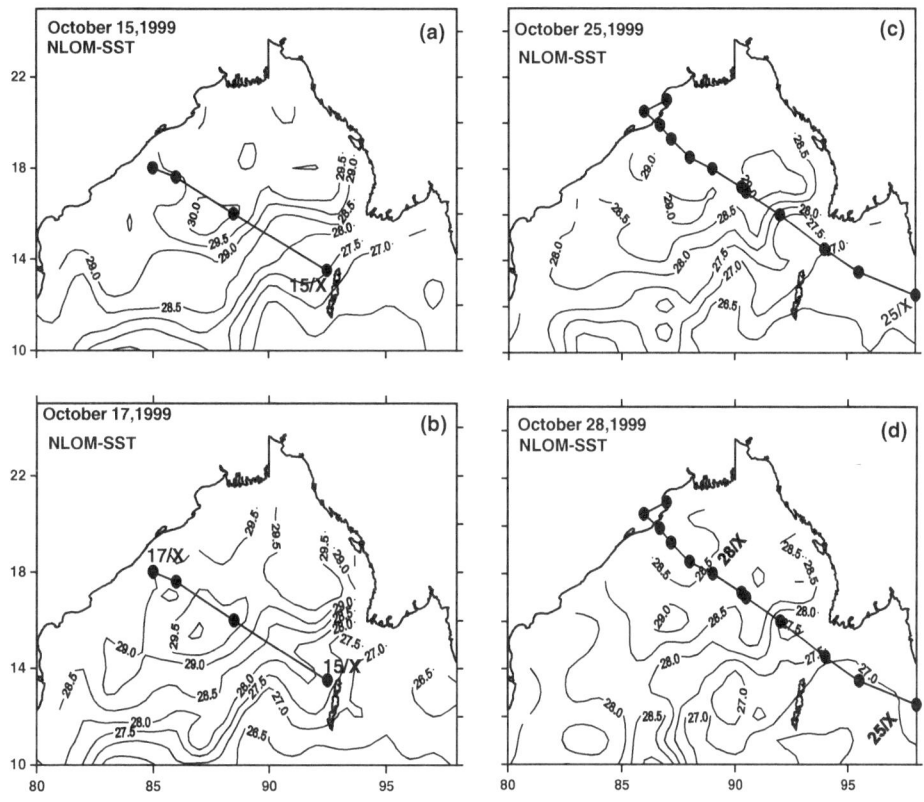

Figure 10
Variation of NLOM model simulated SST during various stages (initial and intense) of (a–b) TC 04B (left panel) and (c–d) TC 05B (right panel).

during the cyclone periods is presented in Figs. 11a–d. At first glance, it appears that the E_{sss} was far lower than the WOA98 SSS for the month of October. However, the variation of E_{sss} followed the TRMM rainfall pattern (not shown here), suggesting some part of the E_{sss} comes as a direct response to the precipitation associated with the cyclones. One would see cells of low salinity (as low as 18–20) beneath the cyclone center and along the track of the cyclone. The day-to-day variation of E_{sss} indicated that the cells of low salinity could be used for predicting the track of the cyclones from remote sensing measurements.

3.4.3 Mixed Layer Depth

The NLOM mixed layer depth was initially shallow (15 m) off the western Andaman Islands where the weather system was at 'depression' stage on 15 October (Figs. 12a–d). However, the MLD was deep (~30 m) ahead (in the direction of) of the cyclone and also on the left side of its track, at the region of maximum winds and

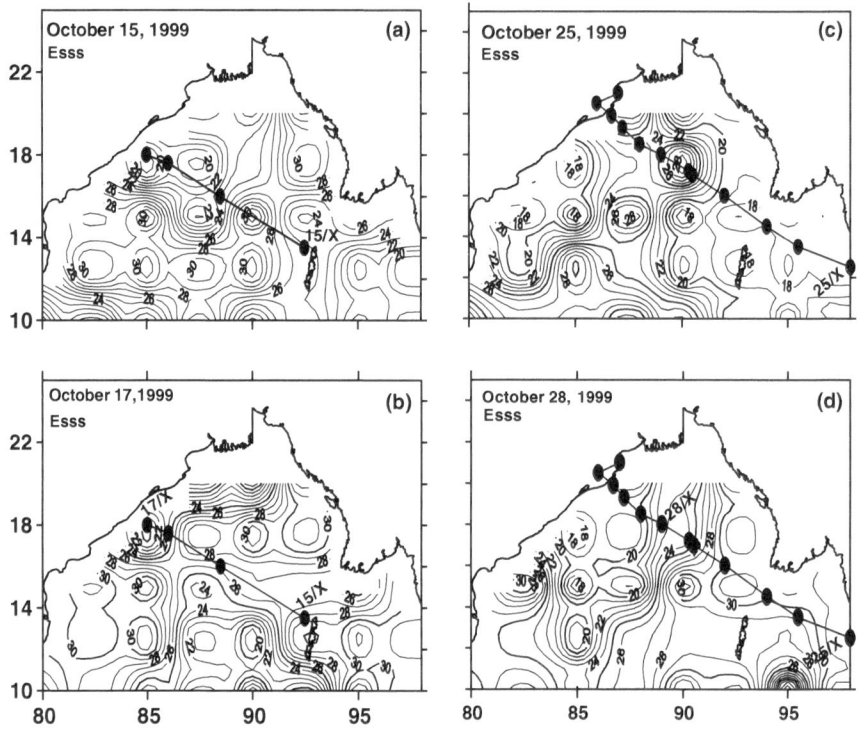

Figure 11
Variation of estimated Sea Surface Salinity (E_{sss}) using the relationship between OLR and climatological SSS during various stages (initial and intense) of TC 04B and 05B.

geostrophic convergence. This trend continued on subsequent days with shallow MLD developing near the Indian coast and a deepening of MLD to 55 m to the left side of the cyclone track between 14 ° and 16 °N. The cyclone 05B also moved over a region of shallow MLD (Figs. 12c-d) with deepening of the MLD to the left of the track. The trend of MLD variation is similar to that observed as in the case of TC 04B. The warm, deep mixed layer ahead of the cyclone provides necessary heat energy for the intensification of the cyclone. As the cyclone center passes over this deep mixed layer, the extraction of heat (due to turbulent air-sea heat fluxes) cools the mixed layer temperature and thus the SST. However, the occurrence of precipitation associated with the cyclone and the presence of low-salinity surface waters (Figs. 11a–d) lead to strong stratification and help maintain warmer SSTs (though cooled to a smaller extent by the cyclone) above 28.0 °C for the sustenance of the cyclones. The salinity stratification and shallow MLD are in turn responsible for the weaker response of SST (observed and estimated as in this study) to the intense cyclones (such as 04B and 05B) as reported through observations (MURTY *et al.*, 1996; SEETARAMAYYA *et al.*, 2001; CHINTHALU *et al.*, 2001). The cyclone-

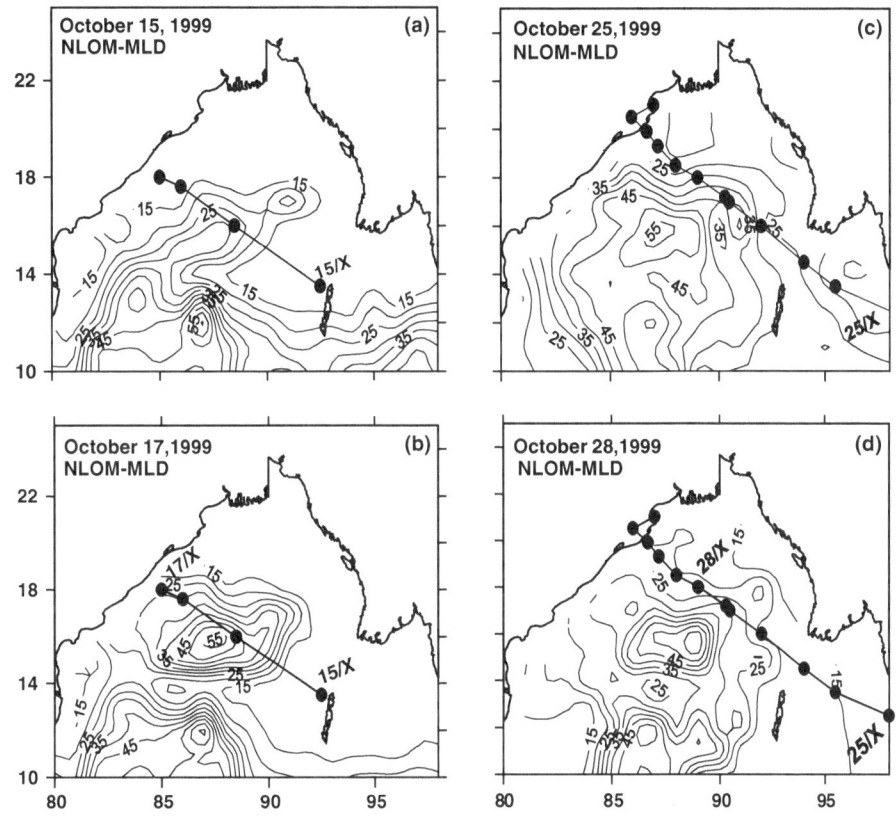

Figure 12
Variation of NLOM model simulated Mixed Layer Depth (MLD, m) during various stages (initial and intense) of (a–b) TC 04B (left panel) and (c–d) TC 05B (right panel).

induced divergence and the stratification lead to the development of a shallow mixed layer at the center of the cyclone with relatively moderate cooling (0.5–0.9 °C) limited to the stratified layer (i.e., mixed layer). This cooling process does not support further deepening of the mixed layer due to strong salinity stratification.

The isolines of MLD are closed across the track of the cyclone with larger MLD to the left of the track. Further, it is seen that the shallow MLD coincides with the lower E_{eol} beneath the cyclone center and the deep MLD with higher E_{eol} to the left side of cyclone track. The OLR shows intense convection ahead of the cyclone and relatively weakened convection to the left of the track. This has caused the estimated SST (using OLR-EOL-SST relationships) and TRMM SST to cool beneath the cyclone center and warm to the left of the cyclone track. When the weather system intensified, lower EOL was observed beneath the center of the cyclone and high on its periphery. These suggest that the cyclone-induced divergence (at the cyclone center) and convergence (at the maximum wind belt) are responsible for the variation of simulated MLD and the associated response in sea-surface conditions (E_{eol}, E_{sst} and E_{sss}, the parameters derived from the OLR).

Thus, it appears that the coupling between the OLR and EOL (through variation in sea-level height, inferred surface circulation, estimated SST and SSS) is useful to predict the path of tropical cyclones or even the monsoon depressions. GOSWAMI (1987) showed that the moisture convergence (or precipitation) on the western sector of a cyclone is responsible for the observed west-northwestward movement of the monsoon depressions that form during the southwest monsoon (June–September) in the northern Bay.

3.5 Upper Ocean Response in the Arabian Sea to the Tropical Cyclone 01A

3.5.1 Sea Surface Temperature

In the Arabian Sea, the estimated SST [$E_{sst} = 2.3055*E_{eol} + 25.8984$ and $r^2 = 0.56$, where the E_{eol} is the estimated EOL] during the period of cyclone is presented in Figs. 13a–c. The day-to-day change of E_{sst} is examined from 21 May to 28 May (shown here only for May 21, 24 and 28, 2001). The superposition of cyclone track on the daily E_{sst} maps indicates that the weather disturbance formed over the warmer SST (30.5–31.0 °C) region in the southeastern Arabian Sea. The extent of warmer waters (> 30.0 °C) on 21 May suggests the presence of Arabian Sea warm pool with core values of SST (31 °C) at 12 °N, 68 °E. The E_{sst} increases from 21 May to 24 May with a core value of 31.5 °C. This warmer SST prevails until 26 May and the position of the weather system is coupled to the warmer SST on these days. With the northward movement of the cyclone on 27 May, the SST cools by 1.5 °C east of 68 °E when compared to the E_{sst} on 24 May and cools by ~0.5 °C from E_{sst} on 26 May. The TRMM SST (Figs. 14 a–c) is closer to the E_{sst}. The TRMM SST effects cooling from 23 May onwards with intense cooling beneath the cyclone center and in the area behind the cyclone. The NLOM SST (Figs. 14d-f) shows warmer SSTs (29 °–30 °C) in the southeastern Arabian Sea from 21 May to 25 May. By 26 May, the NLOM SST exhibits a thermal response to the cyclone and is cooled by 0.5 °C in the area traversed by the cyclone. On 28 May, TRMM SST shows an artifact of abnormal cooling of 5 °C due to heavy precipitation and masks the cooling due to the cyclone. The cell of low SST coincides with the cell of heavy precipitation obtained from TRMM rainfall measurements (figure not shown).

3.5.2 Variation of Sea Surface Salinity

The estimated SSS (E_{sss}) during the cyclone period is obtained using the linear relationship, $E_{sss} = 0.024*OLR + 29.712$; $r^2 = 0.89$ (MURTY *et al.*, 2004) and is presented in Figs. 15a–c. The observed salinity difference (unlike in the Bay of Bengal) was smaller at the center of the cyclone (where the OLR was the lowest). An examination of the E_{sss} indicates that the cyclonic system formed and intensified over the region of low SSS. The minimum E_{sss} of 33.0 was observed on

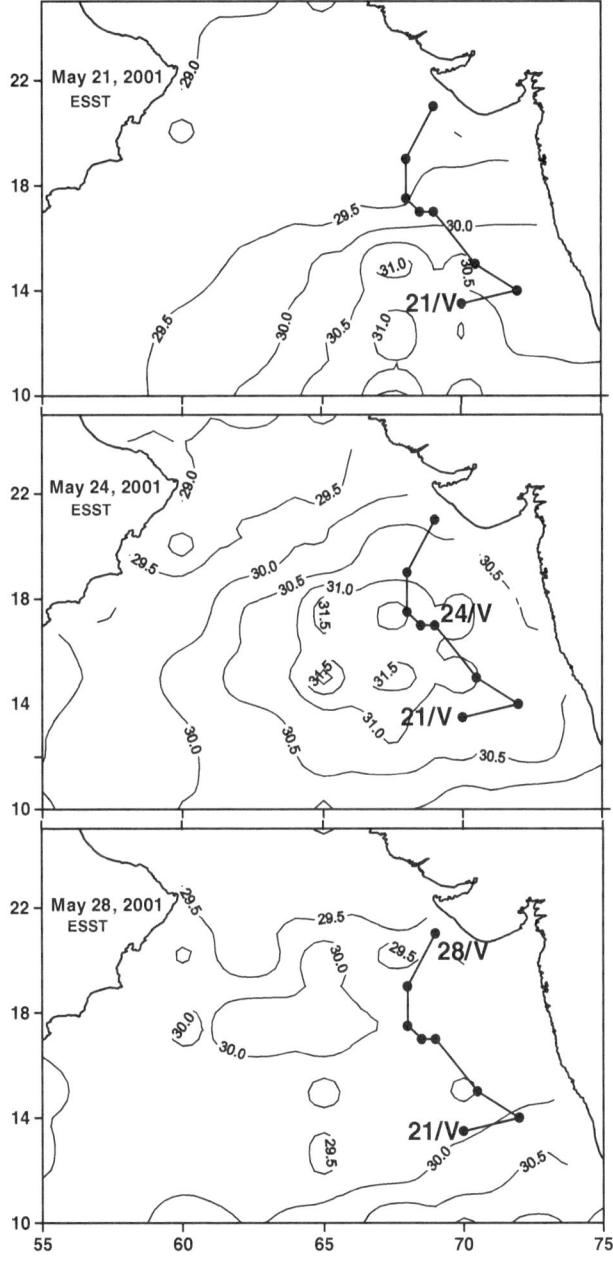

Figure 13
Variation of (a–c) estimated Sea Surface Temperature (E_{sst}) using the relationships between OLR, EOL and climatological SST during various stages (initial, intense and weakening) of TC 01A.

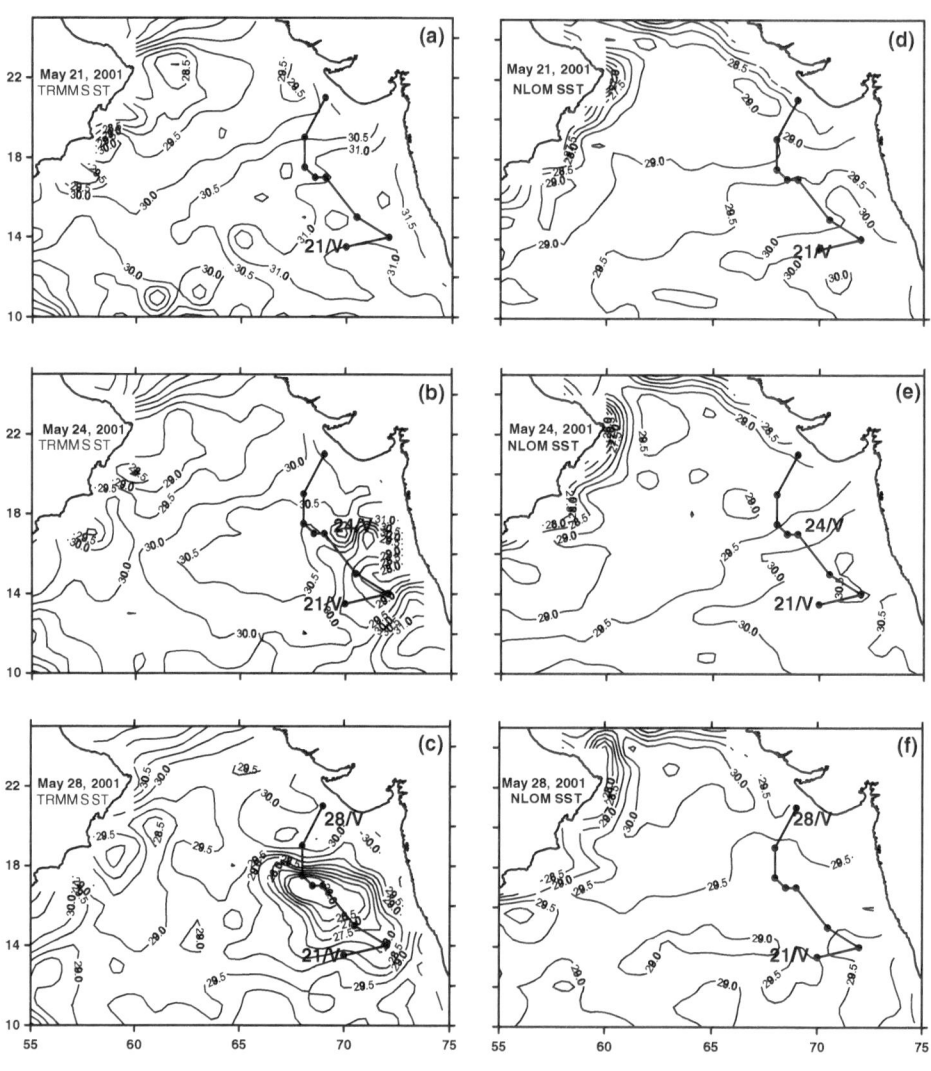

Figure 14
Variation of (a–c) satellite measured TRMM SST and (d–f) NLOM model simulated SST during various
stages (initial, intense and weakening) of TC 01A.

23–24 May when the atmospheric system was fully developed (Fig. 15b). The weather system was nearly stationary at 17 °N, 68 °E from 24 May to 26 May. Subsequently, the weather system moved northward on 27 May over the region of relatively higher E_{sss} and weakened as it moved towards the Indian coast (Fig. 15c). In the southeastern Arabian Sea, the E_{sss} increased when the cyclone moved farther north coinciding with the lower SSTs. It is noted that the center of the cyclone is coupled to the warmer and low saline waters.

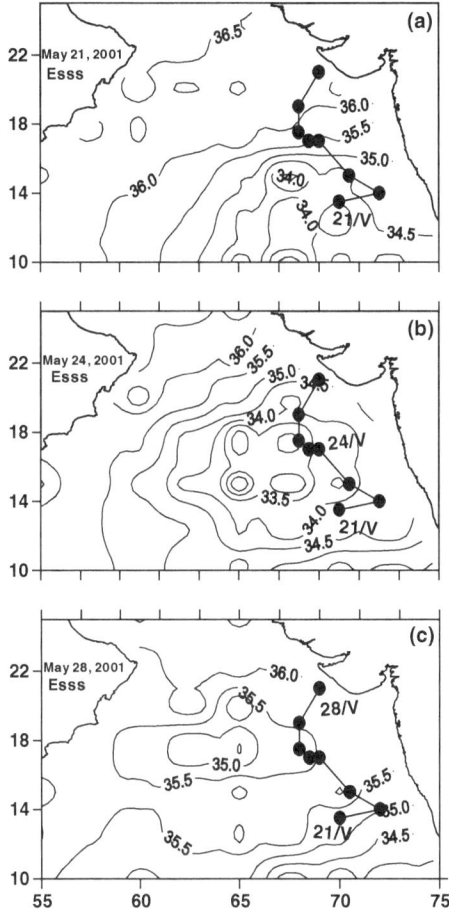

Figure 15
Variation of estimated Sea Surface Salinity (E_{sss}) using the relationship between OLR and climatological SSS during various stages (initial, intense and weakening) of TC 01A.

3.5.3 Variation of Upper Ocean Mixed Layer Depth

The inferred surface circulation from EOL (Figs. 7a–c) correlates well with the NLOM MLD (Figs. 16a–c). In the geostrophic convergence region the MLD is deep, and at the center of the cyclone the MLD is shallow. Moreover, as the cyclone moves northward, the MLD increases in the region behind the cyclone. As the estimated SST cools, the MLD deepens. This suggests that the OLR estimate of SST is realistic. Superposition of cyclone track on the daily NLOM MLD maps shows that the cyclone moves over the shallow MLD (~15 m) region. In accordance with the generally intense warming during May and the associated warmer SSTs from 21 May (Fig. 16a) to 23 May, the MLD was ~12 m in the cyclone-influenced area. As the cyclone intensified, a slight (~3 M) deepening in MLD was noticed on 24 May

(Fig. 16b), with further deepening up to 28 m between 25 May and 28 May
(Fig. 16c). It is observed that the MLD variation is closely related to the E_{eol}
variation. This means that the geostrophic near-surface flow associated with the EOL
governs the MLD distribution. The isolines of MLD run west-east with larger MLD
on the equatorward side. This suggests that the MLD deepening is more biased to the
left of the cyclone track and is due to the convergence of geostrophic flow at the
periphery of the cyclone (SUBRAHMANYAM *et al.*, 2002). The cooling of NLOM SST
on 26 May and 27 May might be a result of entrainment followed by mixing and
deepening of the MLD.

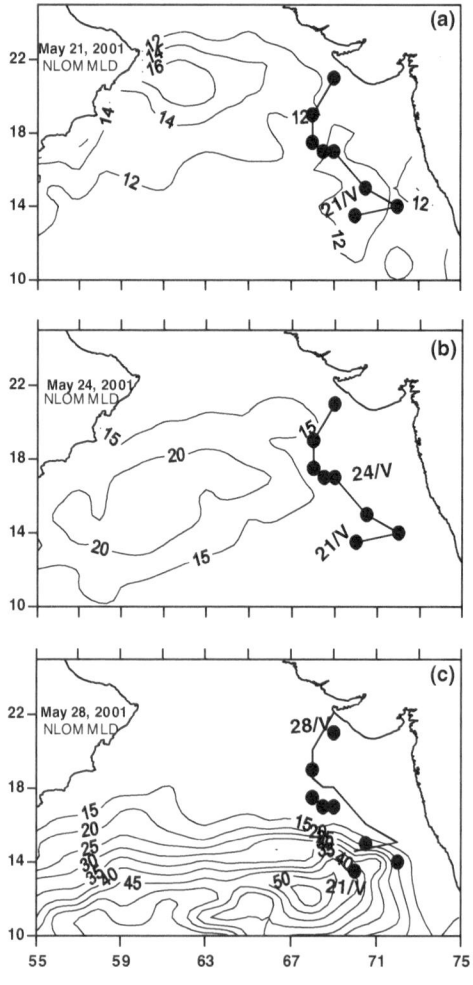

Figure 16
Variation of (a–c) NLOM model simulated Mixed Layer Depth (MLD, m) during various stages (initial,
intense and weakening) of TC 01A.

3.6 Comparison of Estimated SST and SSS with the Time-series Observations during Synoptic Scale Conditions

In this study, dealing with the synoptic weather systems and the associated air-sea coupling, we have applied the algorithms that were derived using the climatological data sets (16 year-mean monthly OLR data, computed monthly EOL using WOA98 temperature and salinity data, and WOA98 SST and SSS data) for the Bay of Bengal and the Arabian Sea (MURTY et al., 2004). In order to assess the extent to which these algorithms suit the synoptic scale studies, we have made use of the short-term time-series temperature and salinity data collected at 3 hourly intervals in the northern Bay of Bengal at 17.5 °N, 89 °E during the Bay of Bengal Monsoon Experiment (BOBMEX) in 1999 (BHAT et al., 2001; VINAYACHANDRAN et al., 2003). The time-series data in the upper 30 m were daily averaged and used to compute the daily time-series of EOL. The INSAT 1-D satellite derived daily OLR data during the BOBMEX period (Phase I: 27 July-5 August, 1999) were used to estimate the SST and SSS using the algorithms (MURTY et al., 2004) for July and August. The SSS is also estimated using the climatological OLR-SSS relationships for July and August (MURTY et al., 2004). It is mentioned that Phase I of BOBMEX-99 period was characterized by intense convection with the occurrence of synoptic weather disturbances (BHAT et al., 2001). The daily OLR arising from the presence of weather disturbances was quite low, compared to the climatological OLR. Figures 17a–b display the comparison of the daily variation of estimated and observed SST and SSS during 27 July-5 August 1999. The values of both the SSTs are closer and their deviations (estimated minus observed SST) give rise to a mean value of –0.01 °C with a standard deviation (SD) of 0.33 °C. One would see that the SST patterns are opposite, as we have not considered other factors such as winds and advection. The estimated SSS values from both the algorithms of SSS, however, are higher than the observed SSS, partly due to nonconsideration of freshwater flux due to evaporation minus precipitation, horizontal advection and mixing in the mixed layer. However, the mean and SD values of SSS deviations are 1.83 ± 2.5 for the OLR-EOL-SSS algorithms and 1.95 ± 1.76 for the OLR-SSS algorithms. We feel that these statistical values are reasonable partly because only 10 days data were used for comparison and partly due to the application of the climatological data based algorithms to the synoptic situation.

4. Conclusions

This study addresses the thermal, salinity and circulation responses at sea surface due to intense tropical cyclones in 1999 and 2001 in the northern tropical Indian Ocean, based on satellite measurements and model simulations, to understand the

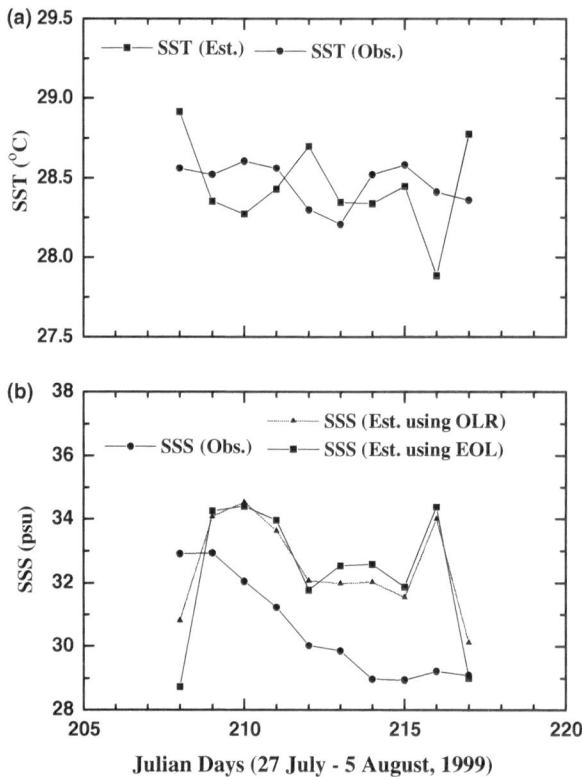

Figure 17
Daily variation of (a) observed SST (dots) and estimated SST (squares) from OLR-EOL-SST algorithms and (b) observed SSS (dots) and estimated SSS from OLR-EOL-SSS algorithms (squares) and from OLR-SSS algorithms (triangles) in the northern Bay of Bengal (17.5°N, 89°E) during 27 July – 5 August, 1999.

air-sea coupling processes. In this study, it is shown that through the relationships between the OLR and EOL, one can estimate the SST, SSS and sea surface elevation which are closely coupled to the oceanic mixed layer depth. It is envisaged that this coupling advances prediction of the path of weather systems in the northern Indian Ocean. The thermal response to the cyclones is documented from the OLR derived SST with the magnitudes of SST cooling falling within the range of reported values. The SST cooling occurred beneath the cyclone center in the Bay of Bengal. However, in the Arabian Sea the SST cooling occurred only a few days after the cyclone had passed over the area. This difference in the air-sea coupling processes is mainly a result of the difference in upper-ocean stratification (related to the EOL parameter); specifically saline stratification in the Bay of Bengal and thermal stratification in the Arabian Sea and the associated mixing processes.

In this study, the salinity response to the cyclones is attempted in terms of estimated SSS variation. In this direction, it is shown that nearly realistic information regarding the daily SSS pattern can be obtained in the vicinity and along the track of cyclones from satellite observations of OLR and would prove helpful until space-borne satellite measurements of salinity are available. The occurrence of cells of low SSS (estimated) beneath the cyclone center (in the Bay of Bengal) suggests intense precipitation associated with the cyclones. Comparison of the OLR derived salinity product with the US Navy's MODAS SSS, both in the Bay of Bengal on 28 October, 1999 and in the Arabian Sea on 24 May 2001, is presented in Figs. 18a–c. The daily MODAS SSS is closer to the monthly WOA98 SSS (Figs. 18b–d), suggesting that the MODAS SSS is good only for longer time scales (weeks or months) and does not depict the surface salinity (SSS) response due to synoptic weather systems.

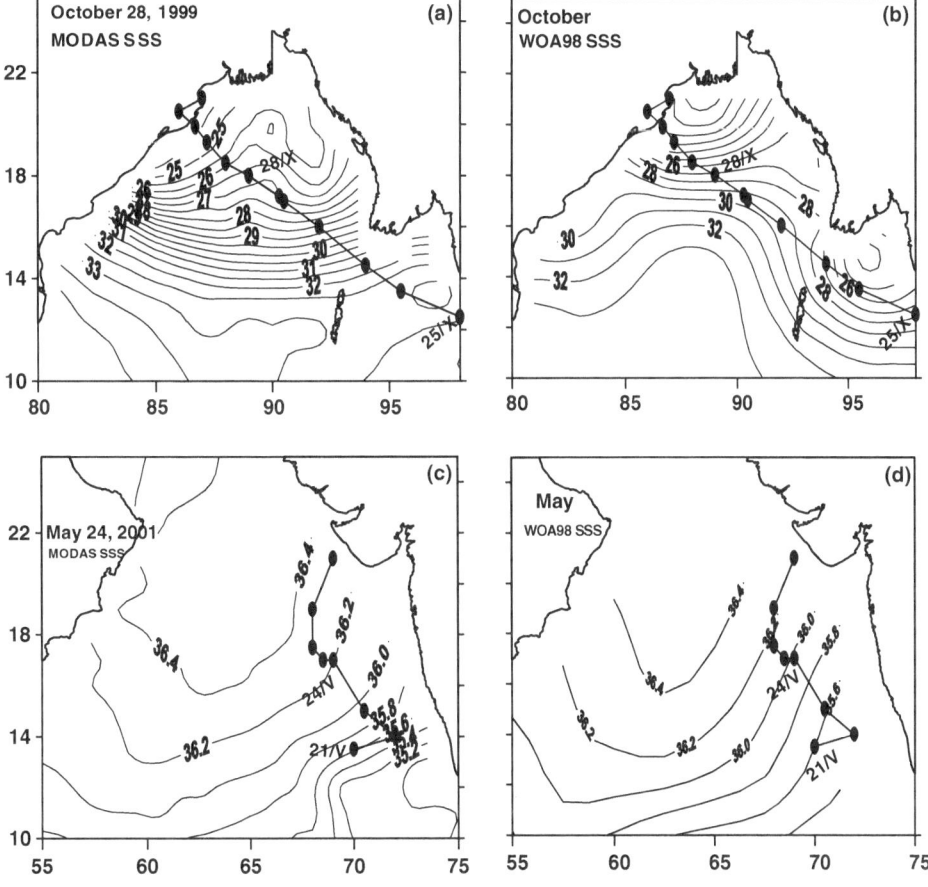

Figure 18
Variation of (a) MODAS SSS and (b) WOA98 SSS in the Bay of Bengal (top panel), (c) MODAS SSS and (d) WOA98 SSS in the Arabian Sea (bottom panel).

Acknowledgements

The interpolated OLR data is provided by the NOAA-CIRES CDC. We thank Dr. Harley Hurlburt and Dr. Birol Kara for providing the NLOM MLD simulations and Dr. Charlie Barron for providing the MODAS sea-surface salinity data. Dr. Mark Bourassa made available the QuikSCAT Scatterometer data on COAPS/FSU web page. TRMM TMI data and images are produced by Remote Sensing Systems and sponsored by NASA's Earth Science Information Partnerships (ESIP): a federation of information sites for Earth Science; and by NASA's TRMM Science Team. Bulusu Subrahmanyam is supported by NASA/JPL grant #961434. The COAPS at the FSU receives its base support from the NASA Physical Oceanography Program and through the Applied Research Center, funded by NOAA Office of Global Programs and also from ONR's Secretary of the Navy Grant awarded to Dr. James J. O'Brien. The BOBMEX programme was supported by the Department of Science and Technology, New Delhi. Dr. V.S.N. MURTY thanks Dr. S.R. Shetye, Director, NIO, India, for his keen interest in this study and also DST, India for funding BOBMEX project. NIO contribution No. 3893.

REFERENCES

ANONYMOUS, (1979), *Tracks of Storms and Depressions in the Bay of Bengal and the Arabian Sea 1877–1970,* India Meteorological Department, New Delhi, Charts 1–186.

ANTONOV, J., LEVIUTS, S., BOYER, T.P., CONKRIGHT, M., O'BREIN, T., STEPHENS, C., and TROTSENKO, B. (1998), *World Ocean Atlas 1998, vol. 3: Temperature of the Indian Ocean,* NOAA Atlas NESDIS 29. U.S. Gov. Printing Office, Washington, D.C., pp. 1–166.

BHAT, G.S., GADGIL, S., HARISH KUMAR, P.V., KALSI, S.R., MADHUSOODANAN, P., MURTY, V.S.N., PRASADA RAO, C.V.K., RAMESH BABU, V., RAO, L.V.G., RAVICHANDRAN, M., REDDY, K.G., SANJEEVA RAO, P., SENGUPTA, D., SIKKA, D.R., SWAIN, J., and VINAYACHANDRAN, P. (2001), *BOBMEX — the Bay of Bengal Monsoon Experiment,* Bull. Amer. Meteor. Soc. *82,* 2217–2243.

BOYER, T.P., LEVIUTS, S., ANTONOV, J., CONKRIGHT, M., O'BRIEN, T., STEPHENS, C., and TROTSENKO, B. (1998), *World Ocean Atlas 1998, vol. 6: Salinity of the Indian Ocean,* NOAA Atlas NESDIS 32. U.S. Gov. Printing Office, Washington, D.C., pp. 1–166.

CHINTHALU, G.R., SEETARAMAYYA, P., RAVINDRAN, M., and MAHAJAN, P.N. (2001), *Response of the Bay of Bengal to Gopalpur and Paradip Super Cyclones during 15–31 October, 1999,* Curr. Sci. *81* (5), 283–291.

FOX, D.N., TEAGUE, W.J., BARRON, C.N., CARNES, M.R., and. LEE, C.M. (2002a), *The Modular Ocean Data Analysis System (MODAS),* J. Atmos. Oceanic Technol. *19,* 240–252.

FOX, D.N., BARRON, C.N., CARNES, M.R., BOODA, M., PEGGION, G., and GURLEY, J.V. (2002b), *The Modular Ocean Data Analysis System,* Oceanography *15,* 22–28.

GILL, A.E., Atmosphere-Ocean Dynamics. *International Geophysical Series 30* (Academic Press 1982) pp. 1–662.

GINIS, I. (2002), Hurricane-ocean interactions, 2002. Tropical cyclone-ocean interactions. Chapter 3. In *Atmosphere – Ocean Interactions* (ed. Perrie, W.) (WIT Press 2002) *Advances in Fluid Mechanics Series,* pp. 33, 83–114.

GOPALAKRISHNA, V.V., MURTY, V.S.N., SARMA, M.S.S., and SASTRY, J.S. (1993), *Thermal Response of Upper Layers of Bay of Bengal to Forcing of a Severe Cyclonic Storm: A Case Study,* Indian J. Mar. Sci. *22,* 8–11.

GOSWAMI, B.N. (1987), *A Mechanism for the West-north-west Movement of Monsoon Depressions*, Nature, *326*, 376–378.

HILBURN, K.A., BOURASSA, M.A., and O'BRIEN, J.J. (2003), *Scatterometer-derived Research-Quality Surface Pressure Fields for the Southern Ocean*, J. Geophys. Res. *108*, doi: 10.1029/2003JC001772, 37–1 to 37–11.

KARA, A.B., ROCHFORD, P.A., and HURLBURT, H.E. (2000), *An Optimal Definition for Ocean Mixed Layer Depth*, J. Geophys. Res. *105*, 16,803–16,821.

KARA, A.B., ROCHFORD, P.A., and HURLBURT, H.E. (2003a), *Mixed Layer Depth Variability Over the Global Ocean*, J. Geophys. Res. *108*, doi:10.1029/2000JC000736, 24–1 to 24–15.

KARA, A.B., WALLCRAFT, A.J. and HURLBURT, H.E. (2003b), *Climatological SST and MLD Predictions from a Global Layered Ocean Model with an Embedded Mixed Layer*, J. Atmos. Oceanic Technol, *20*, 1616–1632.

LAL, M. (2001), *Tropical Cyclones in a Warmer World*, Current Science *80*, 1103–1104.

LEVITUS, S. and BOYER, T.P. (1994a), *World Ocean Atlas 1994, vol. 4, Temperature*, NOAA Atlas, NESDIS, vol. 4, U.S. Department of Commerce, Washington D.C., U.S.A. pp. 1–117.

LEVITUS, S., BURGETT, R., and BOYER, T.P. (1994b), *World Ocean Atlas 1994, vol. 3, Salinity*, NOAA Atlas, NESDIS, vol. 4, U.S. Department of Commerce, Washington D.C., USA, pp 1–99.

MURTY, V.S.N. (1983), The lowering of sea-surface temperature in the east central Arabian Sea associated with a cyclone, Mahasagar — Bull. Natl. Inst.Oceanogr, *16*, 67–71.

MURTY, V.S.N., SARMA, Y.V.B., and RAO, D.P. (1996), *Variability of the Oceanic Boundary Layer Characteristics in the Northern Bay of Bengal during MONTBLEX-90*, Proc. Indian Acad. Sci. (Earth Planet. Sci.), *105*, 41–61.

MURTY, V.S.N., SARMA, M.S.S., TILVI, V. (2000), Seasonal cyclogenesis and the role of the near-surface stratified layer in the Bay of Bengal. In *PORSEC Proceedings*, Vol. I, NIO, Goa, India during 5–8 December, 2000, pp. 453–457.

MURTY, V.S.N., SUBRAHMANYAM, B., SARMA, M.S.S., TILVI, V., and RAMESH BABU, V. (2002), *Estimation of Sea Surface Salinity in the Bay of Bengal using Outgoing Longwave Radiation*, Geophys. Res. Lett. *29*, doi: 10.1029/2001GL014424, 11–1 to 11–4.

MURTY, V.S.N., SUBRAHMANYAM, B., TILVI, V., and O'BRIEN, J.J. (2004), *A new technique for the estimation of sea surface salinity in the tropical Indian ocean from OLR*, J. Geophys. Res., 109, C12006, doi: 10.1029/2003JC001928.

OBASI, G.O.P. (1997), *WMO's Programme on Tropical Cyclone*, Mausam, *48*, 103–112.

PREMKUMAR, K., RAVICHANDRAN, M., KALSI, S.R., SENGUPTA, D., and GADGIL, S. (2000), *First Results from a New Observational System over the Indian Seas*, Curr. Sci. *78*(3), 323–330.

RAO, R.R. (1987), *Further Analysis on the Thermal Response of the Upper Bay of Bengal to the Forcing of Pre-monsoon Cyclonic Storm and Summer Monsoonal Onset during MONEX-79*, Mausam *38*, 147–156.

RAO, Y.R. (2002), *The Bay of Bengal and Tropical Cyclones*, Curr. Sci. *82*, 379–381.

SHARP, R.J. BOURASSA, M.A., and O'BRIEN, J.J. (2002), *Early Detection of Tropical Cyclones Using SeaWinds Derived Vorticity*, Bull. Am. Meteor. Soc. *83*, 879–889.

SHAY, L.K., Oceanic response to tropical cyclones. In *The Oceans: Physical-Chemical Dynamics and Human Impact* (eds. S.K. Mujumdar, E.W. Miller, G.S. Forbes, R.F. Schmalz, and Assad A. Panah) (The Pennsylvania Academy of Science 1994) pp. 1–497.

SHAY, L., GONI, G.J., and BLACK, P.G. (2000), Effects of Warm Oceanic Feature on Hurricane Opal, Mon. Weath. Rev., *128*, 1367–1383.

SEETARAMAYYA, P., NAGAR, S.G., and MULLAN, A.H. (2001), *Response of the North Bay of Bengal (Head Bay) to Monsoon Depression during MONTBLEX-90*, The Global Atmos. and Ocean Sys. 7, 325–345.

STRAMMA, L., CORNILLON, P., and PRICE, J.F. (1986), *Satellite Observations of Sea-surface Cooling by Hurricanes*, J. Geophys. Res. *91*, 5031–5035.

SUBRAHMANYAM, B., RAO, K.H., SRINIVASA RAO, N., MURTY, V.S.N., and SHARP, R.J. (2002), *Influence of a Tropical Cyclone on Chlorophyll-1 Concentration in the Arabian Sea*, Geophys. Res. Lett. *29*, doi:10.1029/2002GL015892, 22–1 to 22–4.

TURK, G. and BANKS, D.C. (1996), *Image-Guided Streamline Placement*, Computer Graphics, SIGGRAPH 96 Conf. Proceedings, pp. 453–460.

VINAYCHANDRAN, P.N., MURTY, V.S.N., and RAMESH BABU, V. (2002), *Observations of Barrier Layer Formation in the Bay of Bengal during Southwest Monsoon*, J. Geophys. Res. *107*, 8018, doi:10.1029/2001JC000831. SRF 19-1 to SRF 19-9.

(Received October 14, 2003; accepted April 29, 2004)

 To access this journal online:
http://www.birkhauser.ch

Pure appl. geophys. 162 (2005) 1673–1688
0033–4553/05/091673–16
DOI 10.1007/s00024-005-2688-5

┃ Pure and Applied Geophysics

Effect of the Mahanadi River on the Development of Storm Surge Along the Orissa Coast of India: A Numerical Study

S.K. Dube,[1,2] P.C. Sinha,[1] A.D. Rao,[1] Indu Jain,[1] and Neetu Agnihotri[1]

Abstract—River-ocean coupled models are described for the evaluation of the interaction between river discharge and surge development along the Orissa coast of India. The models are used to study the effect of fresh water discharge from the Mahanadi River on the surge response along the Orissa coast due to the October 1999 super cyclone which led to severe flooding of the coastal and delta regions of Orissa. The so-called 1999 Paradip cyclone was one of the most severe cyclones; causing extensive damage to property and loss of lives. The present study emphasizes the impact of the Mahanadi River on overall surge development along the Orissa coast. Therefore, we have developed a location specific fine resolution model for the Orissa coast and coupled it with a one–dimensional river model. The numerical experiments are carried out, both with and without inclusion of fresh water discharge from the river. The bathymetry for the model has been taken from the naval hydrographic charts extending from the south of Orissa to the south of west Bengal. A simple drying scheme has also been included in the model in order to avoid the exposure of land near the coast due to strong negative sea-surface elevations. The simulations with river-ocean coupled models show that the discharge of fresh water carried by the river may modify the surge height in the Bay, especially in the western Bay of Bengal where one of the largest river systems of the east coast of India, the Mahanadi River, joins with the Bay of Bengal. Another dynamic effect of this inlet is the potentially deep inland penetration of the surge originating in the Bay. The model results are in good agreement with the available observations/estimates.

Key words: Mahanadi River, Bay of Bengal, tropical cyclone, numerical model, storm surge, fresh water.

1. Introduction

The destruction caused by storm surges associated with severe cyclonic storms along the Indian coastline is of great concern (Dube *et al.*, 1997). The coastal regions of the state of Orissa (Fig. 1) occasionally experiences loss of life and severe damage to property from tropical cyclones originating in the Bay of Bengal (Murty *et al.*, 1986; Dube *et al.*, 1994, 1997, 2000a,b; Das *et al.*, 1983). The coastal districts of Orissa (Fig. 2) have experienced major surges in the past. Storm surges and the rains

---

[1] Centre for Atmospheric Sciences, Indian Institute of Technology, Hauz Khas, New Delhi 110 016, India.
[2] Indian Institute of Technology, Kharagpur, West Bengal 721 302, India.

Figure 1
Bay of Bengal coast of India showing the location of the state of Orissa.

associated with the cyclones are major causes of coastal flooding in the region. Damage can be minimized if the storm surges are forecasted well in advance. RAO (1968) classified the Orissa coast in a category in which surges of around 3–5 m may be expected. The most vulnerable region is the coastal stretch between Puri and Balasore. It has been found that most of the severe cyclones in this region made landfall near Paradip.

DUBE *et al.* (1994) developed a real-time storm surge prediction system. Further improvements in the prediction system are carried out by developing a model on the regional scale for the Andhra coast (RAO *et al.*, 1997). In the present paper we have developed a location specific storm surge model for the Orissa coast (henceforth referred to as the bay model), which is vulnerable to inundation due to storm surges. The model runs on a personal computer. The analysis region extends from 18.25°N to 22°N and 84°E to 90°E (Fig. 2). We use the actual bathymetry from the naval hydrographic chart for the region extending from the south of Orissa to the south of west Bengal and a simple drying scheme has also been included in the model in order to avoid the exposure of land near the coast due to strong negative surges. The coastal boundary in this model has been taken as vertical side-walls through which there is no flux of water. However, the use of an idealized vertical side-wall may lead to a misrepresentation of the surge development.

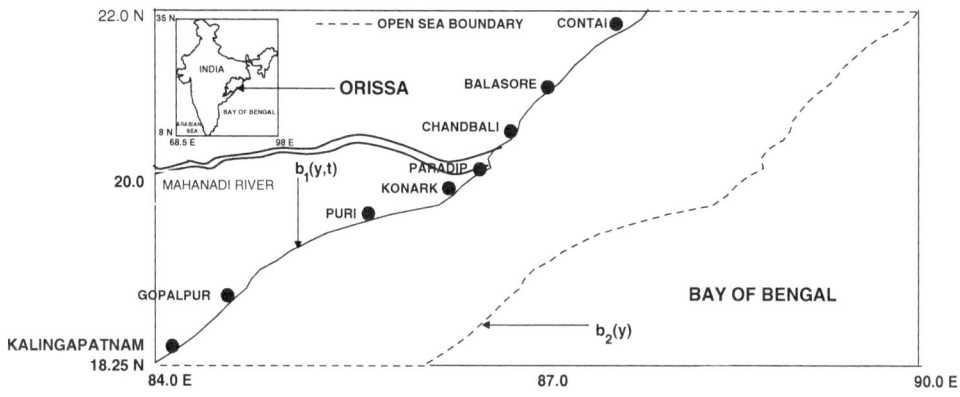

Figure 2
Map showing the Orissa coast and the analyzed region of the model.

Initially, the numerical model does not take into account the effect of any river or inlet joining the sea. This may result in an unrealistic high sea-surface elevation in the region where one of the largest river systems, the Mahanadi River, joins the Bay. Dynamically, this river system may be important for two reasons: (i) variations in the fresh water discharge may modify the surge height in the Bay, and (ii) the presence of such waterway allows a potentially deep inland penetration of the surge originating in the Bay.

Earlier, JOHNS and ALI (1980) developed a river-bay coupled model for the simulation of surges along the Bangladesh coast. In the present work we have modified the bay model by taking into account the effect of the River Mahanadi joining the Bay near Paradip. However, in contrast to the work of Johns and Ali (1980) our method does not depend on the patching together of computational regions with different uniform grid spacing. It is more similar to the work earlier reported by JOHNS et al. (1983) wherein a continuously contracting grid system is used to resolve the surge development along the Orissa coast of India. Another noteworthy difference in the formulation of our model from that of JOHNS and ALI (1980) is the way in which the topography of the Orissa coast is represented. The treatment of the coastal boundary involves a procedure leading to a realistic curvilinear representation of the natural shoreline.

Using this ocean-river coupled model, we have performed numerical experiments to simulate the surge generated by the October 1999 Orissa super cyclone. Comparison of the results of this model with those obtained from the bay model indicates that the surge height is significantly affected in the region where the river joins the Bay. Experiments are also performed to investigate the effect of fresh water discharge on the modification of surge height along the coast, and to estimate the extent of inland penetration of surge water through the river. Hence, simulations using both models, with and without inclusion of the river along the Orissa coast, show that the incorporation of a river in the storm surge model leads to an important change in the prediction detail.

2. Basic Equations

For the formulation of the model a system of rectangular Cartesian coordinates are used in which the origin, O, is within the equilibrium level of the free surface. Ox points towards the east, Oy towards the north and Oz is directed vertically upwards. The displaced position of the free surface is given by $z = \zeta(x, y, t)$ and the position of the sea floor by $z = -h(x,y)$. A western coastal boundary (the east coast of India is situated at $x = b_1(y)$ and an eastern open sea boundary is situated at $x = b_2(y)$. Southern and northern open sea boundaries are at $y = 0$ and $y = L$, respectively. The treatment of the boundaries involves a procedure leading to a realistic curvilinear representation of both the western and the eastern sides of the analysis region (Fig. 2).

The depth-averaged equations of continuity and momentum for the dynamical processes in the western Bay, after neglecting barometric forcing and tide generating forces, are given in the flux form by

$$\frac{\partial \zeta}{\partial t} + \frac{\partial \tilde{u}}{\partial x} + \frac{\partial \tilde{v}}{\partial y} = 0 \tag{1}$$

$$\frac{\partial \tilde{u}}{\partial t} + \frac{\partial}{\partial x}(u\tilde{u}) + \frac{\partial}{\partial y}(v\tilde{u}) - f\tilde{v} = -g(\zeta + h)\frac{\partial \zeta}{\partial x} + \frac{F_s}{\rho} - \frac{c_f \tilde{u}}{(\zeta + h)}(u^2 + v^2)^{\frac{1}{2}} \tag{2}$$

$$\frac{\partial \tilde{v}}{\partial t} + \frac{\partial}{\partial x}(u\tilde{v}) + \frac{\partial}{\partial y}(v\tilde{v}) + f\tilde{u} = -g(\zeta + h)\frac{\partial \zeta}{\partial y} + \frac{G_s}{\rho} - \frac{c_f \tilde{v}}{(\zeta + h)}(u^2 + v^2)^{\frac{1}{2}} \tag{3}$$

where,

$$(\tilde{u}, \tilde{v}) = (\zeta + h)(u, v)$$

u, v :depth-averaged component of velocity in the direction of x, y respectively,
ζ :sea-surface elevation above the mean water level,
h :water depth,
t :time,
ρ :density of the seawater,
f :Coriolis parameter $(= 2\omega \sin \phi)$,
g :acceleration due to gravity,
F_S, G_S :x and y components of the surface wind stress,
c_f :bottom friction coefficient $(= 2.6 \times 10^{-3})$

The surface stresses are parameterized by a conventional quadratic law as follows

$$(F_s, G_s) = c_d \rho_a (u_a^2 + v_a^2)^{\frac{1}{2}}(u_a, v_a)$$

where $c_d = 2.8 \times 10^{-3}$ is the surface drag coefficient, ρ_a is the density of the air and u_a, v_a are the x and y components of the surface wind.

3. Boundary Conditions

The boundary condition at the western coastal boundary which is a vertical side-wall, is given by

$$u - v \frac{\partial b_1}{\partial y} = 0 \quad \text{at} \quad x = b_1(y). \tag{4}$$

At the eastern open sea boundary, a radiation condition is applied similar to that used by JOHNS et al. (1981). This leads to

$$u - v \frac{\partial b_2}{\partial y} - \left(\frac{g}{h}\right)^{\frac{1}{2}} \zeta = 0 \quad \text{at} \quad x = b_2(y). \tag{5}$$

Conceptually similar radiation conditions are applied at the southern and northern open sea boundaries to yield

$$v + \left(\frac{g}{h}\right)^{\frac{1}{2}} \zeta = 0 \quad \text{at} \quad y = 0, \tag{6}$$

$$v - \left(\frac{g}{h}\right)^{\frac{1}{2}} \zeta = 0 \quad \text{at} \quad y = L. \tag{7}$$

Further, a curvilinear representation of the coastal boundaries has been carried out, similar to the work by DUBE et al . (1994).

4. Fine Grid Resolution Adjacent to the Coast

Increased resolution near the coast is important to represent the topography more realistically. Since the location of the first off-shore point influence compu-tation of the surge, more accuracy in prediction can be achieved by increasing the resolution near the coast. This can be done by further transformation of ξ coordinate. This is achieved by defining a new variable η by

$$\eta = \xi + \varepsilon \ln\left(\frac{\xi + \xi_0}{\xi_0}\right), \tag{8}$$

where $\xi = (x - b_1(y))/b(y)$, $b(y) = b_2(y) - b_1(y)$, and ε and ξ_0 are disposable parameters.

By virtue of equation (8) the increments $\Delta\xi$ and $\Delta\eta$ are related by

$$\Delta\xi = \frac{\Delta\eta}{1 + \varepsilon/(\xi + \xi_0)} \tag{9}$$

If we consider a uniform value of $\Delta\eta$ and take $\varepsilon = 0.04$ and $\xi_0 = 0.001$, $\Delta\eta \to 0$, so that

$$(\Delta\xi)_{\xi=0.0} \cong (0.024)\Delta\eta,$$

$$(\Delta\xi)_{\xi=0.5} \cong (0.0926)\Delta\eta,$$

$$(\Delta\xi)_{\xi=1.0} \cong (0.0962)\Delta\eta.$$

Thus, near the coastline, the transformation (8) will lead to a substantial mesh refinement in which the grid increment is reduced to a fraction of its essentially uniform value between $\xi = 0.5$ and $\xi = 1.0$. Therefore, the use of equation (8) in equations (1)–(3) will assist in the incorporation of a more detailed bathymetric specification in the important coastal region. It must, however, be emphasized that the use of such a transformation is not without attendant danger. Equation (8) would be a natural transformation to use if the solution were known to have an almost linear dependence on ζ. This, of course, is not necessarily the case and an injudicious choice of values for ε and ξ_0 could lead to results contaminated by a substantial truncation error.

Writing

$$\frac{\partial\xi}{\partial\eta} = F(\eta) = \frac{1}{1 + \varepsilon/(\xi + \xi_0)} \tag{10}$$

equations (1)–(3) take the form

$$\frac{\partial}{\partial t}(b\zeta) + \frac{1}{F}\frac{\partial}{\partial\eta}(bHU) + \frac{\partial}{\partial y}(bHv) = 0 \tag{11}$$

$$\frac{\partial\tilde{u}}{\partial t} + \frac{1}{F}\frac{\partial}{\partial\eta}(U\tilde{u}) + \frac{\partial}{\partial y}(v\tilde{u}) - f\tilde{v} = -\frac{gH}{F}\frac{\partial\zeta}{\partial\eta} + \frac{bF_S}{\rho} - \frac{c_f\tilde{u}}{H}(u^2 + v^2)^{\frac{1}{2}} \tag{12}$$

$$\frac{\partial\tilde{v}}{\partial t} + \frac{1}{F}\frac{\partial}{\partial\eta}(U\tilde{v}) + \frac{\partial}{\partial y}(v\tilde{v}) + f\tilde{u}$$
$$= -gH\left[\frac{b\partial\zeta}{\partial y} - \left(\frac{\partial b_1}{\partial y} + \xi\frac{\partial b}{\partial y}\right)\frac{1}{F}\frac{\partial\zeta}{\partial\eta}\right] + \frac{bG_s}{\rho} - \frac{c_f\tilde{v}}{H}(u^2 + v^2)^{\frac{1}{2}} \tag{13}$$

where U is the velocity normal to the coastline and is given by

$$U = \frac{1}{b(y)}\left[u - \left(\frac{\partial b_1}{\partial y} + \xi\frac{\partial b}{\partial y}\right)v\right] \tag{14}$$

with $\tilde{u} = Hbu, \tilde{v} = Hbv, H = \zeta + h$.

A conditionally stable semi-explicit finite difference scheme with staggered grid is used for the numerical solution of the model equations. The staggered grid consists of three distinct types of computational points on which the sea-surface elevations and the zonal and meridional components of depth-averaged currents are computed. Following SIELECKI (1968), the computational stability is achieved by satisfying the

CFL (Courant-Friedrich-Lewy) criterion. In the present model this condition is satisfied by limiting the time step of integration to three minutes.

The bottom stress is computed from the depth-averaged currents using the conventional quadratic law with a constant coefficient of 0.0026. This value has been achieved by performing several numerical experiments (JOHNS et al., 1983).

The bathymetry for the model is derived from the Naval Hydrographic Charts and is interpolated at the model grid points by using a cubic spline interpolation scheme. With this procedure, sufficiently accurate and realistic bathymetry is generated. A simple drying scheme has also been included in the model to avoid the exposure of land near the coast due to strong negative surges.

5. The River Model

The Mahanadi River is the sixth largest among the 14 major rivers of India that traverses and spreads over the states of Madhya Pradesh, Bihar, Maharashtra and Orissa before joining the Bay of Bengal. The river is 857 km long with a drainage area of 141,600 km^2 and accounts for 4.3% of the total area of the country. The depth of the river varies from 2–12 m. It extends between 80°30′E and 86°45′E longitudes; 19°20′N and 23°35′N latitudes. RAO (1990) suggested that the monthly stream flow data vary greatly; the largest value being in August and smallest during pre-monsoon months.

The Mahanadi estuarine region is determined by the extent of penetration of saline water, which depends on the season and different phases of the semidiurnal tide. It is observed that the distance of penetration of salt water varies from 25 to 35 km from the mouth, depending on the season. Hence the river channel is modelled by considering its length, ℓ, to be about 30 km upto which saline water intrusion is prevalent, neglecting other tributaries (Fig. 2). The breadth and mean depth are based on the data available from the Survey of India (SOI). Numerical experiments are performed for the river during the month of October. The channel has 36 uniform computational points in the horizontal with the grid increment (Δx) of 857 m and the time step has been taken as 90s. The freshwater discharge, Q, at the head of the river is specified to be 32, 376 m^{3}s^{-1} (ADSORBS, 1993–94). The same depth has been ensured for the common grid point of the river and the bay models.

The river is modelled as a channel with a variable but prescribed rectangular cross section with a flow that is assumed to be unidirectional. It is a 1-D barotropic depth-averaged numerical model. In the western Bay of Bengal the Mahanadi River connects with the main analysis area whose configuration is shown in Fig. 2. Following JOHNS and HAMZAH (1968), the meanders of the river have not been considered in the present model. However, in order to take into account the actual breadth of the river it is assumed that the x-axis is chosen in such a way that it is equidistant from either side of the river banks. The direct action of the meteorological and astronomical influences is ignored. The origin, O, is taken in the equilibrium level of the free surface and is located at the mouth of the river at $x = 0$. The landward end is denoted by $x = \ell$. The axis Ox is directed towards the west and Oz is directed vertically upwards. The elevation of the free-surface above its mean undisturbed level is denoted by $z = \zeta(x, t)$ and the bottom topography by $z = -h(x)$. The depth-averaged velocity, u, in the river then satisfies

$$B\frac{\partial \zeta}{\partial t} + \frac{\partial}{\partial x}(BHu) = 0 \tag{15}$$

and

$$\frac{\partial}{\partial t}(BHu) + \frac{\partial}{\partial x}(BHu^2) = -gBH\frac{\partial \zeta}{\partial x} - Bc_f u|u|, \tag{16}$$

where $B(x)$ is the breadth of the river at x. The bottom stress has been parameterized by a conventional quadric law $\rho c_f u|u|$. At the free surface the applied surface wind stress is taken as zero.

Dynamical conditions in the Mahanadi River are determined from equations (15) and (16). In order to generate the motion in the river the forcing is prescribed from the bay model, in the form of surge height at the river mouth.

5.1 Matching Conditions

The precise way in which the matching conditions between the bay and the river at the river mouth is accomplished, is important because of the transition from a fully two-dimensional bay model to a one-dimensional river model. The appropriate matching conditions at the river-bay junction are

$$\zeta_{bay} = \zeta_{river} \tag{17}$$

$$u_{bay} - v_{bay}\frac{\partial b_1}{\partial y} + u_{river}\left(1 + \left(\frac{\partial b_1}{\partial y}\right)^2\right)^{-1/2} = 0. \tag{18}$$

Equation (17) implies continuity of sea-surface elevation across the section while (18) ensures continuity of the volume flux. Similar matching conditions of surface continuity and volume continuity have also been used by JOHNS and ALI (1980).

5.2 Boundary and Initial Conditions for the River Model

At the river head we have considered two types of conditions. Either

$$u = 0 \tag{19}$$

when no fresh water input is considered, or

$$u = \frac{Q}{BH} \tag{20}$$

when the fresh water discharge is considered. Here, Q denotes the volume of fresh water entering the river per unit time, and it may be a constant or a prescribed function of time.

The matching condition (17) and boundary conditions (4) and (5) take the form

$$bU + u_{river} \left(1 + \left(\frac{\partial b_1}{\partial y} \right)^2 \right)^{-1/2} = 0 \quad \text{at the bay-river junction} \tag{21}$$

$$U = 0 \quad \text{at} \quad \eta = 0 \tag{22}$$

and

$$bU - \left(\frac{g}{h} \right)^{1/2} \zeta = 0 \quad \text{at} \quad \eta = \eta_m \tag{23}$$

where

$$\eta_m = 1 + \varepsilon \ln \left(1 + \frac{1}{\zeta_0} \right). \tag{24}$$

6. Numerical Procedure

The grid points for the river model are on the x-axis which runs along the middle of the river. There are two kinds of grid points for the river. When i is odd the point is ζ- point at which ζ is computed. When i is even, the point is a u-point at which u is computed. The river mouth is represented by a ζ-point while the river head is given by a u-point. The depth h and the breadth B are prescribed at both types of points.

In the bay model the computations are performed on a staggered grid consisting of three distinct types of grid points. When i is even and j is odd, the point is a ζ-point at which elevation is computed. If i is odd and j is odd, the point is a u-point at which both u and U are computed, and finally when both i and j are even, the point is a v-point at which v is computed. Along $\eta = 0$, there will be only u-points and at each of these points $U = 0$ identically, except at the grid point on the mouth of the river where the continuity of volume flux is ensured. We choose m to be even so that $\eta = \eta_m$ consists of both ζ-points and v-points. We also choose n to be odd thus ensuring that there are only ζ-points and u-points along $y = 0$ and $y = L$. The factor $F(\eta)$ is evaluated at each of the discrete η-points by applying a Newton-Raphson iterative procedure. The corresponding discrete values of $F(\eta)$ are derived from (10).

Equations (15) and (16) yield procedures for updating elevation and current in the Mahanadi River. The value of ζ at the head of the river is extrapolated from the interior. The solution developed by this process must match the solution of the bay model. With a forcing applied at the mouth, the river model can be operated independently of the bay model.

6.1 Matching of the River Model with the Bay Model

For the solution process in the river model, the extrapolated value of the elevation from the bay model at the bay-river junction (referred to as the pivot point) provides a forcing for the motion in the river. The use of equations (15) and (16) leads to updated elevations and currents at the ζ- and u-points in the river, except for the current at the head of the river where the u-velocity is given by one of the boundary conditions (19) and (20). The updated elevation at the pivot point from the river model is chosen to update the elevations in the bay model by using condition (17) for the next iteration.

Using the current computed at the first u-point of the river, the boundary values of U may be determined for the bay model. This must be consistent with (21) which is replaced by

$$(U)_{\eta=0} = -\frac{1}{b}\left(u_{\text{river}}\right)_{x=\Delta x}\left(1 + \left(\frac{\partial b_1}{\partial y}\right)^2\right)^{-1/2}. \tag{25}$$

By making use of the previously updated (or initial) elevations in the bay model, this matching procedure leads to updated elevations in the river and, additionally, yields boundary values of U across the mouth of the river to be used in the next updating of the bay model variables.

7. Numerical Experiments

In the present study numerical experiments have been first performed with a location specific storm surge model (bay model) for the Orissa coast having an idealized vertical side-wall. Next, numerical experiments have been conducted with a model in which the River Mahanadi joins the above-mentioned bay model near Paradip. Here, two experiments have been conducted, with and without river discharge from the upstream station, and their effect on the surge development along the coast has been analyzed. For both the models, without river and with river (for the case when the fresh water discharge is excluded), we prescribe an initial state of rest and integrate the governing equations (11)–(13), (15) and (16) ahead in time upto a total of 36 hours.

When fresh water discharge is taken into account in the river-bay model, the preliminary integration of the model equations requires the establishment of initial conditions corresponding to the fresh water discharge. In order to achieve this, we first prescribe an initial state of rest and integrate the governing equations ahead in time upto a total of 150 hours in order to reach a steady-state with only the prescribed fresh water discharge and no wind stress at the sea surface. Defining this steady-state condition as the initial state, the cyclonic wind stress was subsequently applied and the model was run for another 36 hours to obtain the final results. In all our experiments, a time-step of 3 minutes was found to be consistent with the computational stability.

Using both models, with and without the inclusion of river, numerical experiments are performed to simulate the surge heights associated with cyclonic storms which struck the Orissa coast during October, 1999. The track of the cyclone in the analysis region is shown in Fig. 3. India Meteorological Department (IMD) estimated a lowest pressure of 98 hPa and a radius of maximum wind of 40 km at landfall near Paradip. These data were used in the model to carry out numerical experimentation. The wind stress forcing for driving the models has been computed by using the storm model of JELESNIANSKI and TAYLOR (1973).

8. Results and Discussion

A depression formed in the Bay of Bengal near 12°N and 98.5°E at 0600 UTC of 25th October, 1999. It became a cyclonic storm in the early hours of the 26th and was located at 13.5°N and 96.5°E. The cyclone moved further in a northwesterly direction and lay centered at 16°N and 92°E. It became a severe cyclonic storm with a core of hurricane winds on the 27th at 0300 UTC. The cyclone further intensified into a very severe cyclonic storm and lay centered at 17.5°N and 89.5°E on the 28th. The surge prediction was carried out using the track given by IMD after the cyclone crossed a

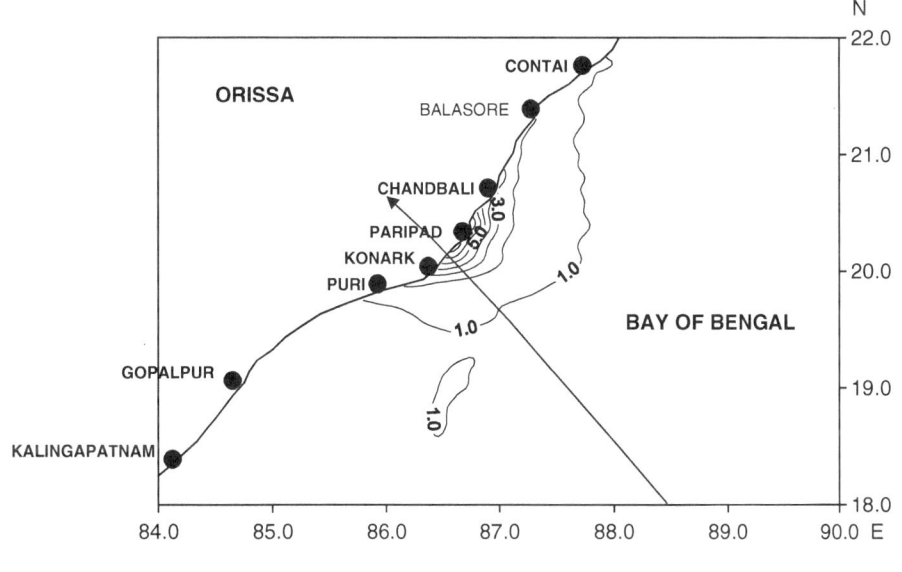

Figure 3
Surge contours (m) associated with 1999 Orissa cyclone and its track.

few km south of Paradip in Orissa. For numerical experiments, the starting point of the cyclone center is at 18°N and 88.5°E on the 29th of October at 0600 UTC.

Figure 3 delineates the computed surge contours along the Orissa coast using the bay model. It may be seen that the maximum surge of 7.7 m occurs to the right of the point of landfall. The coastal region between Konark and Chandbali is affected by a surge of more than 5 m. Post-storm survey reports of IMD also show that the surge was more than 7 m in proximity to Paradip.

In Fig. 4 we provide the distribution of the predicted maximum sea-surface elevation (peak surge envelope) along the Orissa coast using both models. The surge has been computed from the bay-river model both without and with fresh water discharge (henceforth referred to as MWD0 and MWD, respectively). The peak surge envelope for the bay model which does not include the river, henceforth referred to as BM, is also shown in Figure 4 for comparison. It can be seen that the maximum sea-surface elevations along the coast are the same except in the region near Paradip where the River Mahanadi joins the bay. MWD0 and MWD predict a maximum surge of 6.4 m and 6.5 m, respectively, at the river mouth where the predicted peak surge from BM is 6.8 m. Thus, the surge height computed by BM is higher than that obtained from MWD0 and MWD. This may presumably be attributed to the unrealistic accumulation of water near Paradip because of the assumption of an idealized vertical side-wall in BM. In MWD0 and MWD this piling up does not occur since the water is allowed to penetrate through the River Mahanadi thus reducing the maximum sea-surface elevation. A comparison of the results obtained from MWD0 and MWD indicates that the maximum surge predicted by the latter is about 10 cm higher at the river mouth. This may be due to the fact that the initial conditions corresponding to the fresh water discharge lead to a surface elevation above its equilibrium level in the river. We also note that the peak surge of 7.7 m occurs at the same position, P, in all three cases and the inclusion of the river has no impact on the peak surge, as the location of P is about 21 km away from the river mouth.

Fig. 5 presents a comparison of the temporal variation of the surge at the mouth of the Mahanadi River as computed from MWD0, MWD and BM. We note that BM yields, in general, a higher surge with a maximum difference of approximately 7% between 0002 UTC and 0330 UTC on 30th October, 1999. It may be seen that during the resurgence phase BM, MWD0 and MWD predict rapidly falling sea-surface elevations. However, MWD0 produces about 6% lower surge values compared to BM between 0700 UTC and 1800 UTC on 30th October, 1999. We also note that the computed phase difference between MWD0, MWD and BM is insignificant and the peak surge in the three cases occurs at about the same time.

The surface elevations in the Mahanadi River are also studied by integrating the model for 36 hours after steady state is reached in the case of MWD. Figure 6 provides an idea of the inland distance along the river upto which significant surface

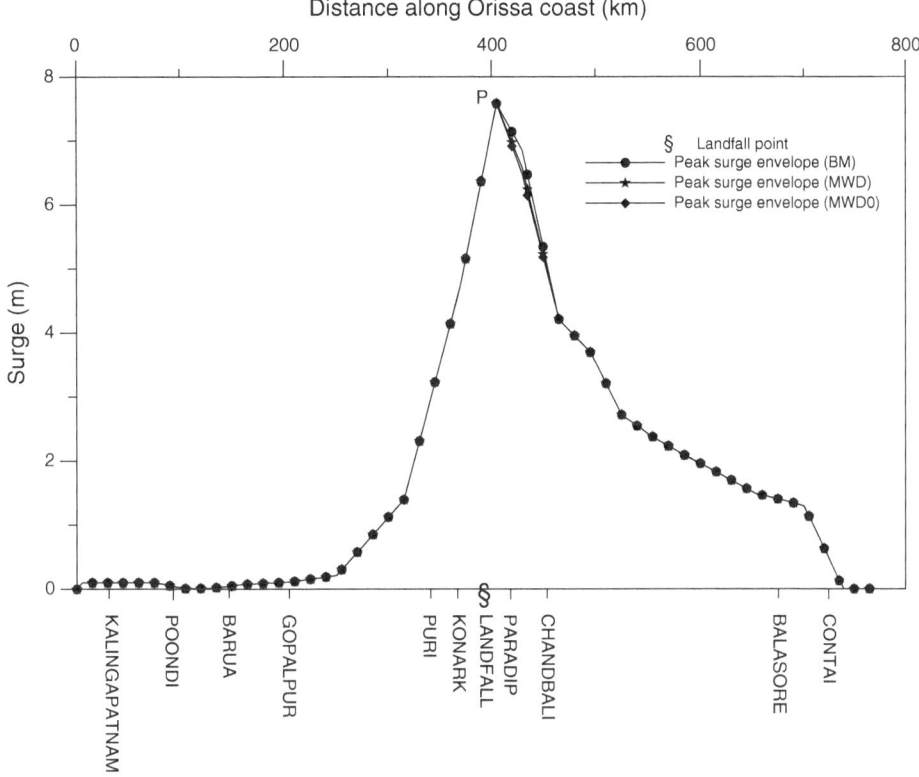

Figure 4
Peak surge envelope along the Orissa coast.

elevations may be expected, using both cases of with and without fresh water discharge at the river head.

It may be seen from Fig. 6 that the difference between the predicted surge response from MWD0 and MWD is insignificant upto a distance of about 8 km from the mouth of the river ($x = 0$). Further landward there is an appreciable difference between the results obtained from MWD0 and MWD with the former predicting lower values of peak surge than the latter all along the river. A maximum difference of 13% is seen at a landward distance of about 21 km. The surface elevation in the river is higher, in general, in the case of MWD as compared to MWD0. This may be due to the initial conditions corresponding to the fresh water discharge leading to a surface elevation above its equilibrium level.

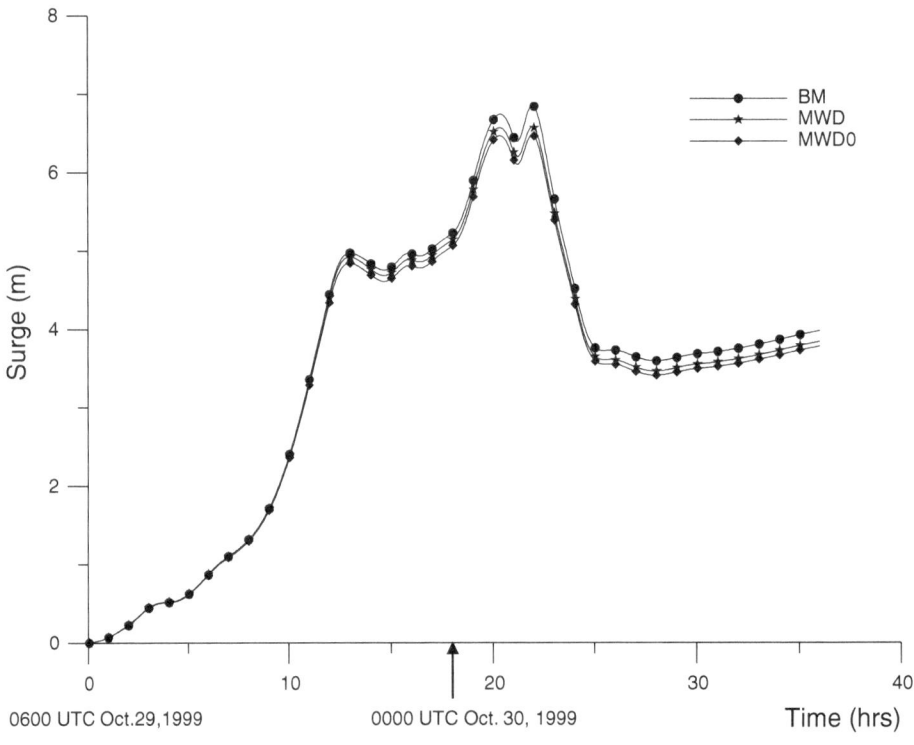

Figure 5
Temporal variation of the predicted surge (m) at the mouth of the Mahanadi River.

9. Conclusions

We have developed depth-averaged storm surge models for simulating the surges along the Orissa coast. Our analysis area includes a representation of the River Mahanadi. Using a forcing wind stress distribution representative of the 1999 Orissa super cyclone, a comparison is made between simulations using models both with and without inclusion of the river along the Orissa coast. The incorporation of a river in our model has shown that surge may penetrate inland, thus leading to a flooding hazard in the inland waterways of Orissa. It has also been shown that the surge response inside the river depends significantly upon the fresh water discharge. Thus, the response of the river on the surge development in the coastal and estuarine regions is appreciable and should be taken into account for any storm surge prediction study.

The model results reported in this study are in agreement with the available peak surge observations/estimates. The results emphasize the suitability of fine resolution location specific models for a reasonable prediction of surges along the Orissa coast.

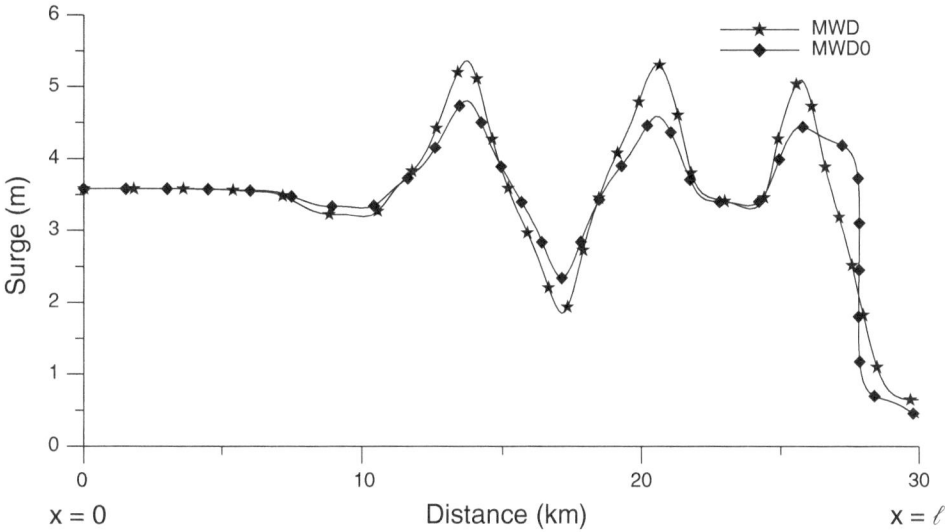

Figure 6
Predicted surface elevation (m) in the River Mahanadi.

In the present study, the cyclonic storm is the sole driving force for the dynamical processes in the sea. Tidal solution has not been used to provide the initial conditions for tide and surge interaction in the bay. Therefore, the nonlinear interaction of surge and the tide has not been considered in this study. Such an interaction may be significant if the occurrence of the surge coincides with that of the high tide. The model may be used for operational prediction of storm surges along the Orissa coast of India.

REFERENCES

ADSORBS/23/1993–94 *Assessment and Development Study of River Basin Series:*, Central Pollution Control Board, Delhi, India.

DAS, P.K., DUBE, S.K., MOHANTY, U.C., SINHA, P.C., and RAO, A.D. (1983), *Numerical Simulation of the Surge Generated by the June 1982 Orissa Cyclone*, Mausam *34*, 359–366.

DUBE, S.K., RAO, A.D., SINHA, P.C., and CHITTIBABU, P. (1994), A *Real Time Storm Surge Prediction System: An Application to East Coast of India*, Proc. Indian Nat. Sci. Acad., *60*, 157–170.

DUBE, S.K., RAO, A.D., SINHA, P.C., MURTY, T.S., and BAHULAYAN, N. (1997), *Storm Surge in the Bay of Bengal and Arabian Sea: The Problem and its Prediction*, Mausam *48*, 283–304.

DUBE, S.K., CHITTIBABU, P., RAO, A.D., SINHA, P.C., and MURTY, T.S. (2000a), *Sea Levels and Coastal Inundation Due to Tropical Cyclones in Indian Coastal Regions of Andhra and Orissa*, Marine Geodesy *23*, 65–74.

DUBE, S.K., CHITTIBABU, P., RAO, A.D., SINHA, P.C., and MURTY, T.S. (2000b), *Extreme Sea Levels Associated with Severe Tropical Cyclones Hitting Orissa Coast of India*, Marine Geodesy *23*, 75–90.

JOHNS, B., and ALI, A. (1980), *The Numerical Modelling of Storm Surges in the Bay of Bengal*, Quart. J. Roy Met. Soc. *106*, 1–18.

JOHNS, B., DUBE, S.K., MOHANTY, U.C., and SINHA, P.C. (1981), *Numerical Simulation of the Surge Generated by the 1977 Andhra Cyclone,* Quart. J. Roy. Met. Soc. *107,* 915–934.

JOHNS, B. and HAMZA, A.M.O. (1968), *Long Standing Waves in a Curved Canal,* J. Fluid Mech. *34,* 759–768.

JOHNS, B., SINHA, P.C., DUBE, S.K., MOHANTY, U.C., and RAO, A.D. (1983), *Simulation of Storm Surges Using a Three-dimensional Numerical Model: An Application to the 1977 Andhra Cyclone,* Quart. J. Roy. Met. Soc. *109,* 211–224.

JELESNIANSKI, C.P. and TAYLOR, A.D. (1973), *NOAA Technical Memorandum,* ERL, WMPO-3, 33 pp.

MURTY, T.S., FLATHER, R.A., and HENRY, R.F. (1986), *The Storm Surge Problem in the Bay of Bengal,* Prog. Oceanog. *16,* 195–233.

RAO, N. S. B. (1968), *On Some Aspects of Local and Tropical Storms in the Indian Area,* Ph.D. Thesis, University of Jadavapur, India.

RAO, P.G. (1990), *Some Stream Flow Characteristics of the Mahanadi Catchment,* The Indian Geograph. J. *65,* 112–122.

RAO, Y.R., CHITTIBABU, P., DUBE, S.K., RAO, A.D., and SINHA, P.C. (1997), *Storm Surge Prediction and Frequency Analysis for Andhra Coast of India,* Mausam *48,* 555–566.

SIELECKI, A. (1968), *An Energy-conserving Difference Scheme for the Storm Surge Equations,* Mon. Weather Rev. *96,* 150–156.

(Received March 2, 2004; accepted May 14, 2004)
Published Online First: May 25 2005

To access this journal online:
http://www.birkhauser.ch

Pure appl. geophys. 162 (2005) 1689–1714
0033–4553/05/091689–26
DOI 10.1007/s00024-005-2689-4

© Birkhäuser Verlag, Basel, 2005

Pure and Applied Geophysics

Numerical Simulation of Mesoscale Circulations in a Region of Contrasting Soil Types

SETHU RAMAN,[1] AARON SIMS,[1,2] ROBB ELLIS,[1] and RYAN BOYLES[1]

Abstract—Mesoscale processes that form due to changes in surface characteristics play a dominant role in the development of the planetary boundary layer structure and the formation of convection. In this study, effects of the Sandhills region of North and South Carolina on mesoscale processes are examined. Climatological analyses indicate increased convective precipitation in this location as compared to the surrounding region. This is believed to be due to enhanced convection induced by horizontal heat flux gradients caused by sharp changes in soil type and hence the heat capacity of the soil. Simulations using a non-hydrostatic mesoscale model (MM5 version 3.3) were made for a non-precipitation case with a 5-km resolution domain centered over the Carolinas from August 15, 2000 to August 18, 2000. The results showed the existence of a mesoscale circulation over the Sandhills region. Differential heating induced by contrasting soil types dividing the Coastal Plain from the central Piedmont causes this circulation. Sea-breeze circulation often combines with the Sandhills circulation to initiate convection in this region. Diurnal variations are handled well by the model indicating that the thermodynamic structure of the atmosphere is well simulated.

Key words: Mesoscale circulations, North Carolina, MM5, soil variation.

1. Introduction

Sandwiched between the Piedmont to the northwest and the Coastal Plain to the southest is an elongated area of sandy rolling hills in the Carolinas of the eastern United States called the Sandhills (Fig. 1). The adjacent Piedmont area contains primarily loam and clay-loam soils. These differing soil types exhibit varying characteristics such as drainage, albedo, and surface evaporation. It is the differences in the characteristics of these soil types that create differential heating of the earth surface. The heat capacity for sand is considerably less than that of clay or loam, so that given the same amount of energy, sandy soil would increase in temperature more than the loam or clay soils.

---

[1]State Climate Office of North Carolina, North Carolina State University, Raleigh, NC, 27695-7236, USA.
[2]Baron Advanced Meteorological Systems, Raleigh, NC 27695, USA.

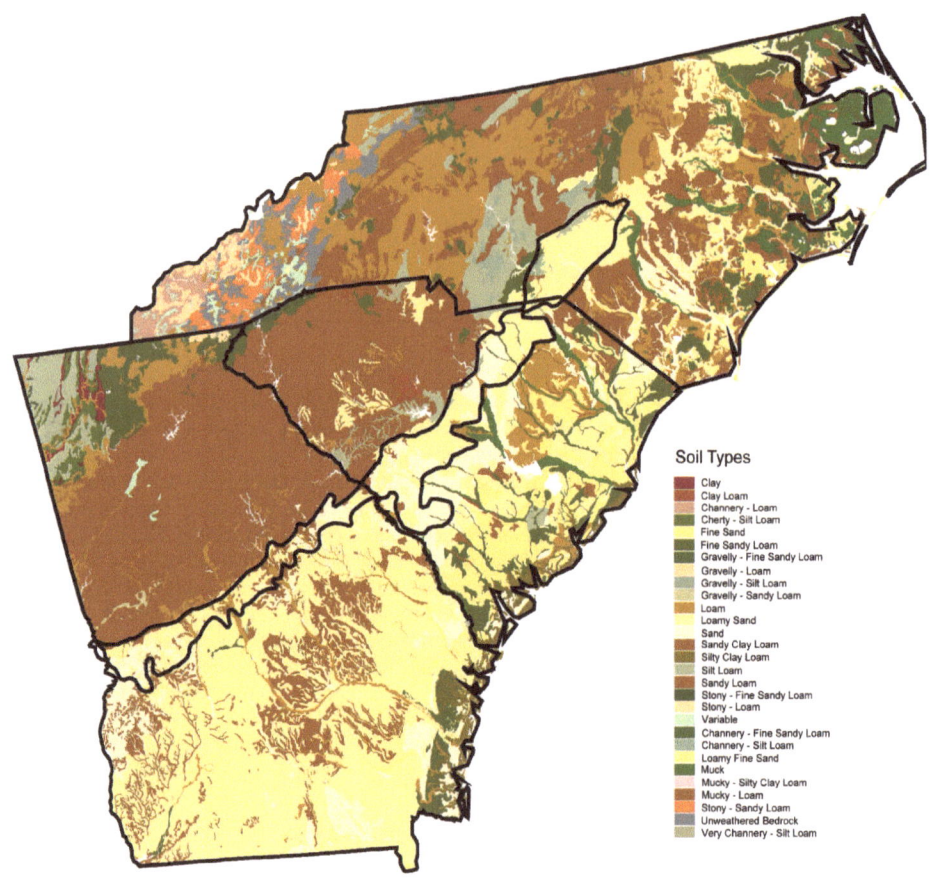

Figure 1
STATSGO-based soil types with Carolina Sandhills overlay.

The sea-breeze effect works in much the same way. The difference in the specific heats of water and land create differential heating. These differential surface heating patterns create mesoscale circulations causing clouds to form. Eventually, with the right large-scale conditions and the availability of moisture, thunderstorms form along the boundaries of these areas. Because surface heating is most intense during the summer months, these mesoscale convective processes are frequent during the months of June, July and August.

An elongated area of low pressure normally found at the transition from the Piedmont to the Coastal Plain during summers is characterized as the "Piedmont Trough" by KOCH and RAY (1997). It is believed that these areas of low-pressure form due to mesoscale convection created by intense heating of the surface. During their study of summertime mesoscale convective boundaries, KOCH and RAY (1997) noticed that this boundary existed for about 40% of the days. KOCH and RAY (1997) explain that these boundaries are autoconvective, or produce convection without the interaction of any other boundaries such as sea-breeze fronts. When coupled with other types of boundaries they were found to be positively influential on convection. Any moisture associated with the coastal circulation or left by thunderstorms brought in by the sea breeze may have a positive effect on increased thunderstorm activity. They also found that the Sandhills region was second only to the sea-breeze front as a producer of the thunderstorms.

Climatological precipitation data for the month of July shown in Figure 2 indicate an increase over the Sandhills region in North Carolina and South Carolina. These data were gathered from over 200 cooperative observer sites in the Carolinas and Georgia for the period, 1960–1999. Also apparent in the figure is the effect of the sea breeze-associated precipitation and its inland extent.

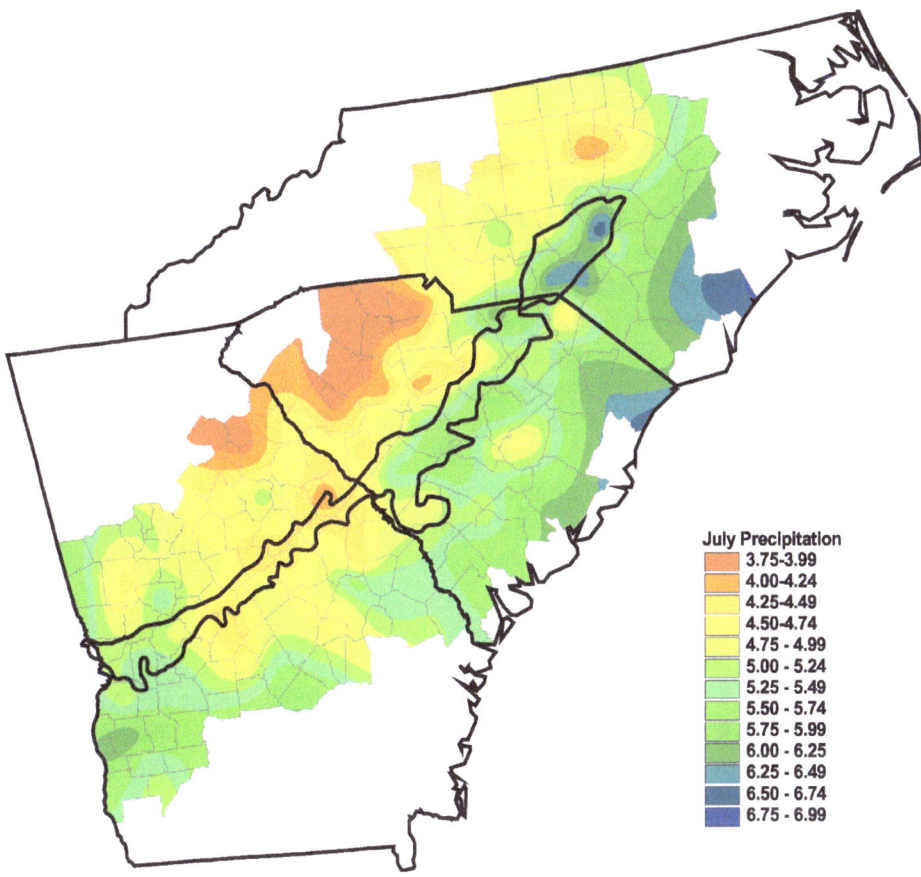

Figure 2
Average of July precipitation in inches for the period 1960–1999.

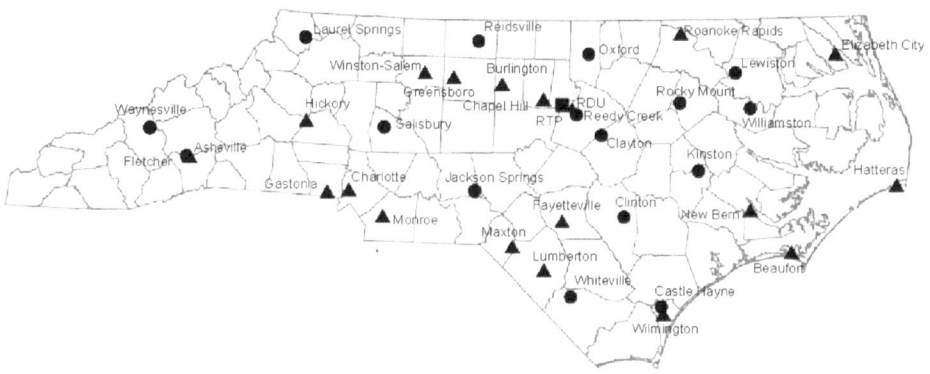

Figure 3
Hourly surface observation stations across North Carolina. Triangle markers indicate ASOS stations. Circle markers indicate ECONet stations. The square marker indicates the location of the RTP SODAR station.

High-resolution observations across North Carolina (NC) provide valuable data to examine the accuracy of the MM5 modeling system. The hourly observation sites included ASOS (Automated Surface Observing System, operated by the National Weather Service) and NC ECONet (NC Environmental and Climate Observing Network, maintained by the State Climate Office of North Carolina) stations.

In the study reported here, numerical simulations of mesoscale processes and boundary layer structure over NC were performed. The combination of complex topography to the west, land-use pattern variations in the Piedmont, and close proximity to the coast in the east can cause significant mesoscale interactions and circulations. Analyses of modeled boundary layer processes and interactions are evaluated and validated with observed data. An examination of simulated horizontal and vertical cross sections in the 5-km domain is made. Additional analyses are performed to estimate model performance on a point-by-point basis. Thirty-six hourly surface observation stations across NC are shown in Figure 3. The relative locations of model grid points to observational sites are shown in Figure 4.

2. *Numerical Model*

Pennsylvania State University (PSU) and the National Center for Atmospheric Research (NCAR) developed the fifth generation mesoscale modeling system (MM5). Its design uses a terrain following a sigma coordinate system and a finite

▶

Figure 4
MM5 model grid for the 5-km domain overlaid on the hourly surface observation stations across North Carolina. Triangle markers indicate ASOS stations. Circle markers indicate AgNet stations. The square marker indicates the location of the SODAR station.

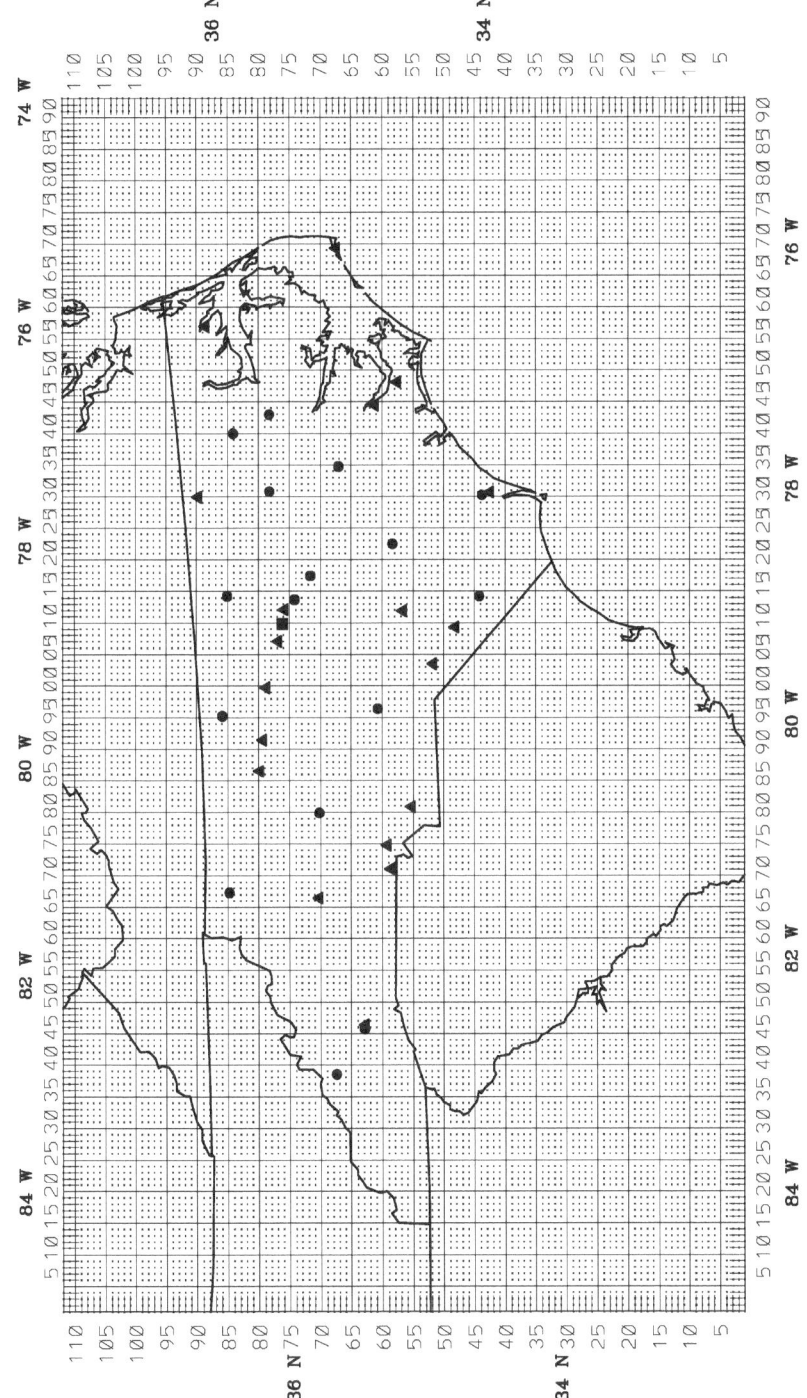

domain for the study of regional and mesoscale atmospheric phenomena. Version 3.3 of MM5 is used for the current study. Model simulations in this study use one-way nested domains. Nested grids employed in this simulation use a 3:1 coarse to fine grid resolution ratio. Boundary values for the coarse domain are derived from the analysis of the archived data set used in these simulations. Lateral boundary conditions are generated for the nested domain from the coarse domain simulation.

A triple nested domain configuration is used. The domain setup is shown in Figure 5. The coarse domain is centered at 36 degrees north and 85 degrees west and has a distance of 45 km between grid points. The number of grid points in the north-south direction is 54 and the east-west direction is 82. The second, intermediate domain has a resolution of 15 km with the number of east-west and north-south grid points totalling 112 and 76, respectively. The innermost domain has a resolution of 5 km. The grid points for this domain are 193 and 112 in the east-west and north-south directions, respectively.

Terrain and land-use data were obtained from the National Center for Atmospheric Research (NCAR). The terrain and land-use data for the innermost nested domain are shown in Figure 6. High-resolution elevation and land-use data were obtained from the United States Geological Survey (USGS). Additional data incorporated into the model also include soil type, vegetation fraction from AVHRR, and annual deep soil temperature from ECMWF analysis (GUTMANN and IGNATOV, 1998; USDA, 1994; ZOBLER, 1986). Soil types for the 5-km domain are shown in Figure 1.

All simulations use National Center for Environmental Prediction (NCEP) Global Data Assimilation System (GDAS) meteorological data obtained from the National Center for Atmospheric Research (NCAR). These data sets are archived at a resolution of 2.5 degrees every 12 hours (00Z and 12Z). These values were then interpolated to the model grid points.

The USGS land-use categories are shown in Table 1. In order to increase the boundary layer resolution, twenty-five of the total thirty-seven vertical levels are below 700 hPa. The cumulus parameterization scheme chosen for all simulations is the Kain-Fritsch (KF) scheme for its simplicity and ability to handle convection in small to medium grids (KAIN and FRITSCH, 1993; FRITSCH and CHAPPELL, 1980). This scheme simplifies the effects of cumulus convection by assuming all clouds in a model grid cell are of the same type and remain consistent while the clouds move through the specified grid cell. Regulation of convection is determined by a parcel's buoyant energy and the time it takes to remove that energy by convection. Moist convection occurs when low level forcing lifts air above the level of free convection (LFC). Vertical wind shear within the cloud layer affects precipitation efficiency. Cumulative effects of environmental compensation of subsidence as a result of updrafts and downdrafts affect the temperature and mixing ratio.

An explicit moisture scheme was also used in this study. MM5 has numerous such schemes including those for dry conditions, stable precipitation, warm rain

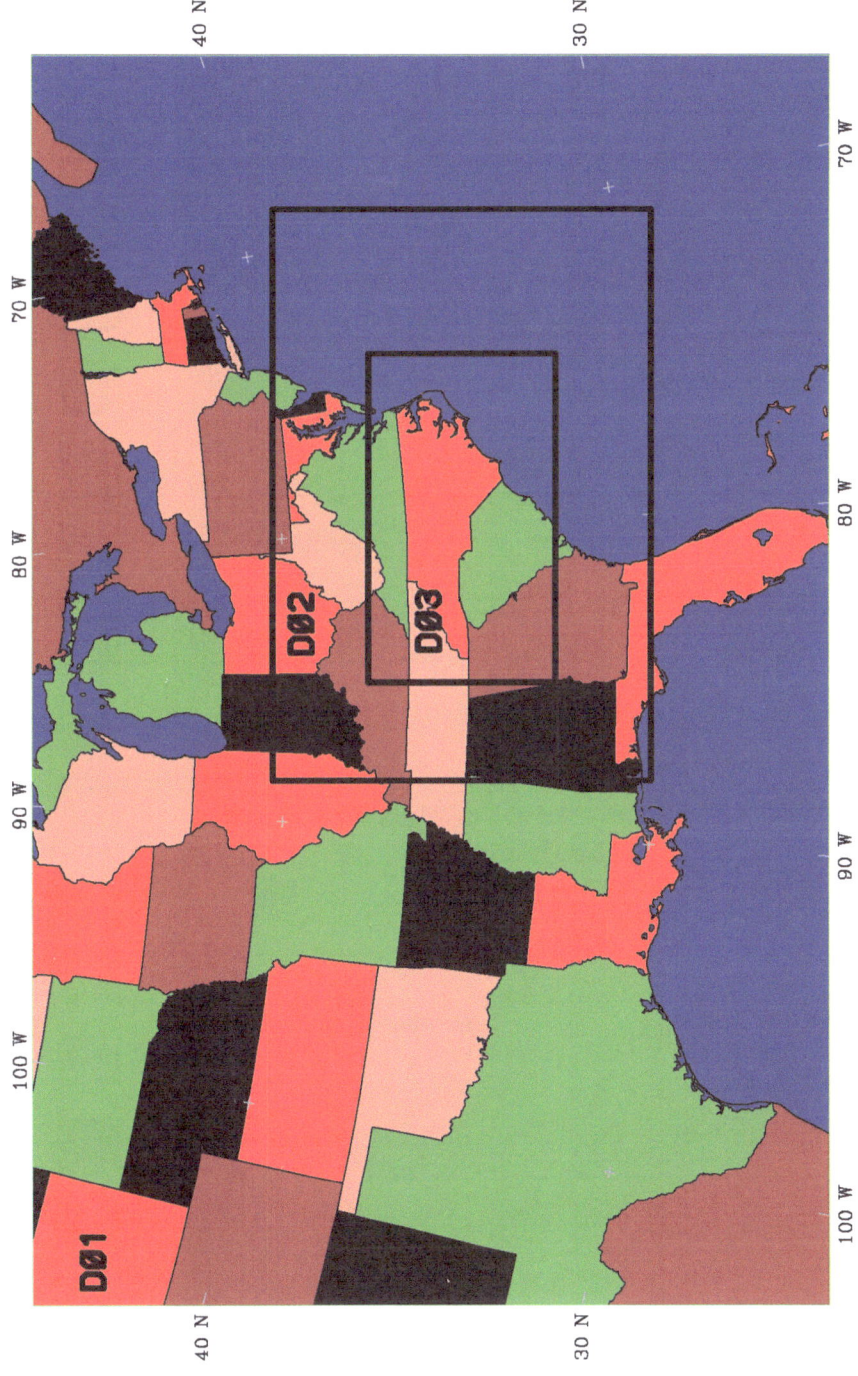

Figure 5
Domain setup centered over North Carolina.

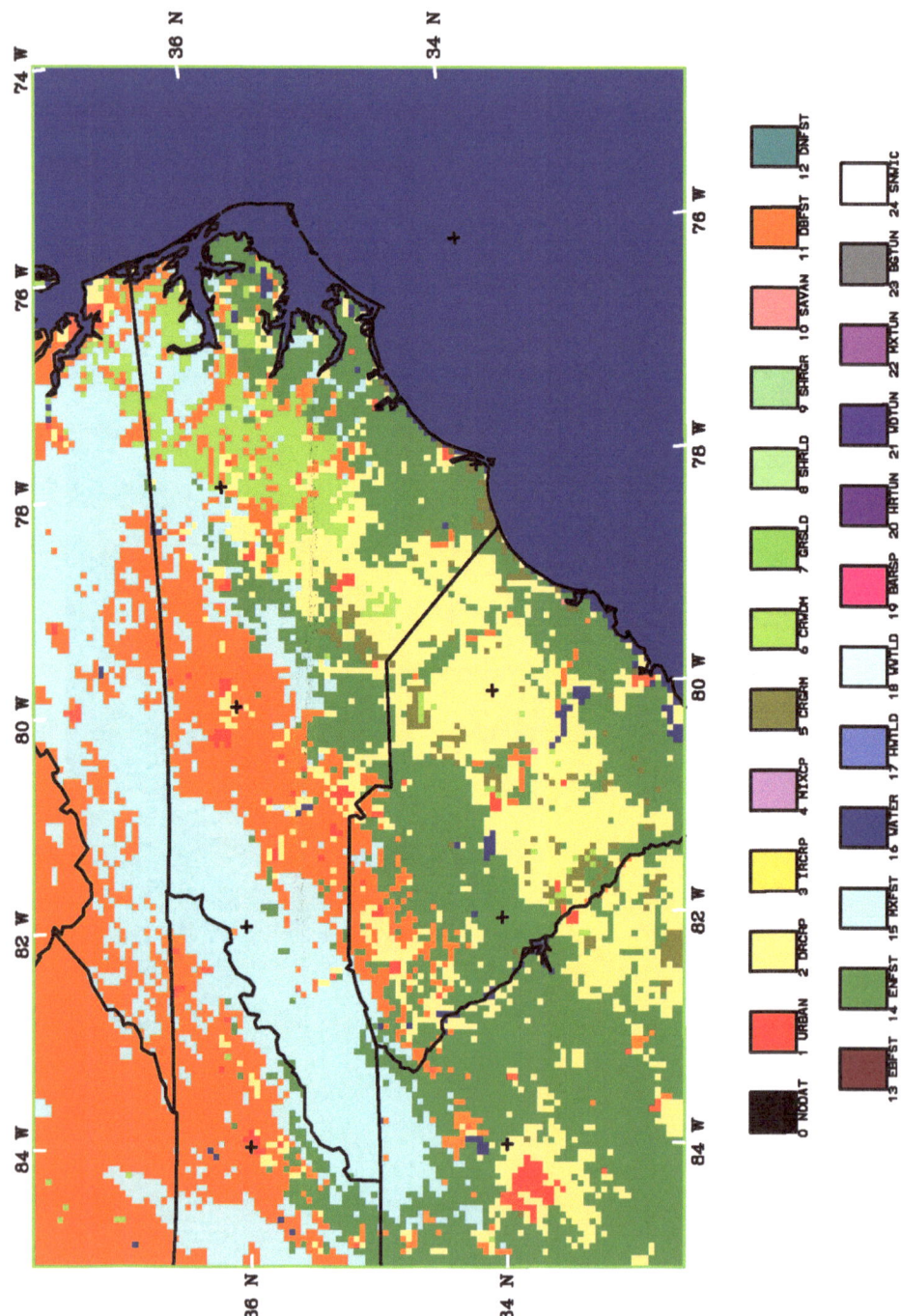

Figure 6
Model dominant vegetation type / land use pattern.

Table 1

Vegetation Type and Land use Category

Category	Vegetation Type / Land Use
1	Urban and Built-Up Land
2	Dryland Cropland and Pasture
3	Irrigated Cropland and Pasture
4	Mixed Dryland/Irrigated Cropland and Pasture
5	Cropland/Grassland Mosaic
6	Cropland/Woodland Mosaic
7	Grassland
8	Shrubland
9	Mixed Shrubland/Grassland
10	Savanna
11	Deciduous Broadleaf Forest
12	Deciduous Needleleaf Forest
13	Evergreen Broadleaf Forest
14	Evergreen Needleleaf Forest
15	Mixed Forest
16	Water Bodies
17	Herbaceous Wetland
18	Wooded Wetland
19	Barren or Sparsely Vegetated
20	Herbaceous Tundra
21	Wooded Tundra
22	Mixed Tundra
23	Bare Ground Tundra
24	Snow or Ice

(KESSLER, 1969), simple ice (DUDHIA, 1989), mixed phase (LIN *et al.*, 1983), Goddard microphysics (TAO and SIMPSON, 1993), Reisner graupel (REISNER *et al.*, 1998), and Schultz microphysics (SCHULTZ, 1995). The simple ice scheme was chosen for its simplicity in these simulations.

The simple ice scheme handles moisture through several key assumptions. There is no super-cooled water in the simulation and ice crystals or snow immediately melt when crossing the zero degree isotherms. The assumption of no super-cooled water is based on the principle of the Bergeron-Findeisen process of ice crystals consuming available water droplets in elevations above the freezing level (DUDHIA 1989).

A cloud radiation scheme is used for these simulations. The scheme allows for longwave and shortwave radiation to interact with the clear air environment and explicit clouds. The longwave radiation calculations are based on the upward and downward fluxes determined by the effective emissivity, while absorption and scattering control calculations for the downward shortwave radiation (DUDHIA, 1989).

The MRF PBL (Medium Range Forecast Planetary Boundary Layer) scheme was chosen for simulations and is described by HONG and PAN (1996). Their effort is based on the boundary layer scheme developed by TROEN and MAHRT (1986). In this scheme, similarity theory is used to represent the surface fluxes. Above the surface layer, mixing is represented by turbulent diffusivities based on bulk similarity and the stability conditions near the top of the surface layer.

3. Synoptic Conditions

The period, August 15, 2000 0000 (1900LST on August 14) through August 18, 2000 0000Z (1900LST on August 17), is a summer case with no precipitation. Throughout this period, synoptic winds were typically light and variable, allowing mesoscale features to influence the local weather. For most of the duration of this case study, high pressure dominated the southeast United States. Mild synoptic events in the case study include the passage of a dry trough of low pressure in western NC between 1200Z (0700LST) on the 16th and 0000Z (1900LST) on the 17th. Also, a dry, mild cold front passed through the domain between 0000Z (1900LST) and 1200Z (0700LST) on August 17 closely following this dry trough.

Cloud cover was minimal over the Carolinas during this case study. Some thin high cirrus clouds may have been present in the inner domain at times. An exception occurs during the last few hours of the period when heavier clouds begin to move into the domain from an approaching synoptic system.

4. Numerical Simulations

4.1 Soil Temperature and Soil Moisture

Observed and modeled soil temperature measurements are compared at a depth of 10 cm. Diurnal variations in soil temperature are handled well by the model in all regions. There appears to be less error in the modeled values during nighttime conditions. Nighttime observed point values of soil temperature are indicated in relation to the contours plotted from model values at 0600Z (0100LST) on August 16 as shown in Figure 7a. Here, the observations (provided in degrees K) match well with the simulated model values. In contrast, the daytime temperature comparison shown at 1800Z (1300LST) on August 16 in Figure 7b indicates large differences between the observed and modeled soil temperatures. In general, the model tends to overestimate the soil temperature at most locations by 3 to 5 degrees C during the daytime, possibly because of approximations and assumptions in the surface energy budget. A more complete statistical analysis of the simulated values is provided by SIMS (2001).

Modeled soil moisture is also compared with point observations from NC ECONet stations with available soil moisture data. The initial model soil moisture on August 15 at 0000Z (1900LST) is shown in Figure 8a. A fairly uniform distribution of approximately 0.33 m^3m^{-3} is apparent. Gradients in the modeled soil moisture develop during the simulation. Observed soil moisture values on August 15 at 1800Z

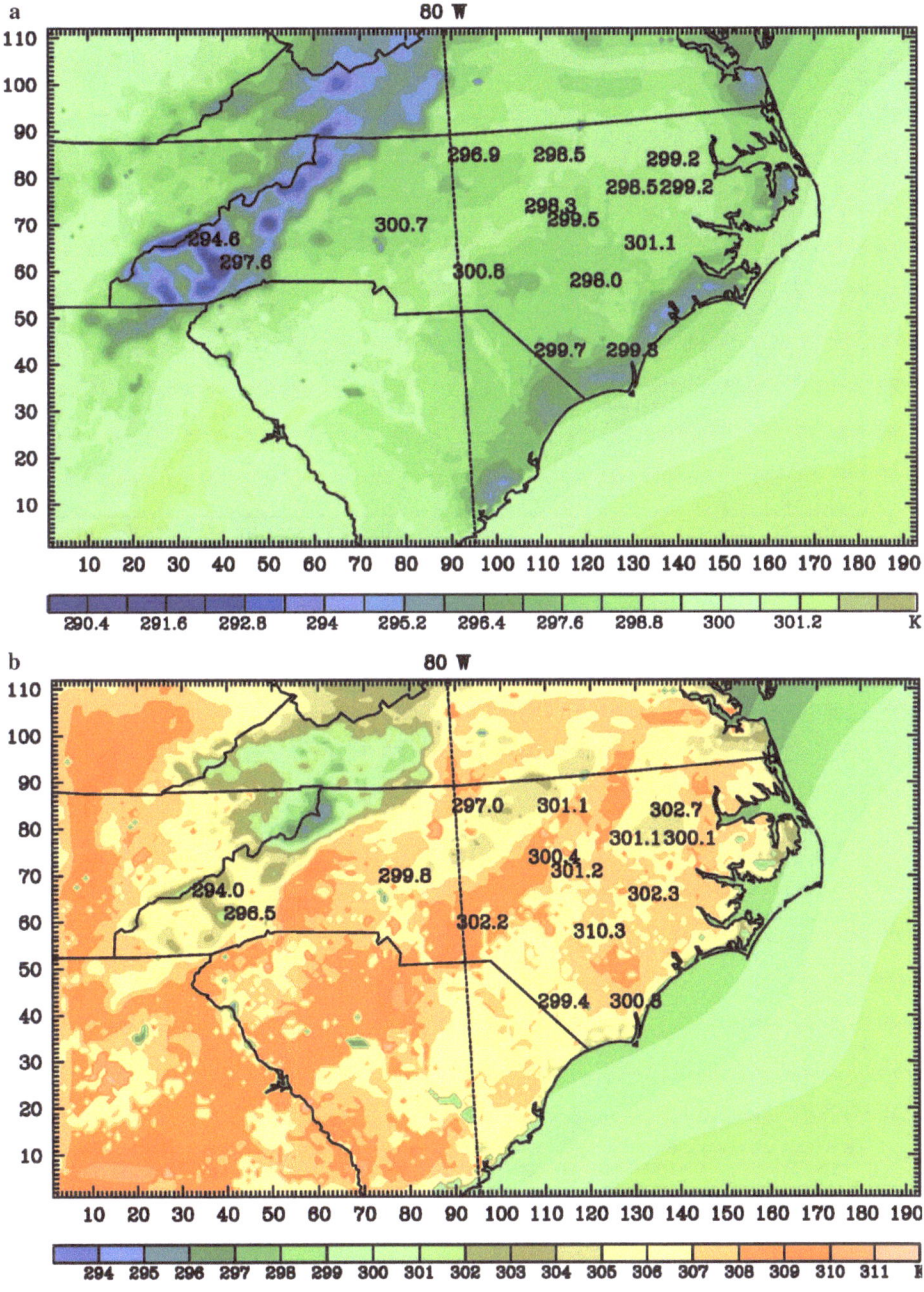

Figure 7a
Modeled soil temperature contours overlaid with actual observed soil temperature values obtained from
NC ECONet stations in North Carolina at 0600Z (0100LST) on August 16, 2000 at 10 cm soil depth.
Figure 7b
As in Figure 5 except at 1800Z (1300LST) on August 16, 2000.

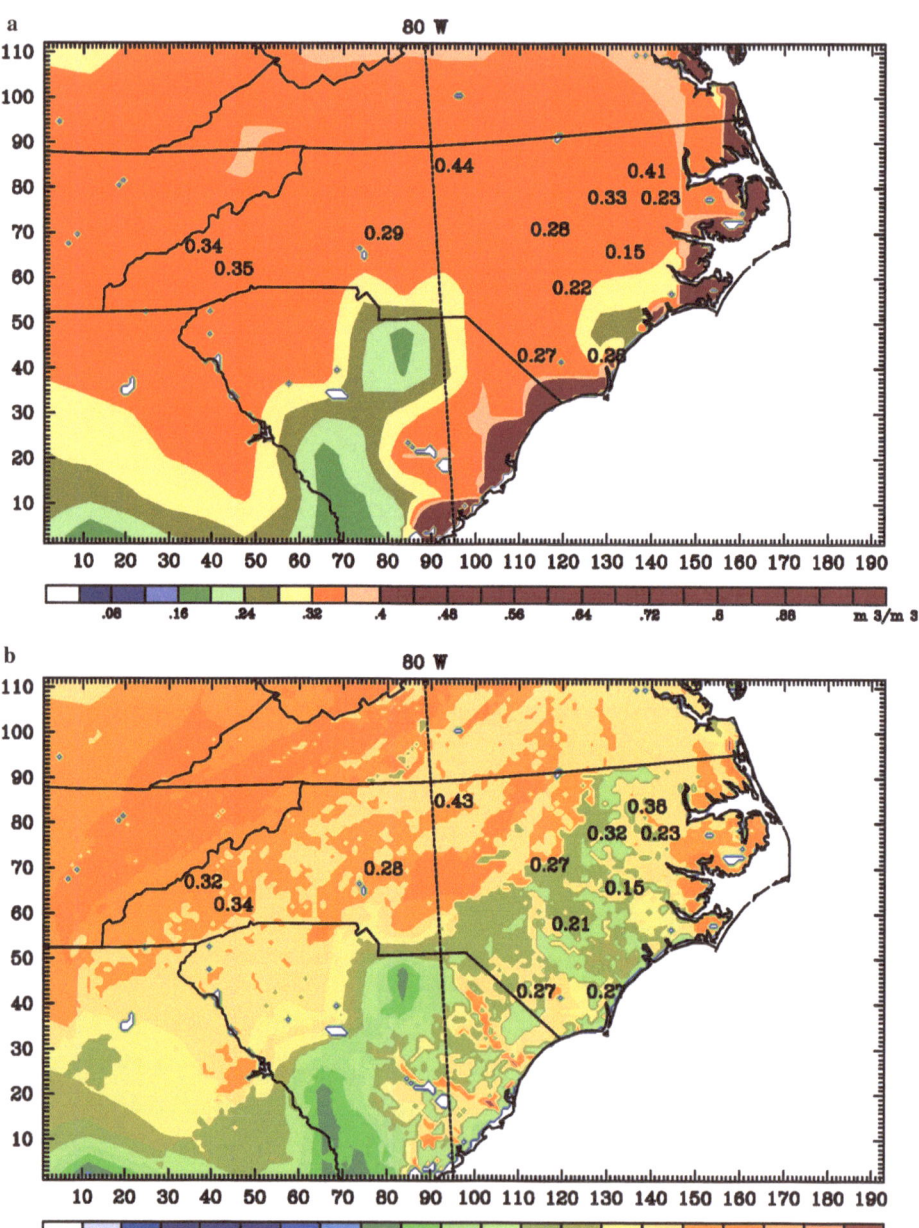

Figure 8a
Initial modeled soil moisture contours overlaid with actual observed soil moisture values obtained from
NC ECONet stations in North Carolina at 0000Z (1900 LST) on August 15, 2000 at 10 cm soil depth. The
initial field is noted to be fairly constant over North Carolina with a value of approximately 0.33 m³/m³.
Figure 8b
Modeled soil moisture contours overlaid with actual observed soil moisture values obtained from NC
ECONet stations in North Carolina at 1800Z (1300 LST) on August 15, 2000 at 10 cm soil depth.

(1300LST) are then compared with model results by overlaying the observed values on the simulated soil moisture contours as shown in Figure 8b. A noticeable gradient in soil moisture is seen at the border of the Coastal Plain and Piedmont regions. This gradient is most likely caused by the difference in soil type and soil texture characteristics. Here, observations match well with the modeled soil moisture values. A drying of the sandy Coastal Plain is evident on August 17 at 1800Z (1300LST), as shown in Figure 8c, causing the soil moisture gradient between the Piedmont and Coastal Plain to increase over the Sandhills area. The modeled soil moisture in the Sandhills region has a value of approximately 0.20 m^3m^{-3}. The area adjacent to and west of the Sandhills still shows a soil moisture value of approximately 0.33 m^3m^{-3}.

In summary, soil moisture observations and model output are in good agreement. A noticeable gradient in soil moisture is seen at the border of the Coastal Plain and Piedmont regions due to the differences in the soil type and texture characteristics.

4.2 Sea Breeze Circulation

Simulated wind patterns for the inner domain are first analyzed using the horizontal wind vectors at the surface. Initial analysis of the wind vectors at 0000Z (1900LST) August 15 show a light northwesterly flow across much of the domain, with the strongest winds occurring offshore and over the northern Piedmont region of NC as indicated in Figure 9a.

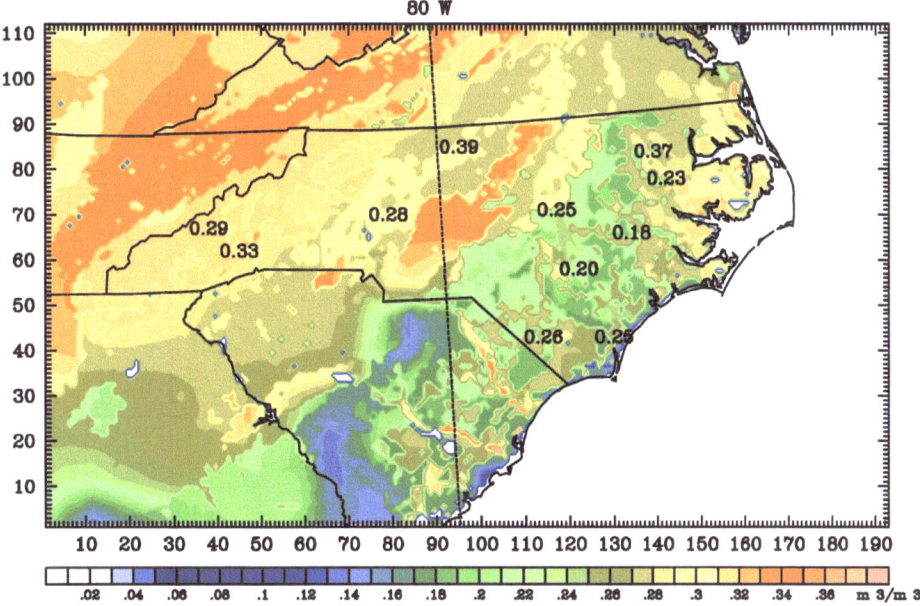

Figure 8c
As in Figure 8b except at 1800Z (1300 LST) on August 17, 2000.

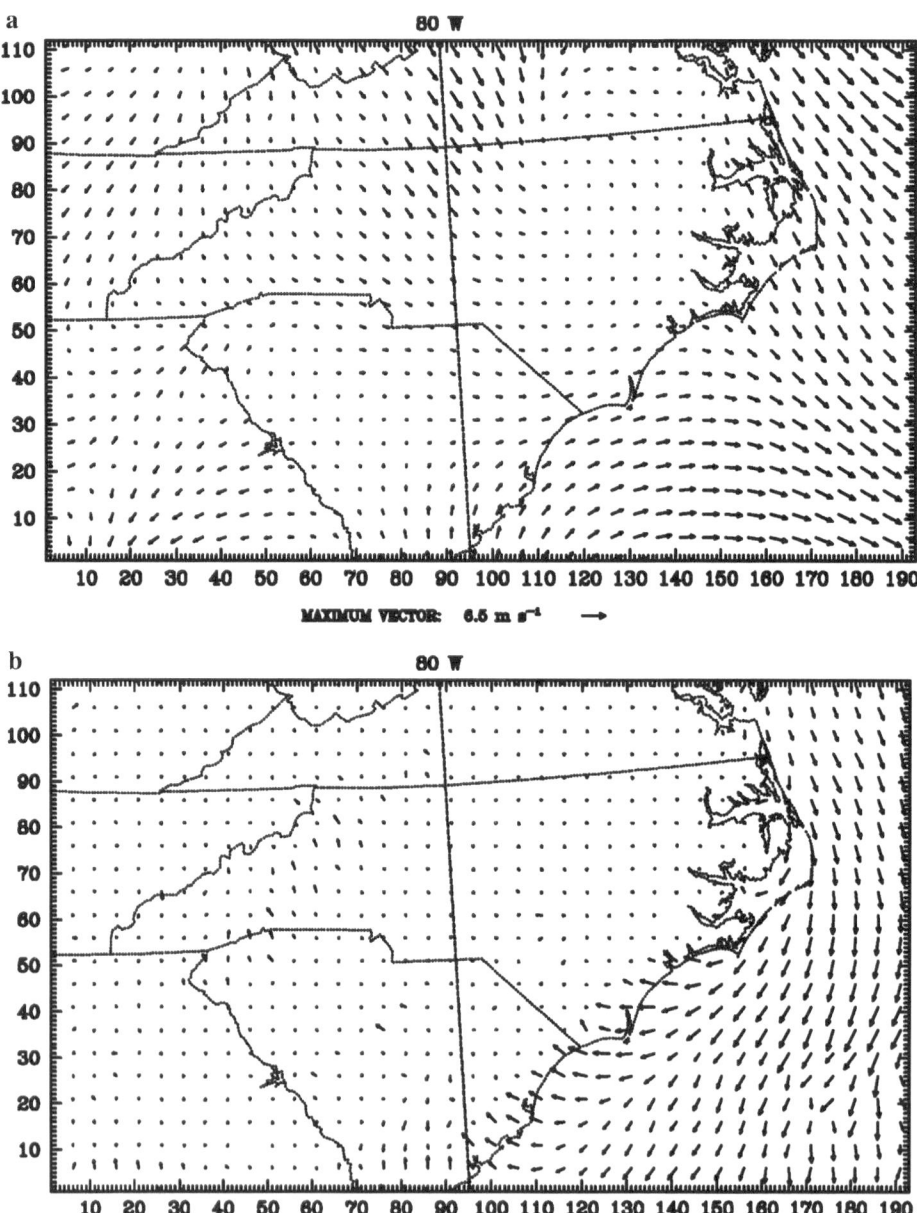

Figure 9a
Horizontal wind vectors at 10 meters above the surface for the 5-km domain on August 15, 2000 at 0000Z
(1900LST).
Figure 9b
As in Figure 9a except at 1900Z (1400LST) on August 17, 2000.

Examination of the sea breeze development helps indicate how well the model simulates diurnal variation and differential heating. At 1900Z (1400LST) on August 15, 16, and 17, there is evidence of the development of a sea breeze indicated by a reversal of simulated surface winds towards the shore. An example of this wind reversal is shown in Figure 9b at 1900Z (1400LST) on August 17. Inland penetration of the sea breeze is approximately 30 km in the southern coastal region. To further determine the extent of the sea breeze, a cross section is taken along the southern coast of North Carolina near Wilmington; the location is indicated by the letter A in Figure 10. This cross section, shown in Figure 11 depicts a well-defined sea breeze circulation at 1800Z (1300LST) on August 15. Similar circulations were modeled on August 16 and 17 as well (not shown). The vertical extent of the simulated sea breeze circulation is approximately 500 m above the surface over the land-water boundary. The onset of the sea breeze was verified (not shown) using surface winds at 10 m height from Wilmington, Whiteville, Castle Hayne, Hatteras, and Beaufort (locations are indicated in Fig. 3).

4.3 Sandhills Circulation

Simulated wind vectors along cross section B across the Sandhills region of the Piedmont are shown in Figure 12. A well-defined circulation is evident in the simulation and extends well inland, to about 90 km over the Sandhills region. This region consists of a narrow band of sandy soil that borders the clay-based soils in the Piedmont as shown in Figure 1. Sharp contrasts in soil characteristics can result in

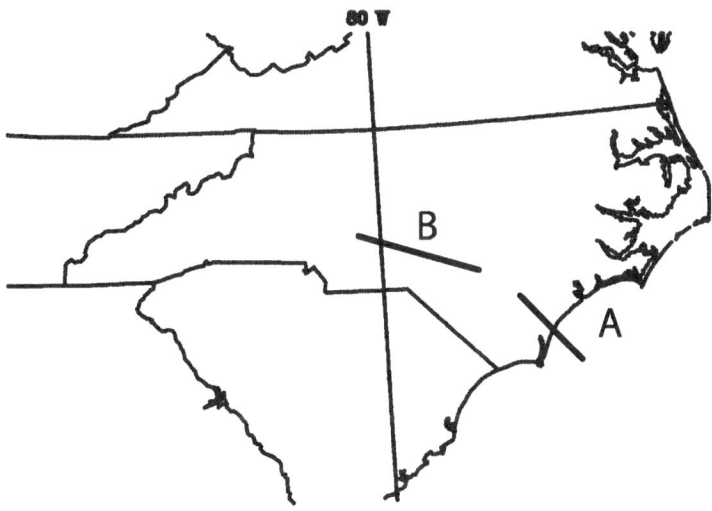

Figure 10
Lines A and B show the locations of the vertical cross sections. A is along the coast in North Carolina (NC) near the city of Wilmington. B is located across the Sandhills of NC.

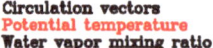

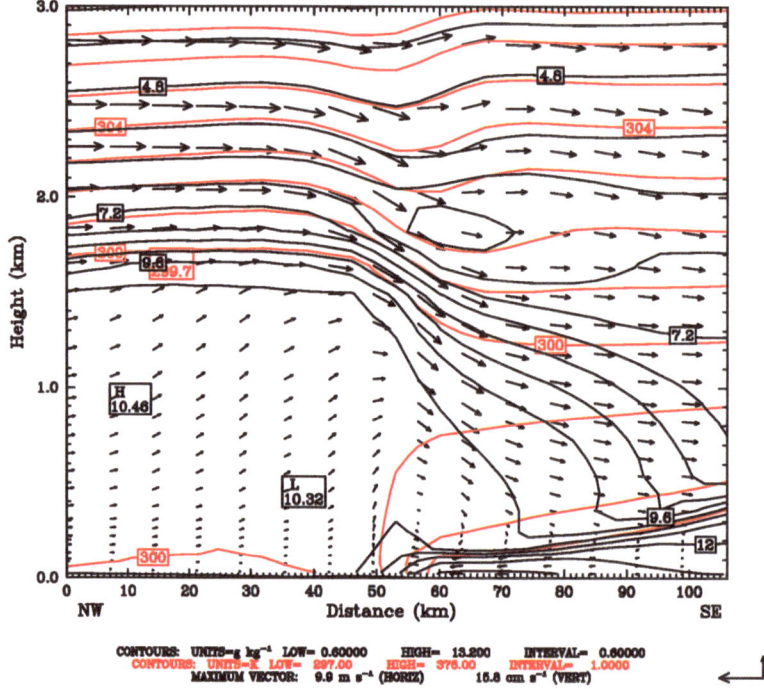

Figure 11
Vertical cross section showing circulation vectors (m/s), potential temperature (K), and water vapor mixing ratio (g/kg) for cross section A at 1800Z (1300LST) on August 15, 2000.

differential heating across this region. There is an area of convergence in the simulated Sandhills region where a strong southerly flow of about 6 m s^{-1} suddenly ceases in a narrow area of near calm winds. This convergence zone extends well into South Carolina and is oriented northeast to southwest.

Locally induced mesoscale circulations during the daytime hours in the Sandhills region at 0000Z (1900 LST) on August 18 is shown in Figure 13. The well developed circulation as seen at 0000Z (1900 LST) on August 18 horizontally extends approximately 70 km along the cross section and vertically extends of 2000 m.

There is a strong surface flow of about 4.6 m s^{-1} associated with this circulation as shown in Figure 13, with a strong upward component on the westward side of the circulation. Evolution of this feature over time produces a westward migration of this circulation feature whereby the center travels 35 km horizontally from east to west over a span of 6 hours. These circulations are not evident at 0600Z (0100 LST) on August 16 as shown in Figure 14. The flow in the early morning hours (0600 LST)

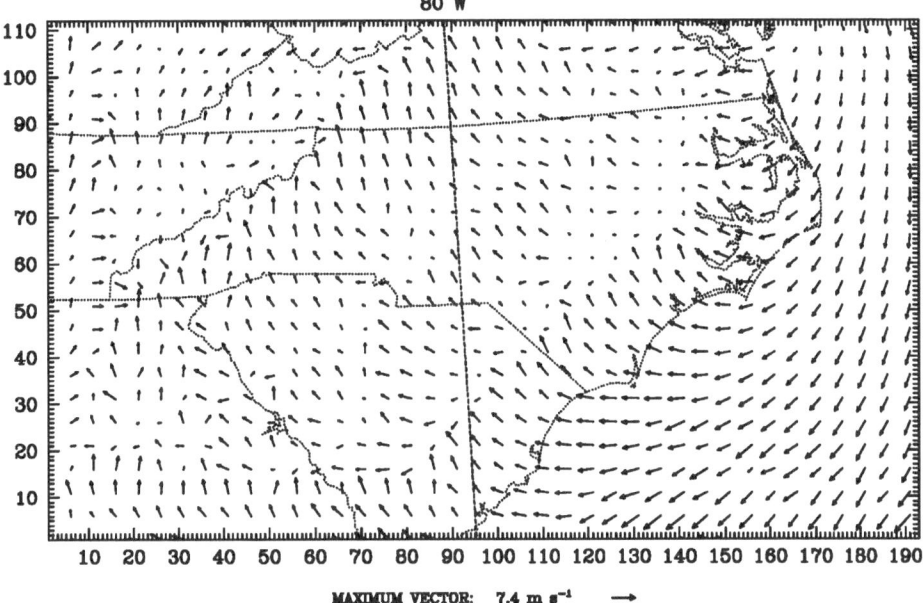

Figure 12
Surface wind vectors showing the formation of a Sandhills convergence zone inland at 0000Z (1900 LST)
on August 18, 2000.

representative of nighttime conditions is fairly uniform and is from the west with a
maximum wind speed of 4.3 m s^{-1}.

4.4 Planetary Boundary Layer Structure

Planetary boundary layer (PBL) heights were examined and compared with
soundings from an upper air station at Greensboro (GSO), North Carolina. Also,
SODAR data at the Environmental Protection Agency (EPA) site at a Research
Triangle Park (RTP) location in North Carolina was obtained. This location is
indicated in Figure 4 by a square marker on the map. Horizontal contour plots
of PBL heights show significant diurnal variation with heights as low as 100 m at
night and over 1500 m during the day. These variations are consistent with the
overall observed variation. The simulated PBL heights at 1200Z (0700 LST) on
August 15 are shown in Figure 15. Here, the simulated stable boundary layer
over the innermost domain indicates an average height of approximately 200–
300 m.

For this simulation period, SODAR observations were available at the Research
Triangle Park in central Piedmont. PBL height variations on August 17 as observed
by SODAR are shown in Figure 16 in which the solid line indicates the growth of the
PBL with time. Modeled PBL heights at 1200Z (0700LST) are shown in Figure 17.

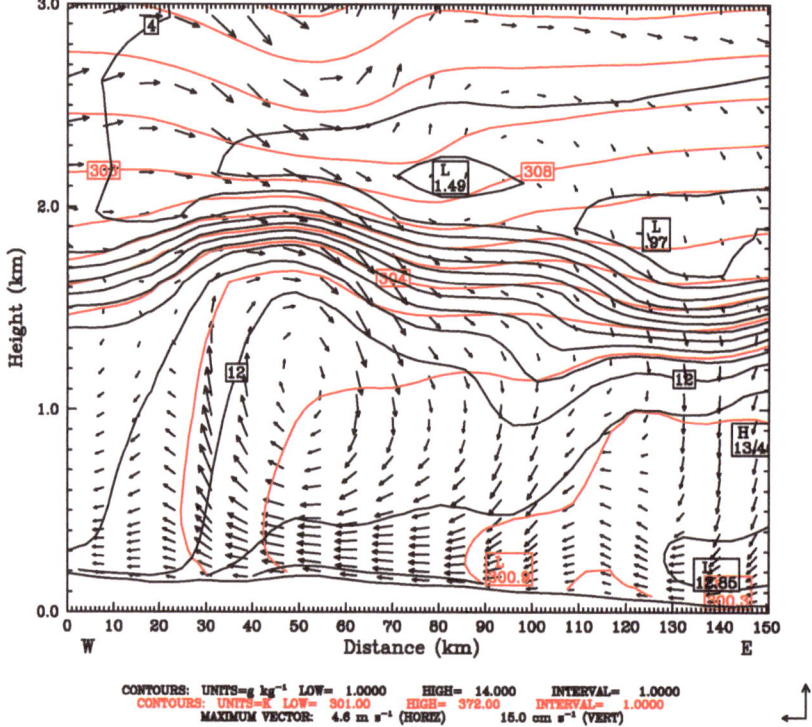

Figure 13
Formation of Sandhills circulation inland in the numerical simulation at 0000Z (1900 LST) on August 18, 2000. Initiation of convection generally takes place over this location.

The modeled PBL height at this time at the SODAR location is 300–400 m, while observations indicate a boundary layer height of about 200 m. Model overestimation was present at 1300Z (0800 LST) as well, where observations indicate a height of about 300 m while the model simulated a height near 400–500 m. Along the shore, modeled PBL growth in the coastal region is about 900 m. By 1400Z (0900LST), modeled heights near the SODAR location approached 800–900m, while observed heights were about 600 m.

The modeled PBL heights at 1500Z (1000LST) are approximately 1200 m to 1300 m at the SODAR location. The extent of the PBL height is not discernable by the SODAR at 1500Z (1000LST) because of range problems as shown in Figure 16. Based on the rate of growth as indicated in Figure 16, PBL heights were most likely at about 1000 m by this time. Along the coast, the model indicates PBL heights of approximately 1400 to 1500 m, whereas heights just offshore of the southern coast are still at 500 to 600 m.

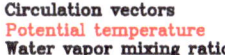

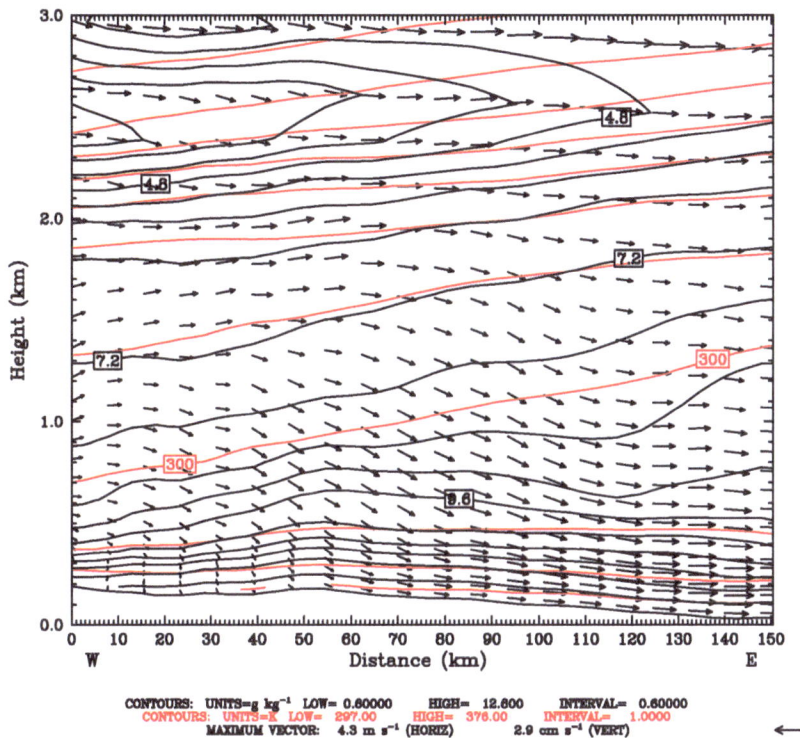

Figure 14

As in Figure 13 except at 0600Z (0100LST) on August 15, 2000. No circulation is seen in the simulation.

4.5 Land Surface Processes

Effects of heterogeneity in land use and soil types were evaluated with respect to the observed and the modeled land surface processes. Surface latent heat fluxes from the model simulation at 1800Z (1300 LST) on August 16 indicate a strong horizontal gradient in the Sandhills region as shown in Figure 18. A comparison of the location of this gradient with surface soil characteristics reveals a strong correlation of this latent heat flux gradient with the land use and soil type. Large latent heat fluxes, on the order of 500 Wm^{-2}, are present in the cropland area in conjunction with the loamy sand soil type. Reduced latent heat flux, about 350 Wm^{-2}, is simulated in the area adjacent to and just west of the Sandhills. The reduced latent heat flux appears to correlate with the finer grained soil and land cover change. Different field capacities for these soil textures combined with developing soil moisture gradients during the simulation may have contributed to these latent heat flux gradients in the Sandhills region. Another noticeable horizontal latent heat flux gradient is along the coast. The

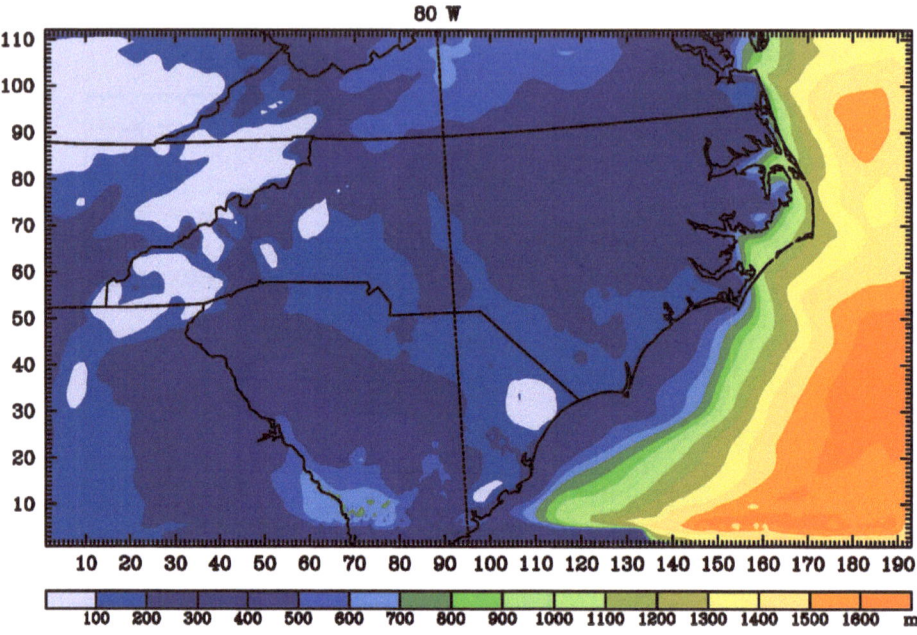

Figure 15
Planetary boundary layer heights for the 5-km domain on August 15, 2000 at 1200Z (0700LST).

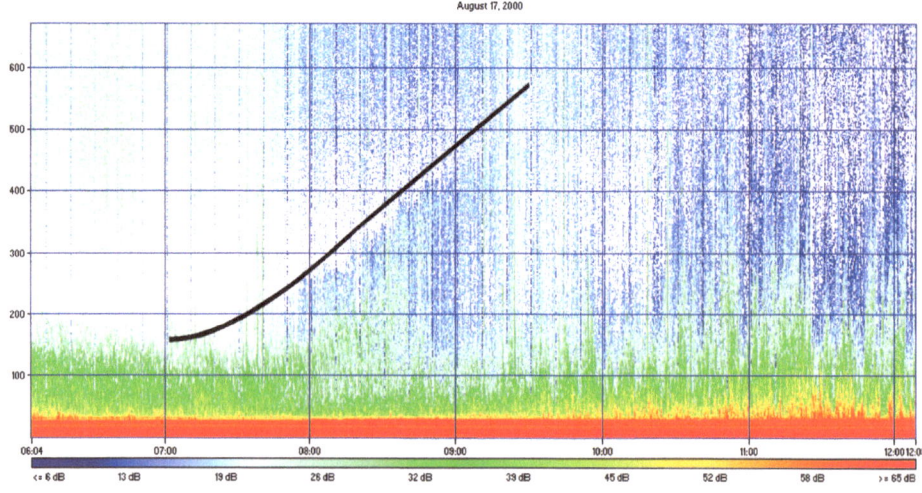

Figure 16
Color enhanced SODAR data obtained from the EPA near Research Triangle Park on August 17, 2000
from 1100Z to 1700Z (0600LST to 1200LST) showing the convective growth of the boundary layer.

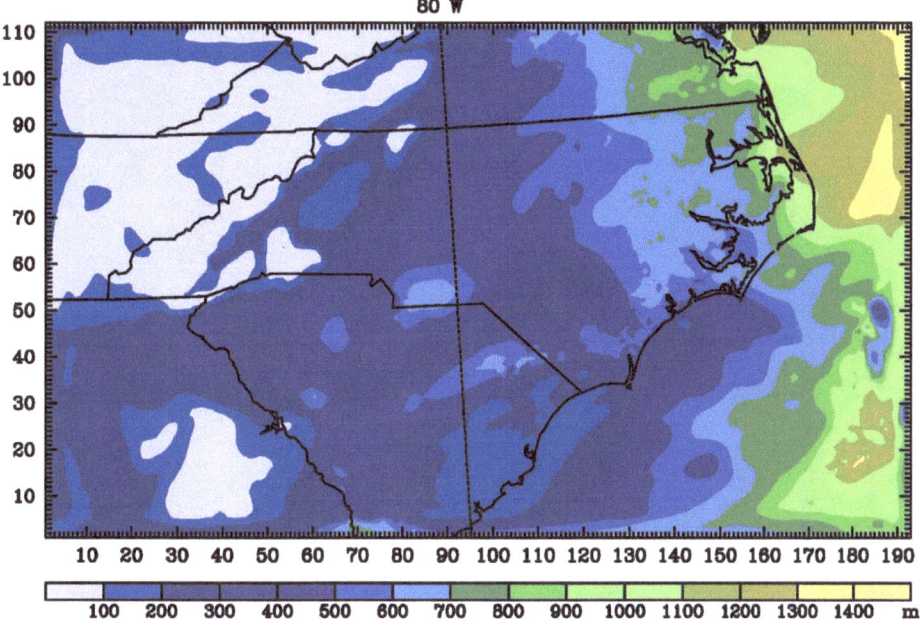

Figure 17
As in Figure 15 except on August 17, 2000 at 1200Z (0700LST).

simulated latent heat flux over land in the coastal region is approximately 400 Wm^{-2} and is larger than the flux over the coastal waters, of about 200 Wm^{-2}. The surface sensible heat flux distribution, shown in Figure 19 corresponds to the latent heat flux patterns at 1800Z (1300LST) on August 16. Sensible heat flux values decrease over the Sandhills region. This is possibly related to larger evapotranspiration occurring in this area as indicated by the larger latent heat fluxes as shown in Figure 18.

5. *Evaluation of Simulated Surface Temperatures*

Results from the time series comparison plots for stations at Wilmington (ILM) and Jackson Springs (JAC) are shown in Figures 20a and 20b, respectively. An overestimation of the nighttime temperatures by the model is apparent for both the locations. The station locations are indicated in Figure 3. This overestimation could be related to radiational cooling, as model performance tends to decrease in the day to night transitional period in which the observed rate of decrease in temperature exceeds the rate of decrease in temperature in the model. The overnight observational temperatures typically drop well below the modeled temperature by as much as 3 to 5 C (or K) at these locations. One possible cause is the error in the initial values of surface temperature and moisture.

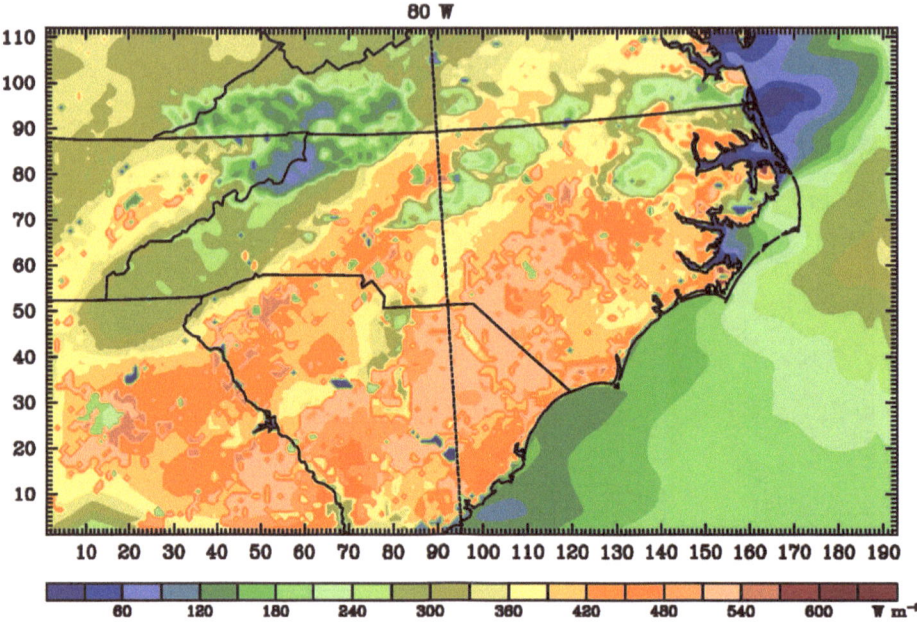

Figure 18
Latent heat flux (W m^{-2}) for the 5-km domain on August 16, 2000 at 1800Z (1300LST). A sharp gradient
in the latent heat flux is seen in the central Piedmont near the Sandhills region.

During the night to day transitional period, the model quickly falls in line with
the observations, handling the maximum temperatures well at most locations. Minor
overestimations and underestimations of maximum temperature occur at roughly
half the stations (SIMS, 2001).

6. Conclusions

During weak synoptic conditions, mesoscale processes can significantly impact
regional weather. Examples of these processes include local surface heat flux
gradients caused by differences in evaporation and transpiration from the earth's
surface. Surface characteristics can also significantly contribute to the development
of the planetary boundary layer and in turn virtually affect cloud and precipitation
patterns. For example, soil moisture and texture and the land use affect surface
characteristics, which in turn affects surface forcings. Correctly treating these land
surface forcings is important for capturing and properly simulating terrain and land-
use induced mesoscale circulations. Performance of a mesoscale model (MM5) was
evaluated using surface observations, and SODAR measurements in North Carolina.
Point-to-point comparisons of surface observations with the closest model grid point

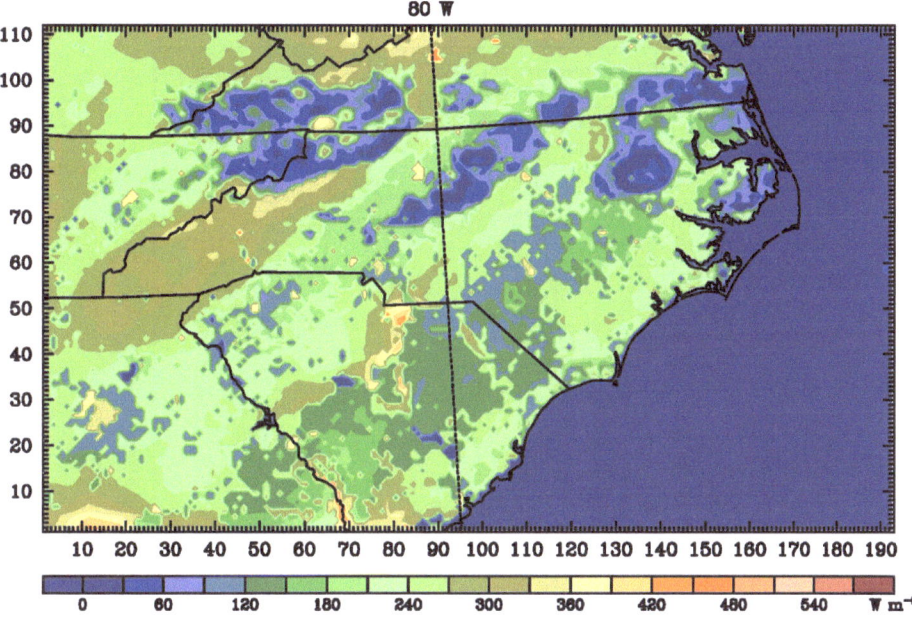

Figure 19

Sensible heat flux (W m^{-2}) for the 5-km domain on August 16, 2000 at 1800Z (1300LST). A sharp gradient in the sensible heat flux is seen in the central Piedmont near the Sandhills region.

values were performed using the 5-km model inner domain. Evaluation and validation of multiple surface parameters were performed.

The general flow patterns over the domain are well simulated by the model. The 5-km domain does simulate more mesoscale variability in the wind fields as compared to the 15-km outer domain. The model overestimates soil temperatures during daytime hours, especially in the Coastal Plain. The model simulates the soil moisture changes reasonably.

Observed nocturnal temperatures are cooler than the model predicted temperatures at the grid points closest to the observation stations. This problem could be due to improper representation, specification, and initialization of surface parameters including soil moisture, land use and texture as well as the surface energy budget. Diurnal variation of the boundary layer structure is well simulated. Nocturnal boundary layer heights appear to be simulated presentably by the model when compared with the soundings and the SODAR observations. The timing of the convective boundary layer growth also matches the observations well. The rate of growth, however, is somewhat overestimated by the model.

Magnitudes of the simulated latent and sensible heat fluxes vary depending on the land cover and soil type. Magnitudes of these heat fluxes depend on the soil moisture

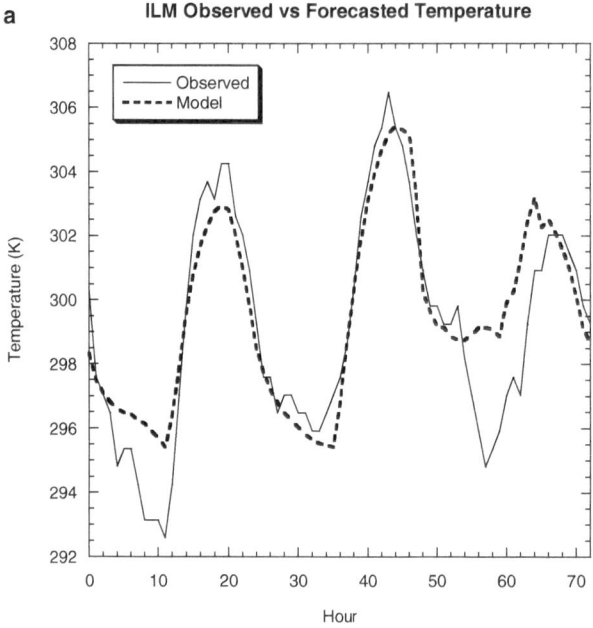

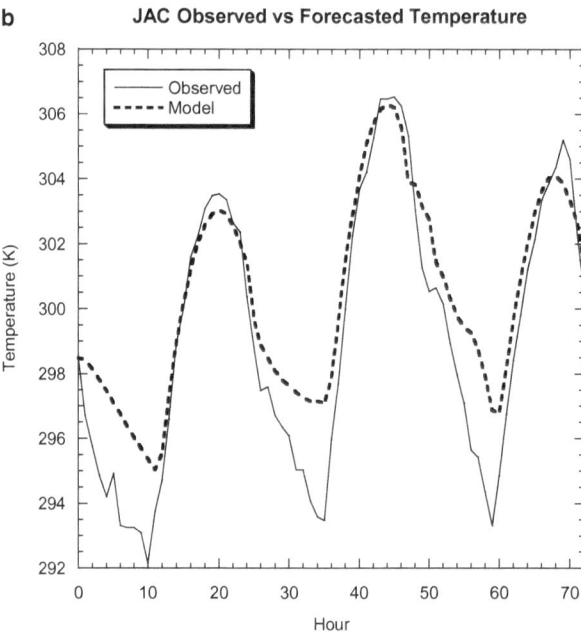

availability in the region and on the field capacity of the soil types. Significant horizontal gradients in the latent and sensible heat fluxes in the Sandhills region contribute to the development of mesoscale circulations observed in this region.

◄

Figure 20a
Observed and modeled air temperature at 2 m above the ground at Wilmington (ILM), NC. Forecast
hours span 72 hours beginning at 0000Z (1900LST) on August 15, 2000 and ending on August 18, 2000 at
0000Z (1900LST). Modeled and observed temperatures are in and out of phase and match well during the
daytime. The model overpredicts nighttime temperatures.
Figure 20b
As in Figure 18a except for the Jackson Springs (JAC) station.

Acknowledgements

This work was supported by the State Climate Office of North Carolina and by the Atmospheric Sciences Division, National Science Foundation under Grant ATM-0233780. North Carolina Super Computing Center and the State Climate Office of North Carolina provided computer resources. Chris Holder, Undergraduate Research Assistant assisted in editing this manuscript. Authors would like to thank the reviewers for several helpful suggestions which enhanced the manuscript.

REFERENCES

DUDHIA, J. (1989), *Numerical Study of Convection Observed during the Winter Monsoon Experiment Using a Mesoscale Two-dimensional Model,* J. Atmos. Sci. *46,* 3077–3107.

FRITSCH, J. M. and CHAPPEL, C. F. (1980), *Numerical Prediction of Convectively Driven Mesoscale Pressure Systems. Part I: Convective Parameterization,* J. Atmos. Sci. *37,* 1722–1733.

GUTMANN, G. J. and Ignatov, A. (1998), *The Derivation of Green Vegetation Fraction from NOAA/ AVHRR Data for Use in Numerical Weather Prediction Models,* Int. J. Remote Sens. *19,* 1533–1543.

HONG, S.-Y. and PAN, H.-L. (1996), *Nonlocal Boundary Layer Vertical Diffusion in a Medium-range Forecast Model,* Mon. Wea. Rev. *124,* 2322–2339.

KAIN, J. S. and FRITSCH, J.M., *Convective parameterization for mesoscale models: The Kain-Fritsch scheme.* In *The Representation of Cumulus Convection in Numerical Models,* (eds. Emanuel, K. A. and Raymond, D.J.) (Amer. Meteor. Soc. Boston 1993), 246 pp.

KESSLER, E. (1969), *On the Distribution and Continuity of Water Substance in Atmospheric Circulation,* Meteorol. Mon. Amer. Meteorol. Soc. *10,* 84.

KOCH, S. E. and RAY, K. A. (1997), *Mesoanalysis of Summertime Convergence Zones in Central and Eastern North Carolina,* Wea. and Forecasting *12,* 56–77.

LIN, Y.-L., FARLEY, R. D., and ORVILLE, H. D. (1983), *Bulk Parameterization of the Snow Field in a Cloud Model,* J. Climate Appl. Meteor. *22,* 1065–1092.

REISNER, J., RASMUSSEN, J., and BRUINTJES, R.T. (1998), *Explicit Forecasting of Supercooled Liquid Water in Winter Storms Using the MM5 Mesoscale Model,* Quart. J. Roy. Meteor. Soc. *124B,* 1071–1107.

SCHULTZ, P. (1995), *An Explicit Cloud Physics Parameterization for Operational Numerical Weather Prediction,* Monthly Wea. Rev. *123,* 3331–3343.

SIMS, A., *Effect of Mesoscale Processes on Boundary Layer Structure and Precipitation Patterns: A Diagnostic Evaluation and Validation of MM5 with North Carolina ECONet observations* (North Carolina State University, Raleigh 2001) 219 pp.

TAO, W.-K. and SIMPSON, J. (1993), *The Goddard Cumulus Ensemble Model, Part 1: Model Description,* Terr. Atmos. Oceanic Sci. *4,* 35–72.

TROEN, I. and MAHRT, L. (1986), *A Simple Model of the Atmospheric Boundary Layer: Sensitivity to Surface Evaporation,* Bound. Layer Meteor. *37,* 129–148.

USDA *State Soil Geographic (STATSGO) Data Base* (U.S Department of Agriculture, Natural Resources Conservation Service, Miscellaneous Publication 1492, 1994).

ZOBLER, L., *A world soil file for global climate modeling* (NASA Tech. Memo. 87802, 1986) [Available from NASA Goddard Space Flight Center, Institute for Space Studies, 2800 Broadway, New York, NY 10025.]

(Received February 10, 2004, accepted June 30, 2004)

To access this journal online:
http://www.birkhauser.ch